Storm Water Pollution Control

Storm Water Pollution Control

Municipal, Industrial, and Construction NPDES Compliance

Roy D. Dodson, P.E.
Dodson & Associates, Inc.

Second Edition

Boston, Massachusetts Burr Ridge, Illinois
Dubuque, Iowa Madison, Wisconsin New York, New York
San Francisco, California St. Louis, Missouri

Library of Congress Cataloging-in-Publication Data

Dodson, Roy D.
Storm water pollution control : municipal, industrial, and construction NPDES compliance / Roy D. Dodson.—2nd ed.
p. cm.
Includes bibliographical references and index.
ISBN 0-07-017388-5
1. Storm sewers—United States. 2. Water—Pollution—United States. 3. Environmental permits—United States. 4. Pollution prevention—United States. 5. Storm sewers—Law and legislation—United States. I. Title.
TD665.D63 1998
363.739′46′0973—dc21 98-25762
CIP

McGraw-Hill

A Division of The McGraw-Hill Companies

7 8 9 IBT/IBT 9 0 9

ISBN 0-07-017388-5

The sponsoring editor for this book was Larry Hager, the editing supervisor was Frank Kotowski, Jr., and the production supervisor was Clare Stanley. It was set in Century Schoolbook by Renee Lipton of McGraw-Hill's Professional Book Group composition unit.

To my wife Pam (who becomes even more beautiful every day), to my children Brad and Alana, and to my mother Vivian E. Dodson. You are God's richest blessings to me.

Contents

Preface

This book will help you understand and comply with the EPA and state requirements for NPDES storm water discharge permits.

Beginning in October 1992, most industrial facilities and construction projects in the United States have been required to obtain permits to discharge storm water according to the provisions of the National Pollutant Discharge Elimination System (NPDES). This requirement affects hundreds of thousands of industrial sites and construction projects. Several hundred large- and medium-size municipalities are also affected. Recent EPA proposals will expand this program to include many more construction sites and smaller municipalities.

There has been tremendous confusion about the requirements of the NPDES program for storm water permits. Construction contractors, engineers, project owners, industry representatives, and public employees have all desperately attempted to obtain reliable information on how to comply with NPDES permit requirements and avoid substantial penalties. This book provides that information.

How I Wrote this book

I began to write this book because of my company's work on a large number of permit applications and storm water pollution prevention plans for industrial and construction clients. The American Society of Civil Engineers also asked me to present a series of continuing education seminars on NPDES storm water permit compliance. As I presented these seminars across the United States, I gathered additional materials for this book. After the first edition was released, I continued to receive helpful comments and suggestions, and collected additional information on new EPA and state requirements.

Describing EPA and state requirements for storm water pollution control, this book naturally relies heavily on EPA publications for its content. The value of this book is that it collects and clearly presents all of the most important material from dozens of EPA documents and other sources, updating the material as necessary to provide a comprehensive yet readable overview of this important topic.

How to Use This Book

Everyone who reads this book should read at least Chapters 1, 2, and 3. These chapters provide basic information about the NPDES storm water program. In order for you to do the best job in complying with a federal government program, it is important to understand the legal and regulatory background of the program. Chapter 1 provides that information. Chapter 2 is important because it reveals how the NPDES storm water program is enforced. However, it contains a great deal of detailed information on penalty assessment that may not be important to those not involved in the negotiation of penalties for Clean Water Act violations. You may wish to peruse Chapter 2 initially, then come back to it as needed for further information.

Chapter 3 is crucially important because it deals with the most fundamental question of all: Do you need a storm water discharge permit?

Chapter 4 is new for the second edition. It provides a complete description of the NPDES storm water discharge permit requirements for municipalities, including the current proposal for future requirements that are likely to affect many more municipalities.

Chapters 5 through 9 deal with industrial storm water discharges. If you are interested only in construction activities, you may skip these chapters.

Chapters 10 through 14 deal with construction storm water discharges. If you are interested only in other industrial activities, you may skip these chapters. However, you should note that obtaining an industrial storm water discharge permit does not allow you to discharge storm water from construction activities at the industrial facility. You may be required to obtain a separate construction storm water discharge permit. Therefore, these chapters may be important to you at some point in the future.

The majority of this book deals with U.S. Environmental Protection Agency (U.S. EPA) requirements under the NPDES program. However, in most states, NPDES permits are actually administered by state agencies. Requirements vary somewhat from one state to another. Therefore, Appendix A discusses the individual requirements for each state, and highlights those areas in which state requirements differ from EPA requirements. Refer to Appendix A for specific information for the state(s) in which you work.

Appendix B is a glossary of terms and acronyms related to storm water quality.

Appendix C provides an overview of the units of measurement used in this book, and the conversion from American to SI units.

Other available resources:

An internet web site has been set up to provide background information and program updates to increase the usefulness of this book:

http: / /www.dodson-hydro.com

This web site also provides the opportunity to submit any suggestions and/or corrections for the book, and to correspond with the author by e-mail or other means.

Acknowledgments

Many people have contributed to the development and completion of this book, including our supportive group of clients, representatives of the U.S. EPA and state agencies, and those who have participated in over 50 ASCE-sponsored continuing education seminars that I have lead since 1992.

I want to particularly acknowledge the efforts of Mr. Taylor Sharpe, of EPA Region 6 in Dallas, and Mr. Harless R. Benthul, an attorney with Gilpin, Paxson & Bersch in Houston, who both reviewed Chapter 2 and made many helpful suggestions. Mr. Kevin Weiss of EPA also reviewed a portion of the first edition of the book in draft form.

I also appreciate the gracious assistance of many EPA regional and state storm water program coordinators and officials who reviewed portions of the book and provided useful comments. Their comments were particularly useful in the completion of Appendix A.

Mrs. Sunny Yung provided diligent and helpful research work to assist in the completion of the second edition. Mark Starks prepared most of the figures. I appreciate my other co-workers at Dodson & Associates, Inc., who "covered for me" while I took the time to write this book.

In spite of all the generous assistance that I received from many sources, I wish to emphasize that any errors or omissions in this book are entirely my responsibility.

Chapter

1

EPA Storm Water Regulations

The Environmental Protection Agency (EPA) regulations on storm water discharges affect thousands of industrial facilities and municipalities in the United States. In complying with these rules, industrial plant managers, industry executives, municipal employees, engineering consultants, and construction contractors may be exposed to EPA water quality requirements for the first time. Often, they do not have a strong background in the legal or regulatory issues related to water pollution control in the United States. Acronyms such as CWA, RCRA, CERCLA, SARA, CZARA, ESA, NHPA, and CFR may be meaningless and confusing to these readers.

This chapter will provide a brief background in the environmental laws and regulations of the United States which pertain to water pollution, particularly storm water discharges. This should make it much easier to understand and apply the more technical information presented in the remainder of the book.

This is the first of three consecutive chapters which provide basic information intended to be of interest to all readers. Chapter 4 provides information on permit requirements for municipal storm water discharges. Chapters 5 through 9 provide information specific to those interested in industrial discharges. Chapters 10 through 14 provide information related to construction storm water discharges only. Appendix A provides additional information on specific state requirements. Appendix B is a glossary of terms and acronyms used in this book.

Legal Basis for the National Storm Water Program

The Environmental Protection Agency developed the *National Storm Water Program (NSWP)* in response to legislation passed by Congress.

The most important item of legislation was the *Federal Clean Water Act of 1972* (Public Law 92-500), which established the *National Pollutant Discharge Elimination System* (*NPDES*). The Clean Water Act (CWA) has been amended several times. One important set of amendments was the *Water Quality Act of 1987* (Public Law 100-4) that established a phased approach for storm water discharge regulation in the United States. Section 1068 of the *Intermodal Surface Transportation Efficiency Act of 1991* (Public Law 102-240) clarified the application of the NSWP to small municipal entities.

Other Related Environmental Legislation

The NSWP also includes references to several other important items of environmental legislation, most of which have to do with hazardous wastes or toxic materials. These include the following:

- *RCRA:* The Resource Conservation and Recovery Act of 1976 (Public Law 94-580) requires a regulatory system for the generation, treatment, storage, and disposal of hazardous wastes.
- *CERCLA:* The Comprehensive Environmental Response, Compensation, and Liability Act of 1980 (Public Law 96-510) establishes a program to mitigate releases of hazardous waste from inactive hazardous waste sites that endanger public health and the environment.
- *SARA:* The Superfund Amendments and Reauthorization Act of 1986 (Public Law 99-499) amended CERCLA.
- *EPCRA:* The Emergency Planning and Community Right-to-Know Act of 1986 is another name for Title III of SARA. Title III was created as a freestanding law establishing requirements regarding emergency planning and reporting on hazardous and toxic chemicals.
- *SMCRA:* The Surface Mining Control and Reclamation Act of 1977 (Public Law 95-87) regulates surface coal mining operations and the acquisition and reclamation of abandoned mines.
- *CZARA:* The Coastal Zone Act Reauthorization Amendments of 1990 (Public Law 101-508) establish coastal management requirements.
- *ESA:* The Endangered Species Act of 1973 (Public Law 93-205) regulates a wide range of activities affecting plants and animals designated as endangered or threatened.
- *NHPA:* The National Historic Preservation Act of 1966 (Public Law 89-665) establishes a program and system of regulations intended to preserve historic places.

- *SDWA:* The Safe Drinking Water Act Amendments of 1996 (Public Law 104-182) require states to establish Source Water Assessment Programs to delineate source water protection areas, to inventory significant contaminants in these areas, and to determine the susceptibility of each public water supply to contamination.

Importance of the Clean Water Act

The Clean Water Act has set the direction of water pollution control in the United States since 1972. It is based on the following principles (Kovalic, 1987):

1. *Waters of the United States:* No one has a right to pollute the navigable waters of the United States.[1] Anyone wishing to discharge pollutants must obtain a permit to do so.
2. *Discharge permits:* Permits shall limit the composition of a discharge and the concentrations of the pollutants in it. Anyone violating the conditions of a permit is subject to fines and imprisonment.
3. *Technology-based controls:* Some permit conditions require specified levels of control based on a consideration of technology and cost, regardless of the receiving water's ability to purify itself naturally. In other words, some levels of control are always presumed to be worth their cost.
4. *Water-quality-based controls:* Any limits or control higher than the minimum federal requirements must be based on the receiving water quality. The only way to impose higher standards than those required under the CWA is to demonstrate that continued protection of the receiving water demands such limits.

Title IV of the CWA gives the EPA the authority and responsibility to issue discharge permits to every point source (discussed in Chap. 3) discharger in the country. Section 402 of the CWA describes the permit system. As of about 1991, the EPA and authorized NPDES states had issued almost 50,000 NPDES permits for industrial process discharges and 16,000 NPDES permits for *publicly owned treatment works* (*POTWs*). As impressive as these numbers are, they are small in comparison to the number of facilities that discharge storm water. For example, the EPA estimates that approximately 150,000 individual industrial facilities are required to obtain storm water discharge permits under the initial requirements (Environmental Protection Agency, 1995).

Technology-based requirements represent the minimum level of control that must be imposed by an NPDES permit. Two technology-based

[1]The term *waters of the United States* has a very broad definition, as explained in Chap. 3.

requirements are appropriate for existing storm water discharges associated with industrial activity:

1. *Best conventional* (pollutant control) *technology* (*BCT*)
2. *Best available technology* (*BAT*) economically achievable .

The BCT standard applies to the control of conventional pollutants, while the BAT standard applies to the control of all toxic pollutants and all pollutants which are neither toxic nor conventional pollutants. Both requirements are consistent with the third principle of the CWA listed above.

Storm water discharges associated with industrial activity are subject to both BAT and BCT technology-based standards for NPDES permits. Permits for most classes of industrial storm water discharges require the development and implementation of site-specific storm water pollution prevention plans. The plans require the identification and implementation of BMPs and pollution prevention measures. EPA has developed numeric effluent limitations for a limited number of industrial storm water discharges. Chapters 5 through 9 discuss industrial storm water discharge permit requirements in detail.

Two technology-based standards have been established for discharges from *municipal separate storm sewer systems* (*MS4s*). The first standard provides that MS4 permits must contain a requirement to effectively prohibit non–storm water discharges into the system. The other standard requires that permits for discharges from MS4s reduce the discharge of pollutants to the *maximum extent practicable* (*MEP*), including management practices, control techniques, systems, design, and engineering methods. EPA has not issued a national guideline defining these standards; instead, specific BMPs and/or measures for implementing these standards are established on a case-by-case basis. Permits for discharges from MS4s have generally required the development and implementation of comprehensive municipal storm water management programs to implement these two standards. Chapter 4 discusses municipal storm water discharge permit requirements in detail.

State water quality standards

In addition to technology-based controls, NPDES permits must include any conditions more stringent than technology-based controls necessary to meet state water quality standards. State water quality standards represent the mechanism by which the fourth principle of the Clean Water Act (discussed earlier) is enacted.

States are required to develop water quality standards for waters of the United States within their boundaries. States are required to review their water quality standards at least once every 3 years and, if appropriate, revise or adopt new standards. The results of this triennial review must be submitted to EPA, and EPA must approve or disapprove any new or revised standards.

The minimum elements that must be included in a state's water quality standards include the use designations for all water bodies in the state, water quality criteria sufficient to protect those use designations, and an antidegradation policy consistent with EPA's water quality standards [40 CFR 131.6]. In their standards states may also include policies generally affecting the standards' application and implementation [40 CFR 131.13]. These policies are subject to EPA review and approval.

Water quality standards establish the "goals" for a water body. In general, states have not had the resources to designate beneficial uses on a segment-by-segment basis for all the states' surface waters. States usually initially designate beneficial uses in a site-specific fashion for a subset of water segments that are potentially threatened by degradation; and then as resources and information become available, states gradually begin to classify the remainder. This allows states to focus limited resources on collecting information to protect the water segments at greatest risk. This approach combined with a default use designation for unclassified waters ensures that all state surface waters have designated beneficial uses and are protected for purposes of the Clean Water Act.

The Clean Water Act states the national goal of achieving "water quality which provides for the protection and propagation of fish, shellfish, and wildlife and...recreation in and on the water," wherever attainable. These national goals are commonly referred to as the *fishable/swimmable goals* of the Clean Water Act. The EPA requires that water quality standards provide for fishable/swimmable uses unless those uses have been shown to be unattainable.

If a state designates a beneficial use which is less stringent than the default fishable/swimmable use, then the state is required to conduct a *use attainability analysis* (*UAA*). This is a "structured scientific assessment of the factors affecting the attainment of a use which may include physical, chemical, biological, and economic factors" [40 CFR 131.3(g)]. The EPA considers a use to be attainable if it can be achieved by using a combination of effluent limitations for point source discharges and cost-effective and reasonable best management practices for non–point source discharges.

Other provisions of the Clean Water Act

The Clean Water Act also contains many other provisions that are beyond the scope of this book, including a huge grant and loan program for the construction of municipal sewage treatment works. Section 404 of the CWA restricts the discharge of fill material into waters of the United States, including wetlands.

Storm water discharge controls under the Clean Water Act

The Clean Water Act originally required control of all point source discharges of pollutants into waters of the United States. However, the appropriate means of regulating storm water point sources within the National Pollutant Discharge Elimination System program has been a matter of serious concern since implementation of the NPDES program in 1972.

Efforts to improve water quality under the NPDES program have traditionally focused on reducing pollutants in discharges of industrial process wastewater and municipal sewage. At the onset of the program in 1972, many sources of industrial process wastewater and municipal sewage were not adequately controlled, and they represented pressing environmental problems. In addition, sewage outfall and industrial process discharges were easily identified as responsible for poor, often drastically degraded water quality conditions. Since enactment of the 1972 Clean Water Act, significant progress has been made in cleaning up industrial process wastewater and municipal sewage. Continuing improvements are expected for these discharges as the NPDES program continues to shift to toxic and water-quality-based pollution control.

In 1973, EPA issued its first storm water regulations. In making these regulations, EPA admitted that storm water discharges fell within the definition of a point source. However, because of the intermittent, variable, and unpredictable nature of storm water discharges, EPA reasoned that the problems caused by storm water discharges were better managed at the local level through nonpoint source controls, such as the imposition of specific management practices to prevent the pollutants from entering the runoff. The EPA also noted that issuing individual NPDES permits for the hundreds of thousands of storm water point sources in the United States would create an overwhelming administrative burden and would divert resources away from control of industrial process wastewater and municipal sewage. Therefore, the first storm water regulations required permits only for particular storm water discharges identified as significant contributors of pollution [38 FR 13530 (May 22, 1973)].

The Natural Resources Defense Council (NRDC) brought suit in the U.S. district court for the District of Columbia, challenging EPA's authority to selectively exempt categories of point sources from permit requirements [*NRDC v. Train,* 396 F. Supp. 1393 (D.D.C. 1975), aff'd.,

NRDC v. Costle, 568 F. 2d 1369 (D.C. Cir. 1977)]. In response to the NRDC suits, a U.S. district court ruled in 1975 that every municipal storm water outfall (a pipe or other conduit that empties into surface waters) must have a permit [*NRDC v. EPA,* 396 F. Supp. 1386, 7 ER Cases 1881 (D.D.C. 1975)]. The district court also ruled that EPA could not exempt discharges identified as point sources from regulation under the NPDES permit program.

The district court also ruled that a more complete permit program than the one proposed by EPA could still be manageable. The court recognized two alternatives for reducing the permit workload:

1. Discretion to define what constitutes a point source.
2. Discretion to use certain administrative devices, such as general permits, to help manage the workload. The court recognized that EPA has wide latitude to rank categories and subcategories of point sources of different importance and treat them differently within a permit program, so long as they were all subject to basic permit requirements.

After the district court decision was appealed, the court of appeals recognized that section 402 provides the agency with flexibility in determining the appropriate scope and form of an NPDES permit. For example, permits may regulate industry practices to lessen point source pollution problems. As a result, the court suggested using area or general permits. In certain cases, it may be appropriate for EPA to require a permittee simply to monitor and report effluent levels [568 F. 2d 1369, 1679 (1977)].

In 1984, EPA again published permit application requirements and deadlines for storm water discharges. However, these regulations were never implemented. The regulations were in litigation when Congress enacted the Water Quality Act (WQA) on February 4, 1987, which directly specified a new national strategy for storm water control.

Despite the lack of a comprehensive permitting program for all storm water discharges prior to the passage of the WQA of 1987, permitting efforts nonetheless proceeded in some areas. Between 1974 and 1982, EPA promulgated effluent limitations guidelines for storm water discharges from nine categories of industrial discharges:

1. Cement manufacturing [40 CFR part 411]
2. Feedlots [40 CFR part 412]
3. Fertilizer manufacturing [40 CFR part 418]
4. Petroleum refining [40 CFR part 419]
5. Phosphate manufacturing [40 CFR part 422]

6. Steam electric power generation [40 CFR part 423]
7. Coal mining [40 CFR part 434]
8. Ore mining and dressing [40 CFR part 440]
9. Asphalt emulsion [40 CFR part 441]

Before 1987, permitting efforts for storm water discharges focused on industrial facilities subject to these effluent limitation guidelines. In addition, some EPA regions and states with authorized state NPDES programs wrote permits for storm water discharges from other industrial facilities. For example, in some states and regions, storm water discharges from industrial facilities are often addressed when NPDES permits for process wastewaters of a facility are reissued.

Water Quality Act of 1987

Congress periodically "reauthorizes" the Clean Water Act. These reauthorizations are generally accompanied by various amendments. In 1987, Congress passed a set of amendments to the Clean Water Act. The 1987 amendments to the CWA are commonly called the *Water Quality Act of 1987*. There are several provisions to this act, including one that established the National Storm Water Program. A two-phase storm water program was established, incorporating a prioritized approach to storm water.

Phase I of the National Storm Water Program

Phase I of the NSWP provides for regulation of the following categories of storm water discharges:

1. *Discharges associated with industrial activity (including construction activity)*: This category includes hundreds of thousands of facilities located in every state and territory. The 9 categories of industrial discharges that already had permits for storm water discharges before February 4, 1987, would continue to require permit coverage.

2. *Discharges from large and medium municipal separate storm sewer systems:* A municipal separate storm sewer system is designed to carry storm water discharges only (not combined with sanitary sewage). A municipal separate storm sewer system includes "municipal streets, catch basins, curbs, gutters, ditches, man-made channels, or storm drains that discharge into the waters of the United States" [55 FR 47989 (Nov. 16, 1990)] An MS4 that serves an urban population of 250,000 or more is considered large. A medium MS4 is one that serves an urban population of 100,000 to 250,000.

3. *Discharges which the director of the NPDES program designates as contributing to a violation of a water quality standard or as a significant contributor of pollutants to the waters of the United States:* This is a general category which was included so that the EPA or state director, after proper study and notification, could require permitting of discharges other than those listed above.

Phase II of the National Storm Water Program

Under phase II of the National Storm Water Program, EPA, in consultation with the states, was required to conduct a study identifying additional sources of storm water contamination and establishing procedures and methods to control these discharges. EPA's Report to Congress (Environmental Protection Agency, 1995) presents the results of the study and discusses the nature and extent of pollutants in the discharges. The report also described various alternative approaches regulating these additional discharges.

EPA issued proposed phase II regulations on January 9, 1998. EPA has committed to finalizing the phase II rules by March 1, 1999. If no further changes are made, most phase II storm water dischargers will have to submit applications for permit coverage under these final rules by August 7, 2001 [60 FR 17950 (Apr. 7, 1995)]. Others will be required to submit applications later [63 FR 01585 (Jan. 9, 1998)].

The proposed phase II rule makes changes in all three areas of storm water regulation: municipal, construction, and industrial. The municipal storm water program would be expanded to include the following:

- All small municipal separate storm sewer systems within urbanized areas (with certain minor exceptions)
- Other small municipal separate storm sewer systems meeting EPA or state criteria for designation (possibly including any system that serves a population center of at least 10,000 people with a population density of at least 1000 people per square mile)
- Any municipal separate storm sewer system contributing substantially to the storm water pollutant loadings of a regulated, physically interconnected municipal separate storm sewer system

All regulated small municipal separate storm sewer systems would be required to develop and implement a storm water management program. Program components would include at least the following measures:

- Public education and outreach
- Public involvement

- Illicit discharge detection and elimination
- Construction site runoff control
- Postconstruction storm water management in new development and redevelopment
- Pollution prevention and good housekeeping of municipal operations

These program components would be implemented through NPDES permits.

The proposed phase II rules will address all construction site activities involving clearing, grading, and excavating land equal to or greater than 1 acre and less than 5 acres, unless requirements are otherwise waived by the NPDES permitting authority.

The proposed phase II rule would also provide relief from NPDES storm water permitting requirements for industrial and other sources that provide a written certification of "no exposure of industrial materials and activities to storm water."

Section 319 of the Clean Water Act

In 1987, section 319 was added to the Clean Water Act to provide a framework for funding state and local efforts to address pollutant sources not addressed by the NPDES program (i.e., nonpoint sources). To obtain funding, states are required to submit nonpoint source assessment reports identifying state waters that could not reasonably be expected to attain or maintain applicable water quality standards or the goals and requirements of the CWA without additional control of nonpoint sources of pollution. States are also required to prepare and submit for EPA approval a statewide nonpoint source management program for controlling nonpoint source water pollution to navigable waters within the state and improving the quality of such waters. State program submittals must identify specific best management practices and measures that the state proposes to implement in the first 4 years after program submission, to reduce pollutant loadings from identified nonpoint sources to levels required to achieve the stated water quality objectives.

State programs funded under section 319 can include both regulatory and nonregulatory state and local approaches. Section 319(b)(2)(B) specifies that a combination of "nonregulatory or regulatory programs for enforcement, technical assistance, financial assistance, education, training, technology transfer, and demonstration projects" may be used, as necessary, to achieve implementation of the BMPs or measures identified in the section 319 submittals.

Although most states have generally emphasized the use of voluntary approaches in their section 319 programs, some states and local governments have implemented regulations and policies to control pollution from urban runoff. States such as Delaware and Florida, as well as local jurisdictions such as the Lower Colorado River Authority, are pursuing storm water management goals through numerical treatment standards for new development. Many states and local governments have enforceable erosion and sediment control regulations.

On a broader scale, nonpoint source pollution is being addressed at the watershed level by such programs as those being implemented by the state of Wisconsin, the Puget Sound Water Quality Authority, and the states that are parties to the Great Lakes Water Quality Agreement. A number of individual states and local communities have adopted legislation or regulations that limit development or require special management practices in areas surrounding water resources of special concern, such as Maryland's Critical Areas Act [63 FR 01585 (Jan. 9, 1998)].

Coastal controls on nonpoint sources

Section 320 of the Clean Water Act established the National Estuary Program, which focused point and nonpoint pollution control on geographically targeted, high-priority estuarine waters. Controls are selected and implemented on a watershed basis. In addition, Congress enacted the Coastal Zone Act Reauthorization Amendments (CZARA) in 1990 to address the impact of nonpoint source pollution on coastal waters.

Section 6217 of CZARA provides that states with approved coastal zone management programs must develop and submit coastal nonpoint pollution control programs to EPA and the National Oceanic and Atmospheric Administration (NOAA) for approval. Failure to submit an approvable program will result in a reduction of federal grants under both the Coastal Zone Management Act and Section 319 of the CWA.

State coastal nonpoint pollution control programs under CZARA must include enforceable policies and mechanisms that ensure implementation of the management measures throughout the coastal management area. Section 6217(g)(5) defines management measures as "economically achievable measures for the control of the addition of pollutants from existing and new categories and classes of nonpoint sources of pollution, which reflect the greatest degree of pollutant reduction achievable through the application of the best available nonpoint pollution control practices, technologies, processes, siting criteria, operating methods, or other alternatives."

To implement CZARA requirements, Congress mandated an approach that relies on technical and economic achievability. This is fundamentally different from an approach based on the cause-and-effect relationships between particular land-use activities and specific water quality problems. Neither the states nor EPA has the money, time, or other resources to create and expeditiously implement a program that depends on establishing cause-and-effect linkages. If the mandated technology-based approach fails to achieve and maintain applicable water quality standards and to protect designated uses, CZARA Section 6217(b)(3) requires additional management measures.

EPA guidance identifies management measures for five major categories of nonpoint source pollution: agriculture, forestry, urbanization, marinas and recreational boating, and hydromodification (Environmental Protection Agency, 1993). The management measures reflect the greatest degree of pollutant reduction that is economically achievable for each of the listed sources. These management measures provide reference standards for the states to use in developing or refining their coastal nonpoint programs. In general, the management measures were written to describe systems designed to reduce the generation of pollutants. A few management measures, however, contain quantitative standards that specify pollutant loading reductions. For example, the New Development Management Measure, which is applicable to construction in urban areas, requires (1) that by design or performance the average annual total suspended solid loadings be reduced by 80 percent and (2) to the extent practicable, that the predevelopment peak runoff rate and average volume be maintained. The management measures approach was adopted to provide state officials flexibility in selecting strategies and management systems and practices that are appropriate for regional or local conditions, provided that equivalent or higher levels of pollutant control are achieved (Environmental Protection Agency, 1993).

Storm water discharges regulated under the existing NPDES program, such as discharges from municipal separate storm sewers serving a population of 100,000 or more and construction activities that disturb 5 acres or more, do not need to be addressed in coastal nonpoint pollution control programs. However, potential new sources, such as urban development adjacent to or surrounding municipal systems serving a population of 100,000 or more, smaller urbanized areas, and construction sites that disturb less than 5 acres, which are identified in management measures under Section 6217 guidance need to be addressed in coastal nonpoint pollution control programs until such discharges are issued an NPDES permit. EPA and NOAA have worked and continue to work together in their activities to ensure that authorities between NPDES and CZARA do not overlap [63 FR 01585 (Jan. 9, 1998)].

Watershed-Based Approach for Water Quality Programs

EPA is promoting an integrated watershed approach for storm water and other discharges that focuses on coordinated public and private sector efforts to address the highest-priority water quality problems within hydrologically defined geographic areas [63 FR 01595 (Jan. 9, 1998)] (Environmental Protection Agency, 1994).

The watershed-based approach is intended to provide a flexible framework for focusing and integrating current efforts and exploring innovative methods for achieving environmental objectives. The approach focuses on four major elements:

1. Identifying specific geographic locations
2. Integrating available authorities to deal with all pollution sources
3. Involving all affected parties in analyzing and creating solutions
4. Measuring effectiveness against clearly established objectives

These key elements are derived from experience gained over the past few years in many states and other EPA efforts such as the Clean Lakes and National Estuaries Programs.

NPDES Storm Water Regulations

When Congress passes a bill and it becomes law through Presidential signature or congressional override, it represents only the beginning of a long process which must take place before the program intended by Congress is fully implemented.

After an item of legislation becomes law, the Government Printing Office incorporates the new law into the *United States Code* (*U.S.C.*), which codifies (categorizes and compiles) all federal legislation. For example, the Clean Water Act and its amendments are codified beginning at 33 U.S.C. 1251.

In connection with new legislation, Congress generally specifies which government agency or department will be responsible for developing and implementing regulations to carry out the legislation. The CWA requires the Environmental Protection Agency to develop and implement regulations under the NPDES program.

The process of developing proposed regulations involves several internal steps within the EPA and coordination with the Office of Management and Budget (OMB). These steps are intended to ensure that the full intent of the congressional legislation is carried out through the regulations and that the resulting rules are as clear and fair as possible, with a minimum adverse impact on the economy.

The *Federal Register*

After the process of preliminary rule making is complete, the EPA publishes the proposed regulation in the *Federal Register,* which is printed every weekday by the Government Printing Office. It contains items published by federal government agencies, including proposed regulations, final regulations, and notices or announcements. The *Federal Register* is usually cited by using the volume number and page number. Each volume represents a separate year. For example, volume 55 represents calendar year 1990. Pages are numbered consecutively throughout the year. In this book, the date of the issue containing the citation is added for easier reference. For example, an item that begins on page 47989 of volume 55 of the *Federal Register* is cited as 55 FR 47989 (Nov. 16, 1990).

Proposed regulations generally have four parts:

1. *Preliminary material,* including the agency name, a summary of the rule, and information on submitting comments.
2. The *preamble,* which provides the background, authorization, and reasoning behind the proposed rule. It can be very useful to review the preamble thoroughly.
3. The *proposed regulation* itself.
4. *Appendixes,* such as application forms, figures, maps, and other supportive material.

A proposed regulation does not have any regulatory standing; it is intended to provide information and give the public an opportunity to comment. It is not binding. After comments have been received on the proposed regulation, the EPA may modify the proposed rule according to the comments.

The EPA also publishes final rules in the *Federal Register.* These have the force and effect of law. Final rules have the same four parts as proposed rules. However, the preamble of a final rule generally contains the EPA's response to comments received on the proposed rule. Instead of stating and responding to each comment individually, the EPA often groups the comments into categories and prints responses to the comments within each category.

In relation to final regulations, the EPA also issues permits for particular types of discharges. A permit which covers an entire class of discharges is called a *general permit.* The EPA publishes announcements of general permits in the *Federal Register,* in order to provide public notice of the general permit.

The EPA has published numerous proposed rules in connection with the NPDES storm water program. The EPA has also published announcements of some general permits in the *Federal Register*. These permits are applicable only in those states in which EPA is directly responsible for issuing permits. (The role of the states in NPDES storm water permitting is described later in this chapter.) However, they also serve as a model for general permits issued by the states.

Code of Federal Regulations

Each July 1, all final rules enacted within the previous year are collected in the *Code of Federal Regulations* (*CFR*). The CFR is divided into "titles" according to subject matter. Most EPA regulations are codified in Title 40 of the CFR, usually abbreviated as simply 40 CFR.

Each CFR title is divided into various sections. The baseline regulations relating to the NPDES program are printed in 40 CFR 122, and the specific requirements for storm water discharge permits are printed in 40 CFR 122.26. However, the EPA general permits themselves are not printed in the CFR.

Role of the court systems in environmental regulations

As indicated by the earlier discussion of district court and appeals court rulings related to storm water permitting, the process of environmental rule making is often strongly affected by litigation. Environmental and industry groups often file suit to challenge proposed or final EPA rules. For example, the Natural Resources Defense Council (NRDC), an environmental organization, filed suit soon after the EPA published the proposed rules for the National Storm Water Program in November 1990, challenging several aspects of the program as being inadequate. Another suit by the American Mining Congress, an industry organization, also challenged the regulations, but for different reasons.

At times, these suits end in settlements which result in revisions to proposed or final EPA rules. Other cases result in court judgments which may uphold the EPA regulations or which may conclude that the EPA regulations must be revised to better reflect the intent of the original legislation.

Role of the states in NPDES permitting

The Clean Water Act allows states to request EPA authorization to administer the NPDES program within their borders. The EPA must approve a state's request to operate the permit program once the EPA determines that the state has adequate legal authorities, procedures, and the ability to administer the program. The EPA is also obligated to adopt standard requirements for state NPDES programs, including guidelines on monitoring, reporting, enforcement, personnel, and funding, and to develop uniform national forms for use by both EPA and approved states. At all times following authorization, state NPDES programs must be consistent with minimum federal requirements, although the programs may always be more stringent. (However, this does not mean that state permits must be as strict as the corresponding EPA permits.)

The 1987 Water Quality Act encouraged state control of NPDES permitting even more by allowing a state to assume the responsibility of only a portion of the NPDES program. At present, more than 40 states have chosen to assume at least some storm water permitting authority. These are called *authorized states.* Within these states, all permit submittals are made to the state agency which administers and enforces the storm water program within the state. Additional nonauthorized states may gain permitting authority. In nonauthorized states, the appropriate EPA regional office is responsible for permitting and permit enforcement.

The EPA requests that authorized states prepare and periodically update state storm water permitting plans. For nonauthorized states, the EPA prepares the state storm water permitting plans. The plans include the following sections:

1. *Municipal separate storm sewer systems:* A list of municipal separate storm sewer systems serving a population of 100,000 or more within the state, a summary of the estimated pollutant loadings for each system, the status of the permits, and an outline of the major components of municipal storm water management programs, including any innovative approaches.
2. *Discharges associated with industrial activity:* The status of the baseline general permits, a list of categories of industrial facilities being considered for industry-specific permits, a description of the procedures used to identify discharges appropriate for individual permits, and a description of how municipal system permits control pollutants in storm water discharges associated with industrial activity.
3. *Water quality standards:* A list of receiving waters where regulated storm water discharges contribute to a violation of a water quality standard, a description of the procedures used to identify these waters, and a plan to evaluate improvements to water quality resulting from the controlling of storm water discharges.
4. *Unregulated storm water discharges:* A list of receiving waters where unregulated storm water discharges contribute to a violation of a water quality standard or significantly contribute pollutants to the waters of the United States and a description of the procedures used to

identify these waters. These unregulated discharges may be considered for designation under Section 402(p)(2)(E) of the CWA as needing a permit.

References

Environmental Protection Agency, *Guidance Specifying Management Measures for Sources of Nonpoint Pollution in Coastal Waters,* EPA 840-B-92-002, January 1993.

———, *Watershed Protection—NPDES Watershed Strategy,* Washington, March 1994.

———, *Storm Water Discharges Potentially Addressed by Phase II of the National Pollutant Discharge Elimination System Storm Water Program,* EPA 833-K-94-002, March 1995.

Government Printing Office: *Federal Register,* Office of the Federal Register, National Archives and Records Administration, Washington.

———: *Code of Federal Regulations,* Office of the Federal Register, National Archives and Records Administration, Washington, July 1, 1992.

———: *United States Code,* Washington, 1988.

Kovalic, Joan M.: *The Clean Water Act of 1987,* Water Environment Federation (formerly Water Pollution Control Federation), Alexandria, VA, 1987.

Chapter

2

Enforcement and Compliance with Storm Water Regulations

As stated in Chap. 1, the National Pollutant Discharge Elimination System (NPDES) permit requirement for storm water discharges began with the Clean Water Act of 1972, as amended by the Water Quality Act of 1987. The Clean Water Act provides severe penalties for those who fail to obtain permit coverage for discharges and for those who do not comply with the terms and conditions of an NPDES permit. This chapter gives some practical advice on dealing with EPA enforcement actions.

This is the second of three consecutive chapters providing basic information for all dischargers.

Agency Enforcement of the Clean Water Act

Section 309 of the Clean Water Act gives the EPA broad enforcement authority. In a state with an approved program (one with NPDES permitting and enforcement authority), the EPA notifies the state whenever a violation comes to the EPA's attention. The approved state must have the same or similar enforcement authority as the EPA has in states without approved programs. EPA retains its authority and exercises oversight of the approved state program. Operating agreements with approved states generally allow that if the state fails to take timely and appropriate action, the EPA can commence enforcement action (Stimson et al., 1993). In states with unapproved programs, the EPA regional offices are the primary enforcement agencies.

Violations may include actions that are inconsistent with provisions of the law itself (e.g., discharges without the required permit) or actions that are inconsistent with conditions of permits issued by the EPA or states under the act. Violations may be cited where the discharge or potential to discharge is to waters of the United States or in some situations to sewer systems. Violations of recordkeeping, reporting, and inspection requirements may also be cited.

EPA brings enforcement actions to require alleged violators to promptly correct the violations and remedy any harm caused by the violations. As part of an enforcement action, EPA also seeks substantial monetary penalties which promote environmental compliance and help protect public health by deterring future violations by the same violator and deterring violations by other members of the regulated community. Penalties help ensure a national level playing field by ensuring that violators do not obtain an unfair economic advantage over competitors who have done whatever was necessary to comply on time. Penalties also encourage companies to adopt pollution prevention and recycling techniques, so that companies minimize their pollutant discharges and reduce their potential liabilities.

The Clean Water Act gives to the EPA three broad categories of enforcement authority:

1. *Civil administrative* authority to issue compliance orders and levy civil penalties for alleged violators of any of the law or permit requirements.
2. *Civil judicial* authority, to cooperate with the U.S. Department of Justice to file a civil action in federal district court for appropriate relief, including temporary or permanent injunctions. The court may also impose civil penalties.
3. *Criminal* authority to prosecute, through the U.S. Department of Justice, three classes of criminal violations: negligent, knowing, and knowing endangerment. Making a false statement, failing to report information, and tampering with monitoring equipment are also prosecuted as criminal offenses.

All Clean Water Act enforcement actions are subject to a 5-year statute of limitations. Therefore, the EPA must initiate enforcement procedures within 5 years of the date of violation.

The law also provides for citizens' suits to enforce the provisions of the Clean Water Act, as described later in this chapter.

Administrative enforcement actions

The EPA's administrative compliance orders are designed simply to give notice of violations and achieve compliance without court involvement. In addition to compliance orders, administrative penalties may be levied through issuance of an administrative complaint. Figure 2.1 illustrates the typical steps in an administrative enforcement procedure.

Class I penalties are limited under Section 309(g)(2)(A) of the Clean Water Act to $11,000 per violation or $27,500 total.[1] To propose a class I penalty, the EPA issues an administrative complaint to the violator, giving notice of the violations alleged and the amount of the penalty proposed. The violator has 30 days to request a hearing and to directly contest any allegation the violator intends to raise as an issue in a hearing. Hearings for class I penalties are conducted by a regional judicial officer acting on the authority of the regional administrator. They are generally less formal than hearings before administrative law judges.

Class II penalties may be up to $11,000 per day with a $137,500 maximum under Section 309(g)(2)(B) of the Act. As with class I actions, the EPA issues an administrative complaint proposing to assess the penalty. Alleged violators are entitled to a formal administrative hearing, but the alleged violator must request the hearing within 30 days and must specifically contest the allegations. An administrative law judge presides over the hearing, which must conform with Section 554 of Title 5, *United States Code.* Class II hearings are courtroom proceedings conducted under 40 CFR part 22.

When an administrative complaint is issued, public notice must be given in the area of the facility where the violations are alleged to have occurred. The public has 30 days to comment on the proposal to assess the penalty in order to establish standing to participate in the action. Public participation in a penalty action can include the right to review the terms of any informal settlement, or to participate in a hearing and present relevant evidence. A person who establishes standing by a timely comment on the complaint may be able to petition the EPA to set aside a proposed settlement in favor of a hearing.

Administrative hearings generally result in the issuance of an order assessing an administrative penalty. There is no authority to order injunctive relief in an administrative penalty action under the Clean Water Act. Any other compliance issues must be taken care of by agreement or by a unilateral administrative order under Section 309(a) of the act, or must be referred for a civil judicial action for injunction.

---

[1]Until January 30, 1997, these penalties were limited to $10,000 per violation or $25,000 total. For violations that occur on or after January 31, 1997, the penalties are 10% higher, as described in EPA's Civil Monetary Penalty Inflation Rule [61 FR 69359; 62 FR 13514]. Under the Debt Collections Improvement Act of 1996, [31 U.S.C. 3701 *et seq.*], the EPA must review all administrative and judicial penalties at least once every 4 years, and it is reasonable to expect these penalty amounts to increase somewhat with each quadrennial review.

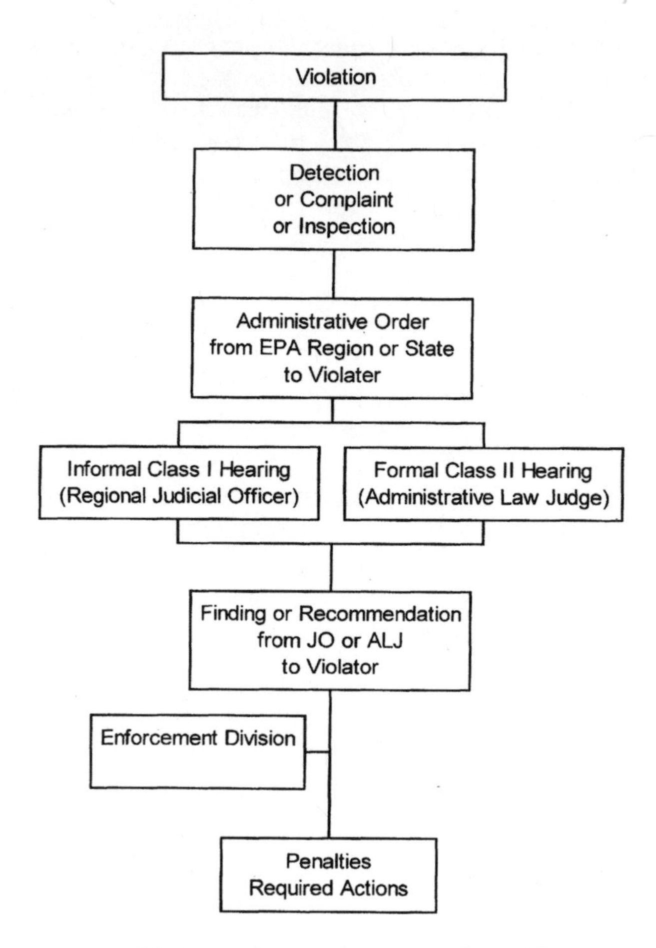

Figure 2.1 Administrative compliance procedures. JO = regional judicial officer; ALJ = administrative law judge.

An order assessing a penalty may be appealed within the EPA. These appeals are ultimately decided by the Environmental Appeals Board (EAB), a three-person panel appointed by the EPA administrator in Washington. The EAB hears appeals of permit and penalty decisions for and in lieu of the EPA administrator.

Within 30 days of the issuance by EPA of a final order assessing a penalty, the violator can petition the federal court system to review the order. For class I penalties, the order is reviewed by a federal district court. For class II penalties, the order is reviewed by a federal court of appeals. Any person who submitted a comment on the proposed penalty and participated in a hearing also has the right to request a federal court review of the final order or to bring a citizen's suit directly against the violator. This could include an environmental group that feels that stronger compliance activities or penalties are required.

Civil enforcement actions

In addition to administrative penalties, the CWA allows the EPA to file civil cases in the district court to enforce the law and assess penalties to violators. Parties that violate the law or permit conditions are subject to court-imposed civil penalties of up to $27,500 per day for each violation. These penalties can accrue until each violation is corrected, with the penalty limited only by the number of violations and the number of days each violation continues, up to the 5-year statute of limitations. Each day that a violation occurs is considered to be a separate violation for the purpose of computing maximum penalties.

Civil judicial actions may be initiated without any administrative orders being issued. As with an administrative penalty action, to successfully prosecute a civil judicial action, the EPA must show only that the law, the applicable EPA regulations, or the terms and conditions of an NPDES permit are being violated. The EPA does not have to prove malice or negligence on the part of the violator. But in a judicial action, the EPA may seek temporary or permanent injunctions in addition to fines, or it may seek court-ordered compliance actions and schedules.

If the violator is a municipality, the state will be named as a defendant as well. The state may be liable for penalties that cannot be paid by the municipality.

Criminal enforcement actions

The most serious violations of the Clean Water Act involve criminal penalties, all of which would be imposed by federal courts. Criminal cases are prosecuted by the Department of Justice, usually after referral from the EPA, although prosecution may result from other sources of information. Criminal penalties may be invoked for negligent violations, knowing violations, knowing endangerment, making false statements, or tampering with monitoring equipment.

These criminal penalties can include jail sentences as well as fines. If the violator is a corporation, the corporation may be fined while the persons holding responsible positions within the corporation may face fines and/or imprisonment. The persons prosecuted may include not only those with hands-on compliance responsibilities but also those who manage or supervise such activities.

Negligent violations are punishable by a fine of not less than $2500 and not more than $25,000 per day of violation and by imprisonment for not more than 1 year. Fines and prison sentences may be twice as high for repeat offenses.

Knowing violations carry penalties of $5000 to $50,000 per day of violation and up to 3 years' imprisonment. Again, courts may double these penalties for repeat violations.

Knowing endangerment occurs when a person commits a knowing violation while also knowing that the discharge places another person in imminent danger of death or serious bodily injury. These serious violations are punishable by fines of up to $250,000 per day and up to 15 years in prison. If the violator is a corporation, the fine may be increased to $1 million per day. Repeat offenses are subject to doubled fines and times of imprisonment.

False statements

The CWA creates a separate class of criminal penalties for parties that file false statements or tamper with monitoring equipment. Violations are punishable by fines of up to $10,000 and jail sentences up to 2 years, with penalties doubled for repeat offenses. The EPA is especially vigorous in prosecuting those who interfere with the flow of information within the NPDES reporting structure. The entire NPDES depends strongly on dischargers obtaining and reporting accurate and complete data. Persons who deliberately withhold or falsify information are a direct threat to the success of such a self-reporting system. Consequently, the EPA is very active in pursuing such cases. The CWA equates withholding of information to falsifying of information.

It is not advisable at any time to falsify or withhold information from the EPA or state regulators. The possible criminal penalties resulting from falsifying or withholding information are likely to be much more severe than the civil penalties resulting from honestly reporting the true facts of a violation, however unfavorable those facts may be.

Determining the actual amount of penalties

The sections above indicate what the *maximum* fine should be for a violation of the CWA. These maximum fines can be quite high. For example, a fine of $27,500 per day will accrue at a rate of about $825,000 per month or more than $10 million per year. Given that the EPA often does not seek enforcement action until a violation has continued for 90 days or more, the potential fine is often $2 million or more before the violator begins to discuss the problem with the EPA.

Simply because the law allows huge penalties does not mean that such penalties are justified in a particular case. The law also requires the EPA (or a civil court) to "take into account the nature, circumstances, extent and gravity of the violation, or violations..." and to consider the violator's "...ability to pay, any prior history of such violations, the degree of culpability, economic benefit or savings (if any) resulting from the violation..." when determining the amount of any penalties [309(g)(3), 33 U.S.C. 1319(g)(3); and 33 U.S.C. 1319].

Whenever the EPA prepares to initiate a civil penalty proceeding, it begins to prepare estimates of the amount of penalties that are appropriate for the violation. These estimates are adjusted as the facts of the case become clearer. It is important to understand that the estimate is not a single number; it is a range that extends from a "floor" value up to a "ceiling" value. The ceiling value would be the maximum statutory penalty, computed by using the most aggressive interpretation of the Clean Water Act and its regulations. This is generally the amount of penalty that EPA will demand in a pleading, or in an administrative or judicial hearing.

The floor value, or the minimum acceptable level of penalty that EPA might be willing to settle for, may be "orders of magnitude" less than the demand amount in civil judicial actions (Environmental Protection Agency, 1995). Therefore, it is extremely important to understand how this "floor" penalty amount is estimated, and the remainder of this section contains a fairly detailed description.

Recomputing the maximum statutory penalty

The first step in computing the minimum acceptable penalty is for EPA staff members to recompute the maximum statutory penalty, using a less aggressive interpretation of the CWA and its regulations. This will generally lead to an amount that is considerably lower than the amount demanded in pleadings and in hearings. *However, this is the amount that EPA will seek to collect through a settlement, if possible.*

The following list provides several examples of types of violations, along with the computed value of the maximum statutory penalty (Environmental Protection Agency, 1995). This list generally reflects a "settlement" rather than a "litigation" approach toward penalty computation:

- Violation of daily maximum limit for pollutant A, on the 5th of January. Maximum statutory penalty: $27,500.
- Violation of daily maximum limit for pollutant A, on the 5th, 10th, and 15th of January. Maximum statutory penalty: $82,500.
- Violation of daily maximum limits for each of pollutants A and B, on the 5th of January. Maximum statutory penalty: $55,000.
- Violation in January of weekly average for pollutant A. Maximum statutory penalty: $27,500 per day, multiplied by 7 days is $192,500.

- Violation in January of monthly average limit for pollutant A. Maximum statutory penalty: $27,500 per day, multiplied by 31 days in January, is $852,500.
- Violation in January of monthly average limit for pollutant A, in which there is evidence that there were no discharges on 4 days (e.g., plant shut down on Sundays). Maximum statutory penalty: $27,500 per day, multiplied by 27 days in January, is $742,500.
- Violation in January of monthly average limits for both pollutants A and B. Maximum statutory penalty: $55,000 per day, multiplied by 31 days in January, is $1,705,000.
- Violation in January of monthly average limit for pollutant A, and of daily maximum limit for pollutant B on January 5th and 15th. Maximum statutory penalty: $852,500 for pollutant A plus $55,000 ($27,500 per day × 2) for pollutant B equals $907,500.
- Violation in January of monthly average limit for pollutant A, and of daily maximum limit for pollutant A on January 5th and 15th. Maximum statutory penalty: $27,500 per day, multiplied by 31 days in January, is $782,500.
- Failure to properly monitor for pollutant A on 4 required days in January.[2] Maximum statutory penalty: $110,000.
- Failure to properly monitor for pollutants A, B, and C on January 15. Maximum statutory penalty: $82,500.
- Failure to monitor for a monthly pollutant parameter. Maximum statutory penalty: $27,500 for each day in which the discharger was required to monitor for that pollutant.
- Failure to submit adequate discharge monitoring report on time (each failure to monitor for a particular pollutant is subject to a separate penalty calculation), based on maximum statutory penalty: $27,500.
- Failure to submit a report or other document on time: $27,500. (Each failure to complete an activity covered by the report in a timely manner is subject to a separate penalty calculation.)

As these examples illustrate, the statutory maximum penalty for violations of an effluent limit for a period longer than 1 day includes a separate penalty for each day in the time period (assuming there was a discharge on each day).

---

[2] For the purposes of calculating penalties, the act of monitoring for a particular pollutant includes the sequence of events starting with the collection of the sample through completion of the analytical testing of the sample. The obligation to report the results of the monitoring is a separate act subject to a separate penalty calculation.

Computing the minimum acceptable penalty

After preparing a revised estimate of the maximum statutory penalty, the EPA begins to "take into account the nature, circumstances, extent and gravity of the violation, or violations..." and to consider the violator's "...ability to pay, any prior history of such violations, the degree of culpability, economic benefit or savings (if any) resulting from the violation..." as required by law [309(g)(3), 33 U.S.C. 1319(g)(3)].

In carrying out these computations, the EPA has developed and documented a policy that is intended to accomplish four important environmental goals:

1. Penalties should help create a level playing field by ensuring that violators do not obtain an economic advantage over their competitors. Therefore, penalties should be large enough to recover the economic benefit of noncompliance.
2. Penalties should be large enough to deter noncompliance. Therefore, they should include an appropriate *gravity amount* in addition to the economic benefit of noncompliance.
3. CWA penalties should be generally consistent across the country. This is desirable as it not only prevents the creation of "pollution havens" in different parts of the nation, but also provides fair and equitable treatment to the regulated community wherever it may operate.
4. Settlement penalties should be based on a logical calculation methodology to promote swift resolution of enforcement actions and the underlying violations. In other words, EPA is generally motivated to reach a settlement, and it is usually in the best interests of everyone involved to do so, if an equitable settlement can be negotiated.

The EPA policy applies to settlement of civil judicial and administrative penalties sought under Section 309 of the CWA, including violations of NPDES permit limits and conditions; discharges without an NPDES permit; and violations of Section 308 information requests. This policy does not apply to actions brought exclusively under Section 311 (oil and hazardous substance spills) or for violations of requirements in Section 404 ("wetlands" cases involving disposal of dredged or fill material). Separate penalty policies apply to these two types of cases. The policy also does not apply to the assessment of penalties by a judge after a hearing on the merits, either administrative or judicial.

Even though the EPA has developed a detailed methodology for computation of minimum penalties, it recognizes that the methodology is not appropriate for all cases. In particular, the EPA policy is written to reflect wastewater discharge violations rather than storm water discharge violations. However, the primary elements of the policy are still applicable to storm water violations, and the policy is used for classification and settlement of storm water cases. It is also important to understand that EPA staff members must receive the advance approval of the assistant administrator before applying a different method. Therefore, the methodology already established by EPA will be used until a different method or policy is approved and promulgated.

The "bottom-line" settlement penalty that EPA computes will determine whether EPA pursues an administrative action or files a civil judicial action. As noted earlier, the maximum penalty that EPA can collect in an administrative hearing is $137,500; if the bottom-line penalty is greater than that amount, then a civil judicial action is required.

The bottom-line settlement penalty is calculated as follows:

Economic Benefit of Noncompliance

+	Gravity of the violation
±	Gravity adjustment factors
−	Litigation considerations
−	Ability to pay
−	Supplemental environmental projects
=	**Minimum penalty acceptable for settlement**

Each component of the penalty calculation is discussed below. A worksheet summarizing the penalty calculation is included.

Economic benefit of noncompliance

In general, the economic benefit component is the most difficult to avoid in any settlement with EPA. The EPA is required to make every effort to calculate and recover the economic benefit of noncompliance. It is assumed that most violations occur because dischargers are unaware of the actions or requirements necessary to comply with the law in a timely manner. No proof of intent or knowledge is required. The fact that money was not spent on CWA compliance represents an *economic benefit* to the discharger which EPA is required to estimate and eliminate, if possible. If it were shown that a person knew of the requirement and refused to spend the money to comply, it is more likely that the violations would be pursued as a criminal action than a civil action.

Commonly delayed and avoided CWA pollution control expenditures include

- Monitoring and reporting (including costs of the sampling and proper laboratory analysis)
- Capital equipment improvements or repairs, including engineering design, purchase, installation, and replacement

- Operation and maintenance expenses (e.g., labor, power, chemicals) and other annual expenses
- One-time acquisitions (such as land purchase)

The EPA has developed a computer program called BEN for calculating the economic benefit from delayed or avoided compliance expenditures (Environmental Protection Agency, 1993). The benefit should be calculated from the first date of noncompliance, but EPA generally does not go back more than 5 years prior to the date when the complaint is filed.

In some cases, the standard EPA methods may not be appropriate for estimating economic benefit or will not capture the full scope of the economic benefit. If the violator decides that its "method of compliance" is to cease operations at the facility, conducting a benefit analysis may be complicated.[3]

Gravity of the violation

Gravity-based penalties represent the "seriousness" or "punitive" portion of penalties over and above the portion representing the economic gain from noncompliance. The EPA is required to make every reasonable effort to calculate and recover a gravity component in addition to the economic benefit component. The gravity component of the penalty is calculated for each month in which there was a violation. The total gravity component for the penalty calculation equals the sum of each monthly gravity component. The monthly gravity formula is

$$\text{Monthly gravity component} = (1 + A + B + C + D) \times \$1000$$

where A = a factor that accounts for the significance of the violation, generally ranging from 0 to 20 per month

B = a factor that accounts for the health and environmental harm caused by the violation, generally ranging from 0 to 50 per month

C = a factor that accounts for the number of effluent limit violations, generally ranging from 0 to 5 per month

D = a factor that accounts for the significance of noneffluent limit violations, generally ranging from 0 to 70 per month

---

[3] In cases where a facility determines that it can comply only by ceasing operations, the EPA will generally consider the savings obtained from the delayed closure costs and the avoided costs of not implementing the appropriate pollution prevention measures during the period of noncompliance.

Therefore, the total range of the monthly gravity component could be from $1000 to as much as $146,000. Note that many storm water discharges are not subject to effluent limit violations. Therefore, some of the gravity factors (especially *C*) may not apply to some storm water discharges.

The four gravity factors *A, B, C,* and *D* are considered for each month in which there were one or more violations. In performing the gravity calculation, the monthly gravity component is calculated from the first date of noncompliance up to when the violations ceased or the date the complaint is expected to be filed; but EPA has the option to start the gravity calculation no more than 5 years prior to the date when the complaint should be filed.

Factor *A—significance of violation.* This factor is based on the degree of exceedance of the most significant effluent limit violation in each month.

If there were no effluent limit violations in a particular month but there were other violations, then gravity factor *A* is assigned a value of 0 in that month's gravity calculation. If the discharger finds an effluent violation, but did not notify the control authority and repeat the sampling afterward as required by 40 CFR 403.12(g)(2), then an appropriate value for gravity factor *D* should be assigned for the notification or monitoring violation(s), as described later in this chapter. For a storm water discharge permit that does not include numeric effluent limits, this factor would probably be given a value of 0.

Toxic pollutants are weighted most heavily in the determination of gravity factor *A*. For the purposes of this computation, a toxic pollutant is any pollutant not specifically listed in connection with Tables 2.2, 2.3, or 2.4 below.

Table 2.1 lists the range of gravity factor values appropriate for different levels of exceedance of the effluent limits for toxic pollutants specified in the discharge permit. All violations are considered, but the violation that yields the highest gravity factor controls the computation. For example, if a particular discharge exceeded the effluent limit by 50% for the

TABLE 2.1 Selection of Gravity Factor *A* for Violations of Effluent Limits for Toxic Pollutants

Monthly average exceedance, %	7-Day average exceedance, %	Daily maximum exceedance, %	Factor *A* value ranges
1–20	1–30	1–50	1–3
21–40	31–60	51–100	1–4
41–100	61–150	101–200	3–7
101–300	151–450	201–600	5–15
301+	451+	601+	10–20

TABLE 2.2 Selection of Gravity Factor *A* for Violations of Effluent Limits for Conventional and Nonconventional Pollutants

Monthly average exceedance, %	7-Day average exceedance, %	Daily maximum exceedance, %	Factor *A* value ranges
1–20	1–30	1–50	0–2
21–40	31–60	51–100	1–3
41–100	61–150	101–200	2–5
101–300	151–450	201–600	3–6
301+	451+	601+	5–15

monthly average, by 60% for the 7-day average, and by 70% for the daily maximum value, the gravity factor would be controlled by the monthly average. A monthly average exceedance of 50% yields a gravity factor of 3 to 7, whereas a 7-day average exceedance of 60% and a daily maximum exceedance of 70% would each yield a gravity factor of only 1 to 4.

Once the range of factors has been determined from the table, the EPA staff members within each regional office may select a particular value for factor *A* within the designated range.

Table 2.2 shows the range of values for gravity factor *A* for conventional and nonconventional pollutants, including *biochemical oxygen demand* (*BOD*), *chemical oxygen demand* (*COD*), total oxygen demand, dissolved oxygen, total organic carbon, *total suspended solids* (*TSS*), *total dissolved solids* (*TDS*), inorganic phosphorous compounds, inorganic nitrogen compounds, oil and grease, calcium, chloride, fluoride, magnesium, sodium, potassium, sulfur, sulfate, total alkalinity, total hardness, aluminum, cobalt, iron, vanadium, and temperature.

As was discussed in connection with toxic pollutants, the gravity factor for conventional and nonconventional pollutants is determined by comparing the maximum gravity factor values determined for different types of exceedance. The highest individual gravity factor is then used.

Tables 2.3 and 2.4 are used to determine the gravity factor *A* values for exceeding numeric effluent limits for fecal coliform and pH, respectively.

After independent determination of the gravity factor *A* for toxic pollutants, conventional and nonconventional pollutants, fecal coliform,

TABLE 2.3 Selection of Gravity Factor *A* for Violations of Fecal Coliform

Exceedance of fecal coliform limit, %	Factor *A* value ranges
0–100	0–5
101–500	2–8
501–5000	4–10
5001+	6–12

TABLE 2.4 Selection of Gravity Factor *A* for Violations of pH Limits

Standard units above or below pH limit	Factor *A* value ranges
0–0.50	0–5
0.51–2.0	2–8
2.01–3.0	4–10
3.01–4.0	6–12
4.01+	8–15

and pH (as applicable), these values should be compared. The final selection of gravity factor *A* will be made from the highest range of values determined by these individual determinations. This determination is repeated for each month in which there was a numeric effluent limit exceedance.

Factor *B*—health and environmental harm. A value for this factor is selected for each month in which one or more violations present actual or potential harm to human health or to the environment. Values are selected by using Table 2.5 based on the type of actual or potential harm that yields the highest factor value.

Factor *C*—number of effluent limit violations. This factor is based on the total number of effluent limit violations each month.

In order to properly quantify the gravity of the violations, all effluent limit violations are considered and evaluated. Violations of different parameters at the same outfall are counted separately, and violations of the same parameter at different outfalls are counted separately. However, unlike the calculation of statutory maximum penalty, each day of exceedance is not counted as a separate violation. For example, the violation of a weekly limit is considered as one violation for the purposes of determining gravity factor *C*, but is considered as 7 days of violation for the purpose of estimating the statutory maximum penalty.

The gravity factor *C* may be adjusted to fit within a typical range of 0 to 5. For example, if all the numeric effluent limits within the permit are exceeded for a particular month, then the gravity factor *C* is given a value of 5 for that month. This is true for permits that have only two numeric effluent limits and for permits that have 15 numeric effluent limits. If about 50% of the limits in the permit are violated in a particular month, then the gravity factor *C* is about 2 or 3. A minimum factor *C* value of 1 is generally appropriate whenever there are violations of two or more different pollutants.

It is important to note that violations of interim limitations in administrative orders are not counted as part of the number of effluent limit violations. However, they are considered later as a part of the "recalcitrance" factor that is used to adjust the gravity component.

TABLE 2.5 Selection of Gravity Factor *B* for Health or Environmental Harm

Type of actual or potential harm	Factor *B* value ranges
Impact on human health (e.g., interference with drinking water supplies, harm or increased risks to subsistence fishing)	10–50
Impact on aquatic environment:	
Water quality-based effluent standard(s) or whole effluent toxicity limit violated	1–10
Fish kill, beach closing, restrictions on use of water body	4–50
Other impact on aquatic environment	2–25

Factor *D—significance of noneffluent limit violations.* This factor has a value ranging from 0 to 70 and is based on the severity and number of the different types of noneffluent limitation requirements violated each month:

- Monitoring and reporting
- Unauthorized discharges
- Permit milestone schedules
- Other types of noneffluent violations

The value for factor *D* for each noneffluent limit violation is selected from Table 2.6 or Table 2.7. Since many storm water discharge permits do not include numeric effluent limitations, Table 2.6 may not be applicable.

TABLE 2.6 Selection of Gravity Factor *D* for Effluent Monitoring and Reporting Violations

Extent of violation	Factor *D* value ranges
Failure to conduct or submit adequate pollutant sampling data for one or more pollutant parameters (but not all parameters)	1–6
Failure to conduct or submit any required pollutant sampling data in a given month but with a reasonable belief that the facility was in compliance with applicable limits	2–6
Failure to conduct or submit any required pollutant sampling data in a given month without a reasonable basis to believe that facility was otherwise in compliance with applicable limits	6–10
Failure to conduct or submit whole effluent toxicity sampling data	4–10
Delay in submitting sampling data	0–5
Failure to sample again after finding a violation [40 CFR 403.12(g)(2)]	2–8
Any other monitoring or reporting violation	0–10

TABLE 2.7 Selection of Gravity Factor *D* for Other Nonefluent Limit Violations

Type and extent of violation	Factor *D* value ranges
Unauthorized discharge, e.g., discharge through an unpermitted outfall, discharge of a wastestream not identified in the permit, sewer overflows, or spill (other than oil or section 311 hazardous substance)	1–20
Violation of permit milestone schedule	1–10
Any other type of noneffluent limit violation	0–10

The failure to submit a monitoring report in a timely manner should generally not be treated as a continuing violation past the month in which the report is due. If the violator fails to submit the report at all, a factor *D* value of 5 or greater may be appropriate for this violation.[4]

The factor *D* value for a given month is the sum of the highest value for each type of noneffluent limit violation. Table 2.6 deals with a single type of violation: a monitoring or reporting violation. Therefore, if more than one value for gravity factor *D* may be determined by using Table 2.6, then only the highest of these individual values is used in the final computation. However, each of the violations listed in Tables 2.6 and 2.7 is a separate type of computation. Therefore, if more than one value for gravity factor *D* may be determined by using Table 2.6 or Table 2.7, then the sum of all these individual values is used in the final computation.

As an example of calculating factor *D* for a given month, assume a discharger did not sample for 4 of the 8 parameters in the permit, the discharge monitoring report was submitted 20 days late, and there were several days of discharge through an unauthorized outfall. From Table 2.6, a gravity factor *D* value of 4 (from a tabulated range of 1 to 6) may be selected based on the failure to conduct sampling for one-half of the parameters. The delay in submitting sampling data is not considered since it produces a gravity factor *D* range of 0 to 5, which is less than the range of 1 to 6 indicated for the other violation.

For the unauthorized discharge, a value of 6 (from a tabulated range of 1 to 20) may be selected from Table 2.7. None of the other types of violations listed in Table 2.7 is applicable to this discharge. Therefore, the total value for gravity factor *D* for this month is computed as the highest value determined from Table 2.6 plus the sum of the values determined from Table 2.7, or $4 + 6 = 10$.

[4]The failure to provide required sampling data on the discharge is a very serious violation as this eliminates the government's ability to perform necessary oversight and allows the discharger to avoid the possible application of gravity factor *A*.

Gravity adjustment

In certain circumstances, the total monthly gravity amount may be adjusted by three factors:

Total Monthly Gravity Amount	
−	Flow reduction adjustment
+	Recalcitrance adjustment
−	Quick-settlement adjustment
−	Environmental auditing adjustment
=	**Adjusted monthly gravity amount**

Flow reduction adjustment for small facilities. The total monthly gravity amount may be reduced based on the flow of the facility. This factor is applicable to direct and indirect discharges, both municipal and nonmunicipal. This gravity reduction does not apply to nonmunicipals if the facility or parent corporation employs more than 100 individuals. Table 2.8 lists the appropriate values for the flow reduction adjustment.

Since the flow reduction factor was obviously developed for wastewater flows, its applicability to storm water discharges is unclear. As noted above, it is not available to private organizations that employ more than 100 people in any event. However, it may be worthwhile for a small organization that has a small site (and therefore probably has relatively small storm water discharge volumes) to try to negotiate a flow reduction factor with EPA representatives.

History of recalcitrance adjustment. The *recalcitrance* adjustment is used to increase the penalty based on a violator's bad faith, or unjustified delay in preventing, mitigating, or remedying the violation. Recalcitrance is also present if a violator failed to comply with an EPA-issued administrative compliance order or a Section 308 information request, or with a prior state or local enforcement order.

This adjustment is applied by multiplying the total gravity component by a percentage between 0 and 150. In administrative penalty

TABLE 2.8 Selection of Flow Reduction Adjustment

Average daily wastewater discharge flow, gal/day	Percentage reduction of total gravity factor
Less than 5,000	50
Between 5,000 and 9,999	40
Between 10,000 and 19,999	30
Between 20,000 and 29,999	20
Between 30,000 and 49,999	10
Between 50,000 and 99,999	5
100,000 and above	0 (i.e., no reduction)

actions, violations of administrative compliance orders are not included in the recalcitrance calculation (because EPA lacks the authority to seek penalties in the administrative forum for violations of administrative compliance orders).

A minimum recalcitrance adjustment of 10% is generally appropriate for each instance in which a violator fails to substantially comply in a timely manner with an *administrative* (compliance) *order* (*AO*), a Section 308 information request, or a state enforcement order. Thus, if a particular discharger violated three AOs, a minimum recalcitrance adjustment of 30% is generally appropriate. If a violator completely fails to comply with an AO or Section 308 request, a recalcitrance adjustment of 20% may be appropriate for that failure, while if there were only minor violations of the AO or request, a recalcitrance adjustment of 5% may be appropriate for that violation.

The efforts of the violator to achieve compliance or minimize the violations after EPA, a state, or pretreatment control authority has initiated an enforcement action (i.e., an administrative or judicial enforcement action) do not constitute good-faith efforts. If such efforts are undertaken before the regulatory EPA initiates an enforcement response, the settlement penalty calculation already includes such efforts through a potentially smaller economic benefit amount, a shorter or less serious gravity component, or a lack of any recalcitrance. The efforts may also be reflected in the environmental auditing adjustment, as described later in this chapter.

The EPA penalty policy assumes all members of the regulated community will make good-faith efforts both to achieve compliance and to remedy violations when they occur; consequently the settlement penalty calculation begins at zero and builds upward, with no reductions for good faith. In contrast, the absence of good-faith efforts provides the basis for increasing the penalty through use of the recalcitrance factor.

Quick-settlement adjustment. To provide an extra incentive for violators to negotiate quickly and reasonably, and in recognition of a violator's cooperation, EPA may reduce the gravity amount by 10% if EPA expects the violator to settle quickly.

Table 2.9 lists the time periods during which a settlement must be reached in order to qualify for this adjustment factor. An administrative consent order resolving the violations must be signed in order to receive the quick-settlement adjustment. The time periods listed in this table begin when either of the following events occurs: The government issues a complaint alleging the violation, or the government first sends the violator a written offer to settle the case.

Environmental auditing adjustment. The EPA has issued its final audit/self-policing policy, which took effect on January 22, 1996 [60 FR

TABLE 2.9 Time Available for Quick-Settlement Adjustment

Type of enforcement action	Time period
Class I administrative	4 months
Class II administrative	6 months
Judicial (civil court)	12 months

66706, Dec. 22, 1995]. This policy strongly encourages facilities to discover and disclose any violations before the violations become the basis for enforcement actions. EPA will not seek gravity-based penalties and will not recommend criminal prosecution for companies that meet all the requirements of the audit/self-policing policy. Therefore, this adjustment can completely offset all gravity penalty amounts. The environmental audit/self-policing policy requires the following:

- The violation must have been discovered through the discharger's own systematic procedures or practices, such as a regularly scheduled environmental audit.[5]
- The violation must have been identified voluntarily, and not as a result of any legally prescribed process.
- The violation must have been disclosed to the EPA promptly (within 10 days or less).
- The violation and disclosure must be independent of any third-party action (i.e., citizens' suits, local government regulations, etc.).
- The violation must be corrected within 60 days, and any environmental harm must be remediated.
- The violation will be prevented from recurring by taking appropriate corrective and preventive actions.
- The discharger cooperates fully with EPA in any investigations.

The policy cannot be used for

- Repeated violations
- Violations that result in serious actual harm
- Violations that may present an imminent and substantial endangerment

[5]If the violation is not discovered through a regularly scheduled environmental audit, but through some other inspection or review, the gravity penalty cannot be reduced by 100%. However, the EPA will still provide a 75% reduction in the gravity penalty if all the other conditions listed in this section are fully met.

Corporations remain criminally liable for violations resulting from conscious disregard of their legal duties, and individuals remain liable for criminal wrongdoing. EPA retains discretion to recover the economic benefit gained as a result of noncompliance, so that companies will not be able to obtain an economic advantage over their competitors by delaying investment in compliance.

Litigation considerations (to decrease preliminary penalty amount)

The EPA and the Department of Justice (if applicable) will evaluate every penalty with respect to the possible success of the case in a trial or hearing. Many enforcement cases may have mitigating factors, weaknesses, or equitable problems that could be expected to persuade a court to assess a penalty less than the statutory maximum amount.

The EPA may not be willing to substantially reduce the amount of a computed bottom-line settlement amount simply because there are some weaknesses in the case, however. Some of these weaknesses may have already been accounted for in the process of computing the settlement amount. For example, the gravity calculation will be less in those circumstances in which the period of violation was brief, the exceedances of the limitations were small, the pollutants were not toxic, or there was no evidence of environmental harm. The economic benefit calculation also will be smaller when the violator has already returned to compliance, because the period of violation will be shorter. Therefore, the EPA would not consider these to be litigation considerations.

Litigation considerations should include factors that have not been accounted for by the computed settlement amount. The goal is to try to arrive at an estimate of what penalty the court might assess at trial given the particular strengths and weaknesses of the case. EPA's considerations of litigation risk are for internal purposes only and will not be revealed.

Case law indicates that a court begins with the statutory maximum as its preliminary penalty figure and then reduces that amount, as appropriate, using only the penalty assessment factors in Section 309(d) or 309(g)(3) of the act. (See *Atlantic States Legal Foundation v. Tyson Foods* [897 F.2d 1128 (11th Cir. 1990)].) In contrast, settlement penalties calculated using the EPA policy described in this chapter represent the bottom-line negotiating position which is built upward from zero, generally ending up with a figure orders of magnitude less than the statutory maximum penalty.

The EPA may consider the following factors in evaluating whether the preliminary settlement penalty exceeds the penalty the EPA would likely obtain at trial:

a. Known problems with the reliability or admissibility of the government's evidence proving liability or supporting a civil penalty
b. The credibility, reliability, and availability of witnesses
c. The informed, expressed opinion of the judge or mediator assigned to the case, after evaluating the merits of the case
d. The record of the judge in any other environmental enforcement case presenting similar issues
e. Statements made by federal, state, or local regulators that may allow the respondent or defendant to credibly argue that it believed it was complying with the federal law under which EPA is seeking penalties
f. The payment by the defendant of civil penalties for the same violations in a case brought by another plaintiff[6]
g. The development of new, relevant case law
h. A blend of troublesome facts and weak legal arguments such that the EPA faces a significant risk of obtaining a nationally significant negative precedent at trial

EPA penalty policy includes specific instructions *not* to consider the following factors as litigation considerations for the purpose of decreasing the amount of penalty:

a. A generalized goal to avoid litigation, or to avoid setting new legal precedents, or to reduce litigation costs.
b. Elements already included or assumed elsewhere in the calculation of penalty. These would include inability to pay, good faith, lack of recalcitrance, or a lack of demonstrated environmental harm. EPA's policy expects good-faith efforts, and does not provide additional rewards beyond those already described in connection with gravity factor *D*.
c. The general reputation or demeanor of the judge.
d. Off-the-record statements by the court, before it has had a chance to evaluate the specific merits of the case. Sometimes, such statements are intended to encourage settlements.

---

[6]If the defendant has previously paid civil penalties for the same violations to another plaintiff, this factor may be used to reduce the amount of the settlement penalty by no more than the amount previously paid for the same violations. [If the previous plaintiff was a state qualified to preempt federal enforcement under EPA's interpretation of Section 309(g)(6), EPA's complaint should not include counts already addressed by a penalty.]

e. The fact that the receiving water is already polluted or that the water can assimilate additional pollution.[7]

f. The failure of a regulatory agency to initiate a timely enforcement action.[8]

g. EPA regions may reduce the preliminary penalty amount for litigation considerations for up to one-third of the adjusted gravity amount without headquarters approval (where such approval would otherwise be required).

In those cases against a municipality or other public entity (such as a sewer authority) in which the entity has failed to comply with the Clean Water Act but nevertheless did make good-faith efforts to comply, the EPA may mitigate the preliminary penalty amount. The EPA has developed detailed tables specifying the degree of reduction in penalties, based on the population of the municipality, the number of months the violations continued, and the estimated economic benefit.

The EPA is under no obligation to reduce the amount of penalties in all municipal cases. This discretionary reduction is primarily intended to apply in cases in which there has been a failure to construct treatment facilities or other capital projects, in spite of good-faith efforts.

Ability to pay (to decrease preliminary penalty amount)

EPA typically does not establish settlement penalties that are clearly beyond the financial capability of the violator, when combined with the cost of the necessary injunctive relief. EPA should not seek a penalty that would seriously jeopardize the violator's ability to continue operations and achieve compliance, unless the violator's behavior has been exceptionally uncooperative, culpable, recalcitrant, or threatening to human health or the environment.

The adjustment for ability to pay may be used to reduce the settlement penalty to the highest amount that the violator can reasonably pay and still comply with the CWA. The violator has the primary burden of establishing the claim of inability to pay, although the agency must establish a prima facie case on the issue in order for a default order to be issued. Generally, the violator must submit the necessary

---

[7] See, e.g., *Natural Resources Defense Council v. Texaco Refining and Mktg.*, 800 F. Supp. 1, 24 (D. Del. 1992).

[8] See *PIRG v. Powell Duffryn*, 913 F.2d 64, 80–81 (3rd Cir. 1990).

information demonstrating actual inability to pay as opposed to unwillingness to pay. Further, the claim of inability to pay a penalty should not be confused with a violator's reluctance to adjust its operations in order to pay the penalty.

If the violator is unwilling to cooperate in demonstrating its inability to pay the penalty, the EPA will consider this adjustment in the penalty calculation. Without the cooperation of the violator, the EPA will generally use the information available on the financial position of the violator. The court can draw a negative inference, however, from the violator's failure to provide requested information to support the claim on its behalf. In some cases, the EPA may consult a financial expert to properly evaluate a violator's claim of inability to pay.

If the violator demonstrates an inability to pay the entire negotiated penalty in one lump sum (usually within 30 days of consent decree entry), the EPA may consider a payment schedule. The penalty could be paid in scheduled installments, with appropriate interest accruing on the delayed payments. The period allowed for such installment payments will generally not extend beyond 3 years.

If a payment schedule will not resolve the violator's ability-to-pay issue, as a last recourse the EPA can reduce the amount it seeks in settlement to a more appropriate amount in situations in which inability to pay can be clearly documented and reasonably quantified.

For municipalities, the EPA will often examine the most recent bond rating (within the past 5 years). If the bond rating is below BBB (Standard & Poor's rating scale) or below Baa (Moody's rating scale), the community may be considered to be in poor financial condition, and a detailed financial evaluation by an appropriate expert may be necessary to determine whether the financial condition affects the ability to pay a penalty.

Supplemental environmental projects (SEPs)

Supplemental environmental projects are defined by EPA as environmentally beneficial projects which a violator undertakes, but is not otherwise legally required to perform, in exchange for favorable penalty consideration in settlement of an enforcement action.

In order for a violator to receive a settlement penalty reduction in exchange for performing such a project, the project must conform with the EPA's SEP policy (Environmental Protection Agency, 1995) or be approved in advance by the assistant administrator. Use of SEPs in a particular case falls entirely within the discretion of EPA and the Department of Justice in judicial cases.

TABLE 2.10 Settlement Penalty Calculation Worksheet

Step	Amount
1. Statutory maximum penalty	________
2. Economic benefit	________
3. Total of monthly gravity amounts	________
4. Economic benefit + gravity (lines 2 + 3)	________
5. Gravity adjustments	________
a. Flow reduction adjustment (0 to 50% of gravity amount)	________
b. Recalcitrance adjustment (0 to 150% of gravity amount)	________
c. Quick-settlement adjustment (0 or 10% of gravity amount)	________
d. Environmental auditing adjustment (0 or 75% or 100% of gravity amount)	________
e. Total gravity adjustments (negative value for net gravity reduction)	________
6. Preliminary penalty amount (line 4 + line 5*e*)	________
7. Reduction for litigation considerations (if any)	________
8. Reduction for ability-to-pay reduction (if any)	________
9. Reduction for supplemental environmental projects (if any)	________
10. Bottom-line cash settlement penalty (line 6 less lines 7, 8, and 9)	

Final penalty computation

Table 2.10 presents a worksheet that may be used for penalty computation, considering all of the factors discussed in this chapter.

Citizens' Enforcement of the Clean Water Act

Citizens' suits

Under Section 505 of the Clean Water Act, citizens' suits can be filed in federal district court by individual citizens and citizens' groups as well as local or state governments. Citizens' suits came into wide use during the 1980s, as environmental and public interest groups learned how to use the *discharge monitoring reports* (*DMRs*) submitted by NPDES permit holders to determine whether treatment standards were being violated. However, several requirements must be met to successfully prosecute an enforcement action (Stimson et al., 1993):

1. *Sixty days' notice:* Before a suit is filed, 60 days' notice must be provided to the alleged violator, to the EPA, and to state authorities about the nature of the threatened suit.

2. *Government priority:* The federal or state government may decide to proceed with its own enforcement action, and this "diligent

prosecution" by these authorities will preclude the citizen from filing suit.

3. *Legal standing:* The citizen must have legal standing to pursue the action. The CWA defines this as a "person or persons having an interest which is or may be adversely affected" by a violation of the act. The courts have generally interpreted this requirement broadly, such that almost any citizen who uses a waterway in any manner or is affected by a discharge would qualify.

4. *Belief of ongoing violations:* Since the violator has had 60 days' notice of the proposed suit, the violation may be corrected before the suit can be filed. If this is the case, then the suit is subject to being dismissed. The citizen may show a good-faith belief that the violations were ongoing—that they were continuing at the time the suit was filed. However, if the violations are corrected at any time—even after the suit is filed—the suit cannot proceed. Some courts have allowed suits for penalties for past violations. Generally speaking, however, section 505 actions are allowed only to pursue an injunction to correct any ongoing condition that violates the act or endangers health or the environment.

If the suit proceeds, and if the federal court decides that the facility was indeed liable for discharge permit violations, the court may fine the violator or issue an injunction to remedy any conditions that constitute or cause violations. Fines are paid to the U.S. Treasury. The act also allows for payment of the complaining party's costs of litigation by the defendant, if the court awards them.

The most common method used by citizens' groups to attack storm water dischargers is to file suit alleging inadequacy of the storm water pollution prevention plan. To file such a suit, just as with the enforcement of permitting requirements by the EPA, no discharge is required to have occurred or to be proved to have occurred. This is a key point, because otherwise the citizens' suit would be difficult to pursue. The courts have held that both the lack of a storm water pollution prevention plan and inadequacies in an existing plan are violations of the Clean Water Act, regardless of whether they are accompanied by any actual discharge. Therefore, every day that such problems exist, there is a continuing violation. As noted above, the presence of a continuing violation is a necessary component of a successful citizens' suit. [See *City of New York v. Anglebrook Ltd. Partnership,* 891 F. Supp. 900 (S.D.N.Y., Jan. 3, 1995), and *NRDC v. Southwest Marine,* 945 F. Supp. 1330 (S.D. Cal. 1996).]

Usually, these citizens' actions end in settlements. The alleged violator pays some fine, but most of the settlement usually goes for programs to benefit the damaged waterway through studies, public education, or cleanup projects (Stimson et al., 1993).

Citizens' petitions to EPA

In addition to citizens' suits, the rules of the NPDES program allow private citizens to petition the EPA or state to rescind general permit coverage for a particular discharger, provided that the citizen is able to show cause why the general permit coverage should be rescinded. If the petition is successful, the EPA or state then requires an individual permit application from the discharger.

Compliance with the National Storm Water Program

The EPA estimates that about 150,000 industrial facilities are included in the phase I regulations and should therefore obtain permit coverage. The rate of compliance for these facilities appears to be following a typical pattern for new environmental regulations. This pattern includes an initial level of compliance achieved by the distribution of information, followed by an increased level of compliance achieved by the application of enforcement authority.

Many of those who have complied with the requirement to submit a *Notice of Intent* (*NOI*) for general permit coverage may not be complying with other requirements of the National Storm Water Program. Since no other backup information must be submitted to EPA to show compliance efforts, it is possible that many industrial and construction facilities have submitted an NOI without taking any other real actions toward storm water quality improvement.

EPA and state representatives have continued to educate regulated industries about the permit program. They generally have chosen not to penalize late applicants. However, if a complaint or an EPA inspection reveals that a discharger has made no attempt to obtain coverage, or to implement permit requirements after filing for coverage, then EPA consistently penalizes these dischargers.

As the National Storm Water Program becomes more mature, and the grace periods for compliance fall into the past, the regulatory agencies are adopting a more aggressive stance.

Any dischargers who become aware of the fact that their facility is required to obtain a discharge permit for storm water should contact EPA or applicable state regulatory authorities and submit a permit application (or Notice of Intent for general permit coverage) as soon as possible, without waiting for the regulatory authorities to initiate contact. This type of positive action toward compliance will reduce the possibility that penalties will be imposed, and will lead to a reduction in penalties if they are imposed.

References

Environmental Protection Agency, *BEN User's Manual,* Office of Enforcement, Washington, December 1993.

———, *Interim Clean Water Act Settlement Penalty Policy* (EPA internal document), March 1, 1995.

———, *Audit Policy Interpretive Guidance* (EPA internal document), Office of Regulatory Affairs, Washington, January 1997.

Government Printing Office: *Federal Register,* Office of the Federal Register, National Archives and Records Administration, Washington.

———: *Code of Federal Regulations,* Office of the Federal Register, National Archives and Records Administration, Washington, July 1 1992.

Stimson, James A., Jeffrey J. Kimmel, and Sara Thurin Rollin, *Guide to Environmental Laws from Premanufacture to Disposal,* Bureau of National Affairs, Washington, 1993.

Chapter

3

Regulated Storm Water Discharges

This is the last of three chapters providing basic information for all dischargers. This chapter describes what types of storm water discharges must have permits. Chapter 4 provides more details on storm water discharges from municipalities. Chapters 5 through 9 deal with industrial discharges. Chapters 10 through 14 deal with construction site discharges.

Regulated Discharges

The primary question that must be answered with regard to storm water permitting is simply this: Do I need a permit at all? There are several questions that must be addressed before this primary question can be answered:

1. Does my facility even have a discharge?
2. What if my facility drains into a *combined sewer system* (*CSS*)?
3. Am I discharging "storm water"?
4. What about "run-on" storm water?
5. Is my discharge a point source?
6. Does my discharge enter a "water of the United States"?
7. Is my discharge exempt from NPDES permit requirements for some reason?

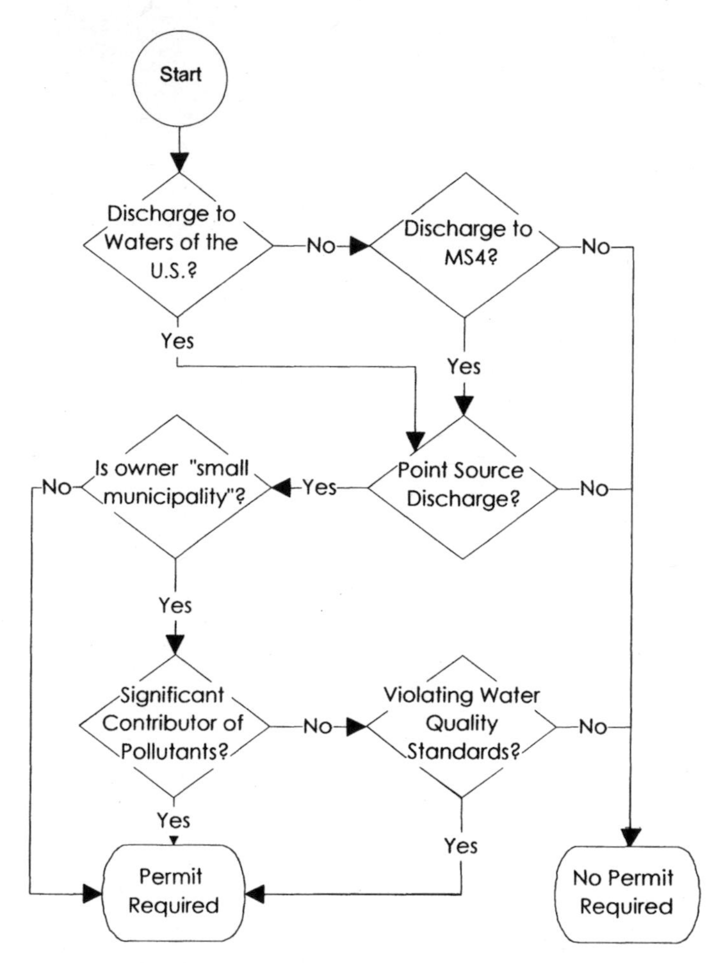

Figure 3.1 Flowchart to identify regulated storm water discharges.

The flowchart in Fig. 3.1 illustrates most of the decisions that must be made in order to determine whether a particular storm water discharge is subject to regulation under the National Storm Water Program. These decisions are described in detail throughout the remainder of this chapter.

Does My Facility Even Have a Discharge?

It is surprising how easy it is to overlook even the most fundamental issue of storm water discharge permitting: If there is no discharge, then there is no need for a discharge permit. It is crucial that we *not* think of an NPDES storm water discharge permit as a permit to *operate* an industrial facility, a construction site, or a municipal drainage system. An NPDES permit is not a permit to operate; it is a permit to *discharge* storm water from such a facility or system.

There is no discharge (and therefore, no discharge permit requirement) until storm water crosses the site property boundary, or enters waters of the United States (defined later in this chapter), or enters a municipal separate storm sewer system (defined in Chap. 4) on the site. Storm water which runs from one part of an industrial or construction site to another part is not a discharge; it is merely an internal transfer of flow, which does not require a permit. Treating an internal flow transfer as a discharge may result in unnecessary effort in sampling, permitting, and measuring flows which do not require a permit.

If the storm water runoff from an industrial facility or construction site flows into a retention basin on site with *no possibility* of discharge, then no permit is required. The storm water would then be eliminated by evaporation or infiltration. The EPA does not regulate discharges to groundwater[1] (such as from an infiltration basin), although some states do.

How large does the retention basin have to be? The answer is not likely to satisfy the engineer or hydrologist: It depends upon the level of risk that the owner of the industrial facility or construction site is willing to accept. If the owner is willing to accept the risk of having an unpermitted storm water discharge every 2 years, then the retention basin should be designed with sufficient capacity for the 2-year storm event. The design capacity of the retention basin is left entirely to the discretion of the permittee. However, every time that the capacity of the retention basin is exceeded, resulting in a discharge, the facility is in violation of the Clean Water Act. The penalties described in Chap. 2 may apply. Therefore, it is prudent to provide sufficient capacity in the retention basin for a large storm event.

What If My Facility Drains into a CSS?

The only types of storm water discharges which are regulated under the National Storm Water Program are discharges into municipal sep-

---

[1]Some recent court cases suggest that EPA may have the right to regulate discharges to groundwater that is hydrologically connected with surface waters. EPA interprets the CWA's NPDES permitting program to regulate discharges to surface water via groundwater where there is a direct and immediate hydrologic connection ("hydrologically connected") between the groundwater and the surface water. However, EPA also believes that this use of NPDES permits is highly dependent on the facts surrounding each permitting situation [63 FR 07858 (Feb. 17, 1998)].

arate storm sewer systems or waters of the United States. Discharges to combined storm sewers are not included, because they are regulated under a separate program. A CSS is a system which carries sanitary sewage under dry-weather conditions, but which is surcharged with runoff under storm conditions. Such systems are present in the older cities of the United States.

Am I Discharging "Storm Water"?

The EPA defines *storm water* as "storm water runoff, snow melt runoff, and surface runoff and drainage" [40 CFR 122.26(b)(13)]. Unless special conditions apply, only these types of discharges are allowed under a storm water discharge permit. Note that the EPA definition includes only discharges that result from precipitation. That is, only water which originates in the form of rainfall, snowfall, or other forms of precipitation can produce storm water discharges.

Water which is carried onto the site in vehicles, is piped or channeled onto the site, or is produced in industrial processes on the site does not fit the definition of storm water. For example, storm water does not include such discharges as water used to flush utility lines or test storage tanks or any form of process wastewater.

As described in later chapters, storm water permits for construction sites and industrial facilities usually allow some types of non–storm water discharges to mix with the storm water that is being permitted. However, it is important to understand that this is an allowance for the convenience of the discharger; these other discharges are still not storm water.

What about "Run-on" Storm Water?

Operators of industrial facilities or construction sites are generally responsible for storm water from upstream adjacent facilities that enters the site or combines with the discharge from a facility submitting a permit application. If such a situation exists, the permit application for the downstream facility needs to address the issue and bring it to the attention of the permitting authority. The permitting authority can then draft appropriate permit conditions to reflect these circumstances. For example, the downstream facility may be required to development management practices or other run-on/runoff controls to segregate or otherwise prevent outside runoff from mixing with on-site storm water discharges. These permit conditions should be acceptable to the owners of the downstream facility, because these conditions will limit the liability for pollutants in such run-on storm water.

Many consider it unfair of the EPA to hold them responsible for obtaining permit coverage for storm water that originates on areas other than their site. However, from the EPA's perspective, this policy is apparently intended to motivate the site owner to at least report the situation to the permitting authority, or to segregate the run-on storm water from the on-site storm water (thereby potentially reducing the possibility that the run-on storm water may pick up additional pollutants from the downstream site). These actions are consistent with the EPA's goal of reducing pollutant discharges.

Is My Discharge a Point Source?

The NPDES permit program regulates only point source discharges. The EPA defines a *point source* as

> any discernible, confined, and discrete conveyance, including but not limited to any pipe, ditch, channel, tunnel, conduit, well, discrete fissure, container, rolling stock, concentrated animal feeding operation, landfill leachate collection system, vessel or other floating craft from which pollutants are or may be discharged. This term does not include return flows from irrigated agriculture or agricultural storm water runoff [40 CFR 122.2].

Since only point sources of storm water are regulated, a permit is not generally necessary for *sheet flow,* which is shallow unconcentrated overland flow. However, the definition of point source is written broadly, so as to include most situations which are commonly referred to as sheet flow. According to the EPA's definition, the discharge must be considered a point source if there is any *discernible* conveyance. A shallow swale or rill, a depression, or a curb cut could fit such a description. The EPA has indicated that it

> intends to embrace the broadest possible definition of point source consistent with the legislative intent of the CWA and court interpretations to include any identifiable conveyance from which pollutants might enter the waters of the United States. In most court cases interpreting the term "point source," the term has been interpreted broadly. For example, the holding in *Sierra Club v. Abston Construction Co., Inc.,* 620 F.2d 41 (5th Cir. 1980) indicates that changing the surface of land or establishing grading patterns on land will result in a point source where the runoff from the site is ultimately discharged to waters of the United States [55 FR 47989 (Nov. 16, 1990)].

A natural channel or rill may form a point source discharge, even if it results only from natural erosion processes and has never been improved in any way at the point of discharge, provided that human activities have increased the amount of impervious cover or have otherwise changed the runoff characteristics of the area drained by the natural channel or rill.

Because of the broad interpretation of point source by the EPA and the courts, it is difficult to argue that an industrial area drains entirely by sheet flow, without any discernible conveyances. The EPA recommends that a permit application be submitted along with information on flow conditions if there is any doubt about whether the flow would be classified as a point source.

There has been considerable confusion regarding the use of the terms *point source* and *non–point source*. This confusion has resulted from the fact that each of these two terms has a legal or technical definition that has changed over a 25-year period. However, the common or popular understanding of these terms has not changed as much, even among members of the legal or technical community. For example, many state storm water pollution control programs operate under a name like "Nonpoint Source Pollution Control Division." Actually, however, such an agency is often involved in NPDES permitting of storm water point source discharges. This contradiction is best explained by remembering that the current, broad definition of *point source* has evolved over the years as a result of various court decisions and regulatory initiatives. At the beginning of that evolutionary process, nearly all storm water discharges were generally thought of as nonpoint sources, and this early understanding is still reflected in popular terminology.

Waters of the United States

As noted above, storm water is regulated only if it discharges to a municipal separate storm sewer system or directly to waters of the United States. There are very few terms which have been debated and litigated as much as the term *waters of the United States*. The definition has grown from the original concept of *navigable waters* to a very broad definition which includes most of the bodies of water and wetlands in the country. The official definition is as follows:

a) All waters which are currently used, were used in the past, or may be susceptible to use in interstate or foreign commerce, including all waters which are subject to the ebb and flow of the tide;
b) All interstate waters, including interstate "wetlands";
c) All other waters such as intra-state lakes, rivers, streams (including intermittent streams), mudflats, sandflats, wetlands, sloughs, prairie potholes, wet meadows, playa lakes, or natural ponds the use, degradation, or destruction of which would affect or could affect interstate or foreign commerce, including any such waters:
    1) Which are or could be used by interstate or foreign travelers for recreational or other purposes;
    2) From which fish or shellfish are or could be taken and sold in interstate or foreign commerce; or
    3) Which are used or could be used for industrial purposes by industries in interstate commerce;

d) All impoundments of waters otherwise defined as waters of the United States under this definition;
e) Tributaries of waters identified in paragraphs a) through d) of this definition;
f) The territorial sea; and
g) Wetlands adjacent to waters (other than waters that are themselves wetlands) identified in paragraphs a) through f) of this definition.

Waste treatment systems, including treatment ponds or lagoons designed to meet the requirements of CWA, are not waters of the United States [40 CFR 122.2].

Many federal government agencies administer programs dealing with waters of the United States. However, for the purposes of the Clean Water Act, the final authority regarding Clean Water Act jurisdiction remains with the EPA.

Tidal bodies of water

The limit of "waters which are subject to the ebb and flow of the tide" is the *high-tide line,* which is the line of intersection of the land with the water's surface at the maximum height reached by a rising tide. The high-tide line may be determined, in the absence of actual data, by a line of oil or scum along shore objects, a more or less continuous deposit of fine shell or debris on the foreshore or berm, other physical markings or characteristics, vegetation lines, tidal gages, or other suitable means that delineate the general height reached by a rising tide. The line encompasses spring high tides and other high tides that occur with periodic frequency but does not include storm surges in which there is a departure from the normal or predicted reach of the tide due to the piling up of water against a coast by strong winds such as those accompanying a hurricane or other intense storm. In other words, the high-tide line is determined by the limits of the astronomic tide. The "territorial sea" extends 3 nautical miles (nmi) offshore [33 CFR 329.12].

Inland bodies of water

In a nontidal water body, the *ordinary high-water mark* is used to define the limits of waters of the United States (except for wetlands). This is that line on the shore established by the fluctuations of water and indicated by physical characteristics such as clear, natural line impressed on the bank, shelving, changes in the character of soil, destruction of terrestrial vegetation, the presence of litter and debris, or other appropriate means that consider the characteristics of the surrounding areas [51 FR 41250, Nov. 13, 1986, as amended at 58 FR 45036, Aug. 25, 1993].

Wetlands

Wetlands are areas that are inundated or saturated by surface water or groundwater at a frequency and duration sufficient to support, and that under normal circumstances do support, a prevalence of vegetation typically adapted for life in saturated soil conditions. Wetlands generally include swamps, marshes, bogs, and similar areas. *Adjacent* wetlands are those that are bordering, contiguous, or neighboring. Wetlands separated from other waters of the United States by artificial dikes or barriers, natural river berms, beach dunes, and the like are adjacent wetlands.

EPA uses the 1987 *Corps of Engineers Wetlands Delineation Manual* (U.S. Army Corps of Engineers, 1987) to identify and delineate wetlands. This document establishes the specific technical criteria that must be satisfied in order for an area to be considered a jurisdictional wetland. Therefore, storm water discharges from a construction activity to jurisdictional wetlands (i.e., waters of the United States) need permit authorization.

Navigable waters of the United States

There is a regulatory distinction between navigable waters of the United States and waters of the United States. *Navigable* waters of the United States are defined in 33 CFR part 329. These are waters that are navigable in the traditional sense where permits are required from the U.S. Army Corps of Engineers for certain work or structures pursuant to Sections 9 and 10 of the Rivers and Harbors Act of 1899. *Waters of the United States* are defined in 33 CFR part 328 (and identically in other locations). These include more than navigable waters of the United States and are the waters where permits are required for the discharge of dredged or fill material pursuant to Section 404 of the Clean Water Act, or for the discharge of other pollutants pursuant to Section 402 of the Clean Water Act.

Precise definitions of *navigable waters of the United States* and *navigability* are ultimately dependent on judicial interpretation and cannot be made conclusively by administrative agencies. However, the Corps of Engineers has developed guidance for making administrative determinations. It is important to note that the Corps of Engineers considers the historical and potential uses of the waterway for interstate commerce as well as the actual present uses. In fact, the Corps may even consider the possibility of "reasonable" improvements that may potentially be made to make the waterway navigable. These improvements do not have to exist, be planned, or even be authorized; it is enough that potentially they could be made. In effect, the Corps is interested in determining whether a waterway is "navigable in law" rather than "navigable in fact."

Interpreting the definition of *waters of the United States*

The definition of the *waters of the United States* may be interpreted in many different ways. One possible method of interpretation may be to review the definitions provided in other federal regulations that use the same terms. For example, many of the same terms are defined for surface mining and reclamation regulations of the U.S. Department of Interior. These definitions are not binding on the EPA. However, they provide some basis for the interpretation of the Clean Water Act.

An *intermittent stream* is (1) a stream or reach of a stream that drains a watershed of at least 1 square mile (mi^2) or (2) a stream or reach of a stream that is below the local water table for at least some part of the year, and obtains its flow from both surface runoff and groundwater discharge [30 CFR 701.5]. The *water table* is the upper surface of a zone of saturation, where the body of groundwater is not confined by an overlying impermeable zone [30 CFR 701.5].

An *ephemeral stream* flows only in direct response to precipitation in the immediate watershed or in response to the melting of a cover of snow and ice, and which has a channel bottom that is always above the local water table [30 CFR 701.5]. A *perennial stream* flows continuously during all the calendar year as a result of groundwater discharge or surface runoff [30 CFR 701.5].

Impoundments are all structures or depressions, either naturally formed or artificially built, that hold water, sediment, slurry, or other liquids or semiliquids [30 CFR 701.5].

Special types of intrastate waters

Sloughs are riparian marshes along slow-moving streams and rivers. Marshes are one of the broadest categories of wetlands and in general harbor the greatest biological diversity. They are characterized by shallow water, little or no peat deposition, and mineral soils. Marshes are dominated by floating-leafed plants (such as water lilies and duckweed) or emergent soft-stemmed aquatic plants (such as cattails, arrowheads, reeds, and sedges). Marshes are frequently or continually inundated with water. Marshes derive most of their water from surface water, including streams, runoff, and overbank flooding; however, they receive inputs from groundwater as well. Environmental conditions in marshes lead to high productivity since the pH of most marshes is generally near a neutral value and nutrients derived from runoff are plentiful.

Prairie potholes are marshlike ponds that have formed in shallow basins caused by glaciation in the Dakotas, Iowa, and the Canadian prairies. Prairie potholes receive runoff from surrounding land uses because of their depressional nature. However, water levels fluctuate seasonally as a result of dependence on precipitation, and prairie potholes may periodically be dry for up to several years. The prairie potholes form the northern staging, resting, and breeding part of the Central Flyway. An estimated 50 to 75 percent of all the waterfowl in North America are hatched in the prairie potholes.

Wet meadows are grasslands with soil that is waterlogged after precipitation events. Wet meadows are often categorized as marshes because their vegetative communities are similar to those of marshes. However, wet meadows are drier than most marshes and are typically found in lower areas on flat landscapes, surrounded by upland meadows or prairie grasses. These systems may also exist as broad transition zones surrounding or leading downgradient to deeper marshes. Because wet meadows depend largely on precipitation for water inputs, they are generally dry during the summer.

Playa lakes are shallow, depressional water bodies in arid regions that receive runoff from surrounding land and occupy a *playa* (Spanish for "beach") in the wet season but dry up in the summer. They are commonly thought of as ephemeral lakes, as they may be dry not only on a seasonal basis, but also on a yearly or multiyear basis, depending on longer-term rainfall patterns. Playas themselves are generally accepted as arid zone basins of greatly varying sizes and origins that, although found above the groundwater table, are subject to ephemeral surface water inundation. Playas may be particularly critical wetland habitat in the intensively farmed, dry plains of New Mexico and Texas. The southern Great Plains area is within the migratory corridor known as the Central Flyway; the birds rely on playas for staging, resting, and breeding areas. Playas can be considered to have four major defining characteristics (Motts, 1970):

1. Occupation of a basin or topographic valley of interior drainage
2. A smooth, barren, extremely flat surface with a low gradient
3. Rare inundation that occurs in low-rainfall regions where evaporation exceeds precipitation
4. Circumference generally more than 2000 to 3000 ft (600 to 900 m) in diameter.

Is My Discharge Exempt from Permit Requirements for Some Reason?

Some types of discharges are exempt from NPDES permit requirements. Generally, these fall into one of the following three categories:

1. Industrial discharges that are already regulated under other programs.

2. Agricultural discharges that have been exempted from NPDES permit requirements. Other programs are also available to deal with many of these agricultural discharges.
3. Discharges from many industrial facilities and construction sites owned by small municipalities. These municipalities were given a temporary deferral of NPDES permit requirements by Congress.

Industries exempt from storm water discharge permits

Certain categories of industries are specifically exempt from regulation by storm water discharge permits:

- *Radioactive wastes:* high-level radioactive wastes regulated under the Atomic Energy Act of 1954
- *Vessels:* sewage discharges from vessels [40 CFR 122.2]
- *Oil and gas operations:* permitted water, gas, or other material injected into wells to aid in the production of oil or gas, so long as groundwaters or surface waters are not degraded [40 CFR 122.3(a)]

Exemptions for agricultural runoff

Return flows from irrigated agriculture or agricultural storm water runoff is exempt from NPDES permit requirements [40 CFR 122.2; 40 CFR 122.3(e) and (f)]. Various programs are underway to control non–point source pollution from agricultural operations. Most of these programs are administered by the U.S. Department of Agriculture.

No permit is required for discharges from non–point source silvicultural activities such as tree nurseries, site preparations, restoration and subsequent cultural treatment, thinning, prescribed burning, pest and fire control, tree-harvesting operations, surface drainage, or road construction or maintenance from which there is natural runoff [40 CFR 122.3(e)]. At 40 CFR 122.27(b)(1), the term *silvicultural point source* is defined to mean any discrete conveyance related to rock crushing, gravel washing, log sorting, or log storage facilities which are operated in connection with silvicultural activities and from which pollutants are discharged into waters of the United States.

Industrial and construction sites owned by small municipalities

Industrial facilities owned or operated by a municipal entity with a total population of less than 100,000 are generally not required to obtain permits to discharge storm water, unless the facility is an airport, power-plant, or uncontrolled sanitary landfill. Construction projects undertaken by small municipalities are also exempt from phase I storm water discharge permit requirements. The broad deferral for small municipalities was granted by Congress in Section 1068 of the Intermodal Surface Transportation Efficiency Act, which was passed in December 1991. The EPA has extended this deferral until the proposed phase II regulations become effective in 2001 [63 FR 01585 (Jan. 9, 1998)].

The EPA uses the Census Bureau's definition of a *municipal entity.* This definition includes not only incorporated places such as cities, towns, boroughs, and villages but also counties and parishes and minor civil divisions such as towns and townships. Special districts such as sanitary sewer districts, flood control districts, school districts, and state agencies are also included. In determining whether a municipal entity has to obtain permit coverage for industrial activities which it owns or operates, the most recent census will be used as follows:

- *Counties:* If the county has a population of 100,000 or more, then its construction activities and the industrial operations which it owns or operates must be permitted.
- *Sewage treatment districts:* Service populations will be used to determine the population of sewage treatment districts which operate publicly owned treatment works (POTWs). Where one sewer district operates a number of plants, the entire service population of the district will be used to determine the applicable population classification of all the treatment works operated by the district. (For example, if a district with a cumulative service population of 150,000 operates two sewage treatment plants, one of which serves 100,000 and the other of which serves 50,000, then both plants will be considered to be a facility that is owned or operated by a municipality with a population of 100,000 or more.)
- *Other special districts:* Flood control districts and other municipal entities with service populations must obtain permits for industrial or construction operations which they own or operate if their service population is 100,000 or more. School districts must obtain storm water discharge permits if their total resident population (not the number of students) is greater than 100,000.
- *State agencies:* The state population will be used to determine the population of most state agencies. Under this approach, the EPA would base the population of facilities operated by a state agency on the entire state population rather than on the population of the local government entity with land-use authority (e.g., city, town, township, county) in which the facility is physically located. This means that most state-owned facilities and projects would be subject to

storm water discharge permit requirements. However, some state agencies with limited service areas (such as a river basin authority) may avoid storm water discharge permit requirements if their total population is less than 100,000.

- *Combined ownership:* Where an industrial facility is owned or operated by more than one municipality, the EPA will use the combined populations of the appropriate municipalities in determining population thresholds.

The effect of these rules is to limit the number of municipalities which can escape the storm water discharge permit requirement. However, the exclusion of industrial facilities and construction projects owned by small municipalities effectively removes thousands of storm water discharges from the National Storm Water Program, at least for the time being. The EPA has specifically extended the deadline for this exemption until at least August 2, 2001, unless further changes are made [60 FR 17950 (Apr. 7, 1995)].

Where a facility is privately owned and operated but has a service contract with a municipality, the facility is not considered to be municipally operated. For example, a privately owned and operated landfill that receives municipal waste pursuant to a contract with a municipality or some other form of reimbursement from a municipality is not exempt from storm water discharge permit requirements.

References

Government Printing Office, *Federal Register,* Office of the Federal Register, National Archives and Records Administration, Washington.

——, *United States Code,* Washington, 1988.

——, *Code of Federal Regulations,* Office of the Federal Register, National Archives and Records Administration, Washington, July 1, 1992.

Motts, W. S., *Introduction to Playa Studies.* Air Force Cambridge Research Laboratory, contract no. AF 19(628)-2486, 1970.

U.S. Army Corps of Engineers, *Corps of Engineers Wetlands Delineation Manual,* Technical Report Y-87-1, Environmental Laboratory, Waterways Experiment Station, Vicksburg, MS, January 1987.

Chapter

4

Municipal Storm Water Permit Requirements

As noted in Chap. 1, the federal Water Quality Act of 1987 required large and medium-size municipalities to obtain storm water discharge permits for their public drainage systems under phase I of the National Storm Water Program. The proposed regulations for phase II require that most small municipalities obtain storm water discharge permits for their public drainage systems as well.

This chapter describes the storm water discharge permit requirements for municipalities. The various types of public drainage systems are discussed, as well as the different permit application requirements and compliance requirements which will apply to different municipal entities under these regulations.

According to the 1990 census, there were 19,289 incorporated cities, 17,796 minor civil divisions, and 3141 counties or equivalents in the United States.

Types of Public Drainage Systems

There are two basic types of public drainage systems: combined sewer systems (CSSs) and municipal separate storm sewer systems (MS4s). A CSS is a system that conveys sanitary sewage to publicly owned treatment works (POTWs) during periods of dry weather, but which also conveys storm water runoff during wet weather. An MS4 is any conveyance or system of conveyances that is owned or operated by a state or local government entity designed for collecting and conveying storm water which is not part of a publicly owned treatment works.[1] A municipal separate storm sewer system includes "municipal streets, catch basins, curbs, gutters, ditches, man-made channels, or storm drains that discharge into the waters of the United States" [55 FR 47989 (Nov. 16, 1990)].

Discharges from CSS are regulated under the NPDES permit program. However, the regulations are not part of the National Storm Water Program. For this reason, CSS requirements are not discussed in this book. This book, and particularly this chapter, focuses on the requirements that apply to MS4 discharges, including the requirements currently applicable to the large and medium-size municipalities regulated under phase I, as well as the smaller municipalities proposed to be regulated under phase II.

Application Requirements for Phase I Municipalities

An MS4 that serves an urban population of 250,000 or more is considered large. A medium-size MS4 is one that serves an urban population of 100,000 to 250,000. Under phase I of the National Storm Water Program, large and medium-size municipalities were required to submit two-part applications for individual storm water discharge permits.

There were 405 urbanized areas of 50,000 people or more and 173 cities of 100,000 people or more identified in the 1990 census. After the elimination of cities served by combined sewer systems, there were 140 cities and 45 counties included in the category of large and medium-size municipalities, as determined by population alone. The EPA and states designated an additional 481 cities, 32 counties, and 60 other government entities (special districts, state DOTs, etc.). Therefore, 658 entities are included in the phase I municipal storm water permitting program, including 521 cities, 77 counties, and 60 others.

Because many of these entities have joined with other adjacent entities to submit joint applications for their storm water systems, a total of 263 municipal permits have been identified for phase I.

The first part of the phase I application required municipalities to provide a map showing storm sewer system outfalls and other discharges, major structural controls, and land-use activities. To accu-

[1] The EPA has proposed to change the definition of MS4 to include public drainage systems owned by the federal government, because there is no essential difference between such systems operated by the federal government and those operated by state and local governments.

rately display this information, field inspection activities were generally required. The Part 1 application also required municipalities to characterize their storm water discharges by compiling existing data on precipitation events, the volume and quality of discharges, and water quality; to conduct a field screening analysis to detect dry-weather flows (e.g., illicit connections, illegal dumping, malfunctioning septic tanks); and to develop a characterization plan that includes sampling locations, frequencies, and equipment for the monitoring required in the Part 2 application.

Part 2 monitoring data were intended to characterize storm water pollutants and provide a basis for estimating annual pollutant loadings and the mean concentration of pollutants in discharges. After the permit application process, a municipality may be subject to additional monitoring requirements as defined by the municipality's permit.

Application Requirements for Phase II Municipalities

The EPA proposes to expand the municipal storm water program to include the following:[2]

- All small municipal separate storm sewer systems within urbanized areas (with certain minor exceptions)
- Other small municipal separate storm sewer systems meeting EPA or state criteria for designation (possibly including any system that serves a population center of at least 10,000 people with a population density of at least 1000 people per square mile)
- Any municipal separate storm sewer system contributing substantially to the storm water pollutant loadings of a regulated, physically interconnected municipal separate storm sewer system.

All regulated small municipal separate storm sewer systems would be required to develop and implement a storm water management program. Program components would include at least the six minimum control measures that define *maximum extent practicable* (*MEP*), as described in the next section. These program components would be implemented through NPDES permits [63 FR 01735 (Jan. 9, 1998)].

[2] The information presented on the phase II requirements is based on the proposed rules as published by EPA on January 9, 1998 [63 FR 01735 (Jan. 9, 1998)]. The final rules, expected in 1999, may be different from the proposed rules.

Statutory Standards for Municipal Separate Storm Sewer Systems

Section 402(p)(3)(B) of the CWA establishes two technology-based standards for permits for discharges from municipal separate storm sewer systems. The first standard provides that MS4 permits must contain a requirement to effectively prohibit non–storm water discharges into the system. In addition, permits for discharges from MS4s must reduce the discharge of pollutants to the maximum extent practicable, including management practices, control techniques, and system, design, and engineering methods.

It is important to note the distinction between the MEP standard and the BCT and BAT standards applicable to many other types of NPDES discharges, including storm water discharges from industrial facilities. It may be assumed that Congress chose not to apply the BCT and BAT standards to municipal systems because they recognized that the lack of practical methods to fully control all access to public drainage systems would inevitably prevent a municipal entity from meeting the same standards that might reasonably be applied to an industrial facility. Industrial facilities can generally maintain more complete control over access to, and discharges into, the storm drainage system on the industrial site. However, by definition, a public drainage system serves public streets and other areas that cannot be fully controlled.

Minimum Control Measures Required to Meet the MEP Standard

EPA has issued preliminary guidelines on the implementation of appropriate pollution control measures to meet the MEP standard [63 FR 01735 (Jan. 9, 1998)]. Municipal dischargers are expected to reduce pollutants to the MEP through implementation of the following minimum control measures:

1. Public education and outreach on storm water impacts
2. Public involvement and participation
3. Illicit discharge detection and elimination
4. Construction site storm water runoff control
5. Postconstruction storm water management in new development and redevelopment
6. Pollution prevention and good housekeeping for municipal operations

The pollutant reductions that represent MEP may be different for each municipality, given the unique storm water concerns that may exist and the differing possible remedies. Therefore, each permittee would determine the specific details in each of the six minimum control measures that represent MEP through an evaluative process. In this process, permittees and permit writers would evaluate the proposed storm water management controls to determine whether reduction of pollutants to the MEP could be achieved with the identified best management practices (BMPs). This evaluative process could consider such factors as the conditions of receiving waters, specific local concerns, and other aspects included in a comprehensive watershed plan. Various municipal entities may choose to cooperate in the development and implementation of the minimum control measures. The following sections provide more detailed information on each of these six minimum control measures, based on EPA guidance [63 FR 01735 (Jan. 9, 1998)].

Public education and outreach on storm water impacts

Education programs can be considered a nonstructural BMP that should be implemented for everyone. Much pollution enters streams, rivers, and lakes through carelessness or ignorance. Many people will adapt new methods or use alternative materials if they are simply informed of techniques that can reduce the impacts on receiving waters.

Public education is essential to the success of a municipal storm water pollution control effort because the actions of the public are so important in determining the level of pollutants present in storm water discharges. The public can reduce pollutant discharges by properly maintaining septic systems to reduce overflows and other discharges, by using lawn and garden chemicals sparingly and carefully, and by properly disposing of used motor oil or household hazardous wastes. However, the public will generally not be sufficiently informed or motivated to carry out these actions without some type of public education and outreach effort. It is estimated that 15% to 20% of household hazardous wastes end up in storm drains or runoff (King County Solid Waste Division, 1990). In addition, EPA estimates that 40% (80 million gal/yr) of "do-it-yourself oil changers" pour onto roads, driveways, or yards or into storm sewers (Environmental Protection Agency, 1993).

Significant amounts of fertilizers and pesticides enter the water from lawn maintenance and landscaping activities. Professional services may apply too much fertilizer and too many pesticides in order to please customers, and homeowners may not know the proper amounts of fertilizer and pesticides to use. Both groups may apply lawn care chemicals too close to water bodies. Local governments can start programs for areawide composting, using yard wastes picked up at the curb. The compost can be sold or given to local gardeners and lawn maintenance services. Homeowners should be informed about the proper use of lawn and garden chemicals.

In addition to educating the public about modifications they can make in their own actions to reduce pollutant discharges, municipalities can reach out to members of the public, asking them to participate in community activities that further reduce pollution. These activities might include roadside litter pickup and storm drain stenciling.

Some public education and outreach efforts should be directed toward targeted groups of commercial, industrial, and institutional entities likely to have significant storm water impacts. The operators of these types of entities (such as restaurants or automotive repair shops) should be given the opportunity to learn about, and understand, the impacts of their discharges on storm water pollution.

A public education program should generally include the following elements:

- A program to promote, publicize, and facilitate public reporting of the presence of illicit discharges or improper disposal of materials into the MS4 (using such methods as a 24-h operational telephone service for reporting, and storm water inlet stenciling)
- A program to promote, publicize, and facilitate the proper management and disposal of used oil and household hazardous wastes
- A program to promote, publicize, and facilitate the proper use, application, and disposal of broadcast chemicals such as pesticides, herbicides, and fertilizers by public, commercial, and private applicators and distributors

Municipalities may enter into partnerships with their states in fulfilling the public education requirement. It may be much more cost-effective to use a state education program rather than have numerous municipalities develop their own. Municipalities may also work with other organizations (e.g., environmental and nonprofit groups and industry) that might be able to assist in fulfilling this requirement. Many of these kinds of organizations already have educational materials, and the groups could work together to educate the public.

Public education can be accomplished by means of a program of general environmental education functions. Businesses and private citizens may be contacted periodically by mail using general public mailouts, business and industry mailouts, newspaper inserts, and utility bill inserts. Broadcast media may be employed through the use of television and radio news reporting, paid commercials, and public service announcements. Partnerships with local businesses may be established to defray all or part of the costs. Many businesses have obtained significant public relations benefits through the sponsorship of, and identification with, environmental improvement programs.

Educational resources may be made available through the development and distribution of educational videos and other materials. Public libraries and school libraries should be included. Local libraries and government agencies, such as the Cooperative Extension Service and the Industrial Extension Service, often have educational materials to use in training.

Personal contact may be established and maintained through the use of a school education program, speaker's bureau, open houses of treatment plants and waste facilities, public employee education, a partnership with business program, citizen advisory groups, telephone hotlines, and walk-in resource centers.

Public involvement and participation

In addition to informing and motivating the public, and enlisting their volunteer participation in community work projects, the municipal storm water program should encourage public involvement in the development, implementation, and assessment of the storm water management program.

Public involvement is an integral part of the municipal storm water program, and should be designed to comply with applicable state and local public notice requirements. The public may become involved through attending public hearings, serving on storm water management panels or advisory committees, serving as trainers for other members of the public, coordinating the effects of the storm water management program with other existing municipal and community activities, or participating in voluntary inspection or monitoring efforts. Public participation should involve all economic and ethnic groups.

People are busy, and it can be difficult to identify enough members of each group who have enough time and interest to participate in the storm water management plan. There will also be conflicting priorities and viewpoints which may not be resolved. However, the potential advantages of public involvement include accelerated implementation of the storm water management plan. This comes about because the concerns of key members of the public may be identified and incorporated into the storm water management plan at an early stage of the development process. The possibility of legal or political challenges by disenfranchised members of the public may also provide a strong motivation to involve these potential opponents in the process and thereby reduce the need for delays due to last-minute objections.

Illicit discharge detection and elimination

As noted previously in this chapter, one of the statutory requirements of the federal Water Quality Act of 1987 is the elimination of illicit discharges from municipal separate storm sewer systems. A municipal storm water management program should include a program for detecting and eliminating illicit discharges.

Discharges from storm water drainage systems often include wastes and wastewater from non–storm water sources. EPA's Nationwide Urban Runoff Program (NURP) indicated that many storm water outfalls still discharge during substantial dry periods. Pollutant levels in these dry-weather flows were shown to be high enough to significantly degrade receiving water quality. Results from a 1987 study conducted in Sacramento, California, revealed that slightly less than one-half of the water discharged from a municipal separate storm sewer system was not directly attributable to precipitation runoff (Environmental Protection Agency, 1993).

A significant portion of these dry-weather flows results from illicit and/or inappropriate discharges and connections to the municipal separate storm sewer system. Illicit discharges enter the system through either direct connections (e.g., wastewater piping either mistakenly or deliberately connected to the storm drains) or indirect connections (e.g., infiltration into the storm drain system or spills collected by drain inlets).

Virtually all municipalities provide, either directly or indirectly (e.g., through contractors), for residential collection of sanitary wastewater, garbage, and storm water. Thus it is appropriate that municipalities take a leadership role in ensuring that residents have the opportunity to properly dispose of household hazardous wastes and used motor vehicle fluids in an environmentally responsible manner. Public education can reduce the amount of waste material produced and inform residents regarding improper disposal methods. However, municipalities must then have a response to the question: "Where can I dispose of my household hazardous wastes and used motor vehicle fluids?"

An adequate long-term program will include periodic collection events and should ensure that a publicly available "drop-off" location (not necessarily owned or operated by the municipal entity) is available on a more regular basis and during hours when working people can drop off waste (i.e., occasional long weekday hours, or weekends). The long-term program could incorporate programs, or program components, run by the private and/or public sector (e.g., automobile parts and discount stores with auto centers accepting used oil from do-it-yourself oil changers).

Detection and elimination of illicit discharges always begin with adequate knowledge of the public drainage system. This knowledge should be recorded in the form of system maps and related documents showing the location of major pipes, outfalls, and topography. The map should identify areas of concentrated activities likely to be a source of storm water pollution, if the data already exist.

Another necessary step toward the elimination of illicit discharges is the legal prohibition of such discharges, to the full extent of available government authority. This prohibition would be in the form of enforceable ordinances or regulations.

The final major step toward the detection and elimination of illicit discharges is to inform key individuals about their responsibilities to properly dispose of wastes. These key individuals may include public employees, business owners and managers, and members of the general public. This information could be disseminated by public education efforts and by targeted outreach efforts to specific groups. The public should be encouraged to report illicit discharges and should be given a convenient way of doing so. Recycling programs should be developed or supported for various types of potential illicit discharges, including used motor oil, antifreeze, pesticides, herbicides, and fertilizers.

Some types of non–storm water discharges such as the following are relatively innocuous, and could be allowable unless they contribute substantial quantities of pollutants:

- Water line flushing
- Landscape irrigation
- Diverted stream flows
- Rising groundwaters
- Uncontaminated groundwater infiltration [as defined at 40 CFR 35.2005(20)] to separate storm sewers
- Uncontaminated pumped groundwater
- Discharges from potable water sources
- Foundation drains
- Air conditioning condensation
- Irrigation water
- Springs
- Water from crawl space pumps
- Footing drains

- Lawn watering
- Individual residential car washing
- Flows from riparian habitats and wetlands
- Dechlorinated swimming pool discharges
- Street wash water
- Discharges or flows from fire fighting

Since most municipalities have made some provision for the prevention and control of spills through the formation of emergency response teams, no special actions related to spills should be taken as a result of storm water discharge permit requirements.

Construction site storm water runoff control

Chapters 10 through 14 of this book describe construction storm water discharge permit requirements in detail. These permit requirements apply to construction sites that are required to obtain permit coverage for storm water discharges from their site.

Over a short time, storm water discharges from construction site activity can contribute more pollutants, including sediment, to a receiving stream than had been deposited over several decades. Storm water runoff from construction sites can include pollutants other than sediment, such as phosphorus and nitrogen from fertilizer, pesticides, petroleum derivatives, construction chemicals, and solid wastes that may become mobilized when land surfaces are disturbed.

Generally, properly implemented construction site ordinances are effective in reducing these pollutants. The program would need to include, at a minimum,

- Requirements for construction site owners or operators to implement appropriate BMPs, such as silt fences, temporary detention ponds, and hay bales
- Provisions for preconstruction review of site management plans
- Procedures for receipt and consideration of information provided by the public
- Regular inspections during construction
- Penalties to ensure compliance

The storm water management program also would need to ensure control of other waste at construction sites that could adversely impact water quality. This waste could include discarded building materials, concrete truck washout, and sanitary waste.

In many areas, however, the effectiveness of ordinances in reducing pollutants is limited due to inadequate enforcement or incomplete compliance with such local ordinances by construction site discharges of storm water. Not all construction site owners or operators properly maintain BMPs. For example, sediment traps and sediment basins may fill up, and silt fencing may break or be overtopped.

Postconstruction storm water management in new development and redevelopment

Most of the storm water management plan components described thus far are nonstructural. That is, they are implemented primarily through the actions of public employees, business owners or employees, or general citizens. However, there is a valid role for structural measures to reduce storm water pollution in certain circumstances. Because of the difficulty and expense involved in retrofitting structural storm water pollution control measures onto existing drainage systems and developments, the most reasonable way to implement structural storm water pollution controls is by incorporating them into new construction projects. In this way, the pollution control measures become a part of the project after completion. A complete storm water management plan will address redevelopment as well as new development. However, minor remodeling of existing structures would not be considered "redevelopment."

Examples of structural BMPs that may be applied to new development or redevelopment include storage practices (wet ponds and extended-detention outlet structures), filtration practices (grassed swales, sand filters, and filter strips), and infiltration practices (infiltration basins, infiltration trenches, and porous pavement). Because of constant improvements in these types of control measures, the municipal entity should allow flexibility in the requirements for implementation of these BMPs. The remainder of this section provides further details. Most of this section is based on information from the WATERSHEDSS on-line resource on the Internet (Osmond et al., 1995).

Not all BMPs for control of pollutant discharges from new construction and redevelopment are necessarily structural BMPs. Nonstructural BMPs may also be incorporated. These may include requirements to limit growth to identified areas, protect sensitive areas such as wetlands and riparian areas, minimize imperviousness, maintain open space, and minimize disturbance of soils and vegetation.

One of the major obstacles to the implementation of long-term storm water pollution control measures in developed areas has been the lack of an effective method for ensuring the adequate long-term operation and maintenance of BMPs. The EPA has announced its intention to develop and publish guidelines in this area [63 FR 01735 (Jan. 9, 1998)].

The primary guideline for the selection of storm water pollution control measures for new development and redevelopment is that the water quality effects of the development should not be significantly different from the water quality effects of the same site before development.

Buffers, setbacks, and easements. Buffer zones are strips of vegetation, either natural or planted, around water bodies. Such vegetated zones help reduce the impact of runoff by trapping sediment and sediment-bound pollutants, encouraging infiltration, and slowing and spreading storm water flows over a wide area.

Setbacks, which establish restrictions on development activities within a specified distance of a stream bank or other water resource through zoning or other mechanisms, can prevent or minimize erosion and gully formation, thus minimizing sedimentation and associated nutrient enrichment downstream. Setbacks are distinct from, although potentially overlapping with, such BMPs as riparian zones, floodplain preservation, and wetland preservation, because the areas involved are defined differently and as the primary goals of these other BMPs include other pollutant removal, preserving natural sediment and nutrient removal functions, and water quantity issues in addition to minimizing erosion and providing sedimentation.

Easements can be created to prevent development on land areas around water bodies. Although easements have not been widely used to protect water resources, they can provide an alternative method of gaining control of strategic land. Easements may be negotiated or purchased from landowners and passed on to future owners as part of the deed to the property. Long-term protection of land adjacent to water resources could include purchase of the land or an easement, purchase of development rights, or some other type of agreement to limit development. Such "green belts" around waterways can be used to protect the water and provide parks and recreational areas for residents.

Buffer zones, setbacks, and easements are most effective at reducing stream bank erosion and providing sediment removal when used as part of a BMP system including structures to diffuse concentrated storm water flows from upgradient development; measures to directly repair eroded stream banks, such as live stakes, fascines, cribwalls, gabions, and revetments; and a means of reducing the artificially heightened erosivity of stream flows at their source, such as routing runoff in grass swales, using detention ponds, and providing discharge spreader swales.

Buffer zones, setbacks, and easements can be established in already-developed corridors, with provisions for shifting some site uses upgradient or improving buffer characteristics. Generally a local unit of government or one of its arms, such as a planning board or zoning commission or the soil and water conservation board, will establish such restrictions and will be responsible for management of these areas. Local governments can pass ordinances and rules to require buffer zones and setbacks around receiving waters. State agencies can work with local governments to obtain such protections by, e.g., offering matching funds for projects that will establish protections on targeted, nonattaining water bodies or, for the long term, through statewide comprehensive growth management requirements.

Dry detention basins. Dry detention basins temporarily detain a portion of storm water runoff for a specified length of time, releasing the storm water slowly to reduce flooding and remove a limited amount of pollutants. The basins are referred to as *dry detention* because these devices dry out between rain events. Pollutants are removed by allowing particulates and solids to settle out of the water. Overall pollutant removal in dry detention devices is low to moderate. Important reasons for use of dry detention basins are to reduce peak storm water discharges, control floods, and prevent downstream channel scouring.

Dry detention basins release storm water through a controlled outlet over a specified time based on design criteria. Extended-detention basins drain more slowly or may retain a permanent pool of water. However, these basins often release water too slowly to empty the basin before the next storm. Since the basin is partially full, only a portion of the design runoff volume from the next storm is detained, and the remainder is bypassed directly into the stream. With little or no detention, few pollutants are removed from the runoff. Such failures can be prevented through adequate design and maintenance to keep the inlets and outlets open.

Design of dry detention basins includes locating proper sites for construction of the basin, calculating the appropriate detention time, treatment of the expected range in volumes of storm water from storms, and maintenance procedures and schedules. The storm water should be held for at least 24 h for maximum pollutant removal. Soils should be permeable to allow the water to drain from these basins between storms, and the water table should be more than 2 ft below the bottom of the basin (to avoid a permanent pool of water in the basin during wet weather). A forebay is a section of the basin separated from the main part of the basin by a wall or dike and receiving the incoming storm water. Forebays help capture debris and sand deposits, which accumulate quickly, and thereby ease routine cleaning.

Dry detention basins are capable of removing significant amounts of pollutants and have proved effective at reducing peak storm flows. An appreciable body of knowledge has been accumulated on the design and maintenance of these structures. Detention basins can serve small to rather large developments and are usually readily incorporated into the design of the overall development. Existing dry basins built to control storm water peak flows can be modified to provide extended detention for storm water, thus improving pollutant removal.

Dry basins can be unsightly, especially if floating and other debris accumulate in them. Basins should be located where they are not easily seen or where they can be concealed with landscaping. Many dry basins end up with permanent pools of water because runoff from previous storms has not either flowed out or infiltrated before another storm occurs. The standing water can be a nuisance and an eyesore to residents.

Because they take up large areas, dry detention basins are generally not best suited for high-density residential developments. Sites must allow easy access for equipment to maintain and clean the basin and remove sediment. The appearance of some dry detention basins has been improved by planting hardy wildflowers in the bottom. Resident acceptance of a "wildflower basin" is much higher than of an unadorned open basin.

Maintenance of dry detention basins is both essential and costly. General objectives of maintenance are to prevent clogging, prevent standing water, and prevent the growth of weeds and wetland plants. This requires frequent unclogging of the outlet and mowing. Normal maintenance costs can range from 3% to 5% of construction costs on an annual basis (Schueler, 1987). Cleaning out sediment, which is expensive, will be necessary in 10 to 20 years' time. Cleaning involves digging out the accumulated sediment, mud, sand, and debris with construction backhoe or other earthmoving equipment.

Infiltration devices. *Infiltration* refers to the process of water entering into the soil, which indicates the predominant means by which these devices evacuate their treatment volume. A number of devices are used to treat storm water that employ infiltration to remove pollutants and to recharge or replenish the groundwater. Infiltration devices include infiltration basins, infiltration trenches, and dry wells. Properly designed infiltration devices can closely reproduce the water balance that existed predevelopment, providing groundwater recharge, control of peak flows from storm water, and protection of stream banks from erosion due to high flows. A significant advantage of infiltration is that in areas with a high percentage of impervious surface, infiltration is one of the few means to provide significant groundwater recharge.

Infiltration devices can remove pollutants very effectively through adsorption onto soil particles and biological and chemical conversion in the soil. Infiltration basins with long detention times and grass bottoms enhance pollutant removal by allowing more time for settling and because the vegetation increases settling and adsorption of sediment and adsorbed pollutants. Although infiltration is a simple concept, infiltration devices must be carefully designed and maintained if they are to work properly. Poorly installed or improperly located devices fail easily. It is critical that infiltration devices be used only where the soil is porous and can absorb the required quality of storm water. Maintenance needs for infiltration devices are higher than those for other devices, partly because of the need for frequent inspection. Nuisance problems can occur, especially with insect breeding, odors, and soggy ground.

Pollutant removal capability for infiltration basins that exfiltrate the entire amount of captured storm water is shown in Table 4.1. Other infiltration devices which exfiltrate only part of the captured storm water (some of the storm water is discharged to receiving waters on the surface) have lower removal effectiveness.

Some infiltration devices (infiltration trenches, dry wells, and catch basins) can be constructed under parking lots and roads, taking very little land from other uses. Other infiltration devices take up considerable areas, depending on their size and the drainage area served. Locating smaller infiltration devices is fairly easy so that large downstream devices can be replaced with a number of small structures

TABLE 4.1 Pollutant Removal Efficiencies

BMP	Nutrients	Sediment	Metals	BOD and COD	Oil and grease	Bacteria
Dry detention basin	Low	High	Moderate	Moderate	Low	High
Infiltration devices	High	Very high	Very high	Very high	High	Very high
Sand filters	Moderate	Very high	Very high	Moderate	High	Moderate
Oil and grease trap	None	Low	Low	Low	High	Low
Vegetative practices	Low	Moderate	Moderate	Low	Moderate	Low
Constructed wetlands	High	Very high	High	Moderate	Very high	High
Wet ponds	Moderate to high	High	Moderate to high	Moderate	High	High

SOURCE: Compiled from Schueler, 1987; Schueler et al., 1992; Environmental Protection Agency, 1990; Phillips, 1992; Birch et al., 1992; and others.

upstream and still achieve the same control of storm water. Infiltration devices require permeable soils and reasonably deep water tables. Smaller infiltration devices such as dry wells and basins can be located near buildings to capture the runoff from roofs and other impervious surfaces.

Infiltration devices help replenish the groundwater and reduce both storm water peak flows and volume. Pollutant removal can be very high for many pollutants. Because they take up little land area and are not highly visible, many underground infiltration devices can be located close to residential and commercial areas.

Infiltration techniques work only where the soils are permeable enough that the water can exit the storage basin and enter the soil. These devices have a high failure rate. Infiltration devices must have sediment removed before the storm water enters the device, to prevent clogging of the soil. The water table must be at least 2 ft under the bottom of the device.

Maintenance requirements include regular inspections, cleaning of inlets, mowing, and possible use of observation wells to maintain proper operation. Infiltration basins and sediment removal devices used to prevent clogging of other infiltration devices must have the sediment removed regularly. If an infiltration device becomes clogged, it may need to be completely rebuilt.

Underground oil and grease catch basins. A number of devices are used to remove oil and grease from storm water. One type, commonly known as *oil-water separators,* are mechanical devices manufactured by various industrial equipment manufacturers and usually installed at industrial sites.

Another type of oil and grease removal device is the oil and grease trap catch basin (or oil and grit separator). These catch basins are underground devices used to remove oils, grease, other floating substances and sediment from storm water before the pollutants enter the storm sewer system. The basins are usually placed to catch the oil and fuel that leak from automobiles and trucks in parking lots, service stations, and loading areas.

A popular design for the oil and grease trap catch basin uses three chambers to pool the storm water, to allow the particulates to settle, and to remove the oil. As the water flows through the three chambers, oils and grease separate either to the surface or sediments, and are skimmed off and held in the catch basin. The storm water then passes on to the storm sewer or into another storm water pollution control device. Because these devices are relatively small and inexpensive, they can be placed throughout a drainage system to capture coarse sediments, floating wastes, and accidental or illegal spills of hazardous wastes. Oil and grease trap catch basins can reduce maintenance of infiltration systems, detention basins, and other storm water devices.

Since these catch basins detain storm water only for short periods, they do not remove other pollutants as effectively as facilities that retain runoff for longer periods. However, these basins can be effectively used as a first stage of treatment to remove oil and sediment from storm water before it enters another, larger storm water pollution control device, such as a wet pond. Pollutant removal varies depending on the basin volume, flow velocity, and depth of baffles and elbows in the chamber design.

Oil and grease trap catch basins can be installed in most areas. Drainage areas flowing into the catch basin must be no larger than 2 ac (0.8 ha), and the catch basin must be large enough to handle dry-weather flows that enter the basin. These catch basins can be installed in almost any soil or terrain, which allows their use near or at the impervious surfaces contributing heavily to the storm water runoff. Little land area is taken up by catch basins, as they require only enough area for proper maintenance.

Since these devices are underground, there should be few complaints concerning appearance. These catch basins can be used very effectively as part of a system of storm water controls to remove oily pollutants and coarse sediment before they enter another storm water control device. Also, small catch basins can be distributed over a large drainage area, which may prove more advantageous than constructing a single large structure downstream.

Pollutant removal is low for contaminants other than oil, grease, and coarse sediment. The accumulated sediment must be removed or cleaned out frequently (at least twice a year) to prevent sediment-bound pollutants from being stirred up and washed out in subsequent storms. Sediment removal removes the oil and grease because these pollutants eventually bind to the sediment. The underground chamber is difficult to maintain because of its enclosed, underground design. Odors are sometimes a problem.

Oil and grease skimmers. Oil and grease skimmers consist of an open sedimentation basin with a skimmer plate extending below the ponding control elevation at the outlet. Storm water velocities are reduced in this sump, dropping out coarse sediment and separating oils and greases and floatables, which are retained in the basin by the skimmer as the storm water discharges to a larger detention device or off-site. The skimmer plate should extend sufficiently below the lowest discharge level to preclude siphoning of the water surface by the discharge.

These sediment sump/skimmers are often designed to be larger than the underground chambers, have longer detention times, and thus remove more of the sediments and oils and greases. They are also simpler and more easily maintained and allow for photodegradation of hydrocarbons in addition to settling.

Pollutant removal varies depending on the basin volume, flow velocity, and the depth of baffles and elbows in the chamber design. Pollutant removal is low for contaminants other than oil, grease, and coarse sediment. The accumulated sediment must be removed or cleaned out frequently to prevent sediment-bound pollutants from being stirred up and washed out in subsequent storms. Sediment removal removes the oil and grease because these pollutants eventually bind to the sediment. Odors are sometimes a problem. Wastes removed from these systems should be tested to determine proper disposal methods. The wastes may be hazardous; therefore, maintenance costs should be budgeted to include disposal at a proper site.

Porous pavement. Porous pavement is an alternative to conventional pavement that is intended to reduce imperviousness and consequently minimize surface runoff. Porous pavement follows one of two basic designs. First, it may be comprised of asphalt or concrete that lacks the finer sediment found in conventional cement. This formulation is usually laid over a thick base of granular material (Urbonas and Stahre, 1993). Second, porous pavement may be formed with modular, interlocking open-cell cement blocks laid over a base of coarse gravel (Urbonas and Stahre, 1993). A geotextile fabric underlying the gravel prevents the migration of soil upward into the gravel bed. Both designs typically include a reservoir of coarse aggregate stone beneath the pavement for storm water storage prior to exfiltration into surrounding soils (Schueler et al., 1992). Use of porous pavement requires permeable soils with a deep water table. Traffic must be restricted to exclude heavy vehicles. Its use is not advisable in areas expecting high levels of off-site sediment input, including chemicals and sand used in snow removal operations.

The porous pavement itself functions less as a treatment BMP and more as a conveyance BMP to the other necessary component of the design, the underlying aggregate chamber, which functions as an infiltration device. As with other infiltration devices, treatment is provided by adsorption, filtration, and microbial decomposition in the subsoil surrounding the aggregate chamber, as well as by particulate filtration within the chamber. Operating systems have been shown to have high removal rates for sediment, nutrients, organic matter, and trace metals. These rates are largely due to the reduction of mass loadings of these pollutants through transfer to groundwater (Schueler et al., 1992).

The big disadvantage of porous pavement is that sites have a high failure rate, due to clogging from improper construction, accumulated sediment and oil, or resurfacing (Schueler et al., 1992). Excessive sediment will cause the pavement to rapidly seal and become ineffective (Urbonas and Stahre, 1993). The modular, interlocking, open-cell concrete block type tends to remain effective for considerably longer than asphalt or concrete porous pavement. Porous pavement must be maintained frequently to continue functioning. Quarterly vacuum sweeping and/or jet hosing is needed to maintain porosity, and this may constitute 1% to 2% percent of the initial construction costs (Schueler et al., 1992).

Positive attributes include the diversion of potentially large volumes of surface runoff to groundwater recharge, providing both water quality and quantity benefits. While more expensive than conventional pavement, it can also eliminate the need for more involved storm water drainage, conveyance, and treatment systems, offering a valuable option for spatially constrained urban sites. Porous pavement may be most beneficial in watersheds with high percentages of impervious surface and high volumes of runoff. Its use typically is recommended for lightly trafficked satellite parking areas and access roads. Increased infiltration at the source (parking lots, etc.) will reduce both the volume of runoff and the delivery of associated pollutants to water bodies.

Sand filters. Sand filters are a type of storm water control device used to treat storm water runoff from large buildings, access roads, and parking lots. As the name implies, sand filters work by filtering storm water through beds of sand. Small sand filters are installed underground in trenches or precast concrete boxes. Large sand filters are aboveground, self-contained sand beds that can treat storm water from drainage areas as large as 5 ac (2 ha).

To date, the city of Austin, Texas, and the state of Florida have made use of the large, aboveground versions of sand filters, while the underground sand filters have been installed in Florida, Maryland, Delaware, and the District of Columbia. Both aboveground and underground versions use some form of pretreatment to remove sediment, floating debris, and oil and grease to protect the filter. After the storm water passes through the pretreatment device, it flows onto a sand filter bed. As the storm water flows through the filter bed, sediment particles and pollutants adsorbed to the sediment particles are captured in the upper few inches of sand.

The underground versions fit in very well in urban areas and on sites with restricted space. Depending on the design, underground sand filters are practically invisible to casual observers and generally receive few complaints from residents. Maintenance of these sand filters is simple and done manually. Aboveground sand filters are often considered to be eyesores by residents. Thus, these sand filters are best used where they cannot be seen or where hedges or other visual barriers can be installed. Because of the construction techniques used to build aboveground sand filters, large filters are proportionately less expensive than small filters. Construction costs can be kept lower if lightweight equipment is used for maintenance, which reduces the structural reinforcing needed in the filter.

Pollutant removal for sand filters varies depending on the site and climate. Overall removal for sediment and trace metals is better than removal of more soluble pollutants because the filter functions by simply straining small particles out of the storm water.

Sand filters remove pollutants by settling out particles in the pretreatment devices and by straining out particles in the filter. Underground sand filters built in two-chambered precast concrete boxes cannot handle large drainage areas. Moderate to large parking lots should be the largest areas drained to underground sand filters.

Sand filters can be installed underground in urban settings and be kept out of sight, or aboveground for large drainage areas. Sand filters can provide effective reduction of the more common urban pollutants in storm water. Sand filters have demonstrated long lifetimes and consistent pollutant removal when properly maintained. Maintenance for sand filters is simple and inexpensive. Mosquito breeding is usually not a problem, even in underground settling chambers that hold pools of water for long periods. Shaver (1992) reports that oil and grease in the storm water form a sheen on the water which prevents mosquito growth.

Sand filters are more expensive to construct than infiltration trenches. If heavy equipment is to be used for maintenance, construction costs are significantly higher. Sand filters on the land surface are considered unattractive. No storm water detention is provided by sand filters. Sand filters have only limited pollutant removal for a number of pollutants.

Sand filters require frequent but simple maintenance. Maintenance for smaller underground filters is usually best done manually. Normal maintenance requirements include raking of the sand surface and disposal of accumulated trash. The upper few inches of dirty sand must be removed and replaced with clean sand when the filter clogs. The pretreatment devices must be cleaned to remove sediment and debris.

Filter strips. Filter strips typically are bands of close-growing vegetation, usually grass, planted between pollutant source areas and a receiving water. They also can be used as outlet or pretreatment devices for other storm water control practices. Filter strips can include shrubs or woody plants that help to stabilize the grass strip, or they can be composed entirely of trees and other natural vegetation. Such strips or buffers are used primarily in residential areas around streams or ponds. Filter strips do not provide enough runoff storage or infiltration to significantly reduce peak discharges or the volume of storm runoff. For this reason, a filter strip should be viewed as only one component in a storm water management system. At some sites, filter strips may help reduce the size and cost of downstream control facilities.

Filter strips reduce pollutants such as sediment, organic matter, and many trace metals by the filtering action of the vegetation, infiltration of pollutant-carrying water, and sediment deposition. Although studies indicate highly varying effectivenesses, trees in strips can be more effective than grass strips alone because of the trees' greater uptake and long-term retention of plant nutrients. Properly constructed forested and grassed filter strips can be expected to remove more than 60% of the particulates and perhaps as much as 40% of the plant nutrients in urban runoff.

Grassed swales. Grassed swales are earthen channels covered with a dense growth of a hardy grass such as tall fescue or reed canary grass. Swales are used primarily in single-family residential developments, at the outlets of road culverts, and as highway medians. Because swales have a limited capacity to convey runoff from large or intense storms, they often lead into concrete-lined channels or other stable storm water control structures. Swales may provide some reduction in storm water pollution through infiltration of runoff water into the soil, filtering of sediment or other solid particles, and slowing the velocity and peak flow rates of runoff. These processes can be enhanced by adding small (4- to 10-in-high) dams across the swale bottom, thereby increasing detention time.

Pollutants are removed from surface flow by the filtering action of the grass, sediment deposition, and/or infiltration into the soil. The pollutant-removing effectiveness of swales has been assessed as moderate to negligible depending on many factors, including the quantity of flow, the slope of the swale, the density and height of the grass, and the permeability of the underlying soil. Research on grassed swales has found varying levels of pollutant removal ranging from 30% to 90% reduction in solids and 0% to 40% reductions in total phosphorus loads.

Vegetative practices remove pollutants by encouraging infiltration into the ground, reducing runoff velocity and allowing particles to settle, and absorbing some pollutants. To be effective, vegetative practices require flat areas that are large in relation to the drainage area, and deep water tables. Swales should have as little slope as possible to maximize infiltration and reduce velocities. Filter strips should not be used where slopes exceed 15%; best performance occurs where the slope is

5% or less. The height of grass in filter strips and swales can affect the pollutant removal. Taller grass will slow velocities more, but grass cut to a short length may take up more plant nutrients.

Vegetative practices are inexpensive, and generally easy to maintain with common procedures such as mowing and trimming. Vegetation is usually pleasing to residents. Filter strips and grassed swales are easily located and constructed. Vegetation is highly effective in preventing erosion and thus controlling sediment in storm water runoff.

Vegetative practices remove only small amounts of pollutants. These practices do little to control peak storm flows or reduce storm water volumes.

Maintenance of vegetation includes periodic inspection, mowing, fertilizer application, and repair of washed-out areas and bare spots. Filter strip maintenance basically involves normal grass- or shrub-growing activities such as mowing, trimming, removing clippings, or replanting when necessary. Strips that are used for sediment removal may require periodic regrading and reseeding of their upslope edge because deposited sediment can kill grass and change the elevation of the edge such that uniform flow through the strip can no longer be obtained. Swale maintenance basically involves normal grass-growing activities such as mowing and resodding when necessary.

Constructed wetlands. Interest has steadily increased in the United States over the last two decades in the use of natural physical, biological, and chemical aquatic processes for the treatment of polluted waters. This interest has been driven by growing recognition of the natural treatment functions performed by wetlands and aquatic plants, by the escalating costs of conventional treatment methods, and by a growing appreciation for the potential ancillary benefits provided by such systems. Aquatic treatment systems have been divided into natural wetlands, constructed wetlands, and aquatic plant systems (Environmental Protection Agency, 1988). Of the three types, constructed wetlands have received the greatest attention for treatment of storm water pollution.

Constructed wetlands typically are not intended to replace all the functions of natural wetlands, but to serve, as do other water quality BMPs, to minimize point source and non–point source pollution prior to its entry into streams, natural wetlands, and other receiving waters. Constructed wetlands which are meant to provide habitat, water quantity, aesthetic, and other functions as well as water quality functions—termed *created, restored,* or *mitigation wetlands* (Hammer, 1994)—typically call for different design considerations from those used solely for water quality improvement, and such systems are not addressed here. In fact, debate continues over the advisability of intentionally combining primary pollution control and habitat functions in the same constructed facilities. Nonetheless, constructed wetlands can provide many of the water quality improvement functions of natural wetlands with the advantage of control over location, design, and management to optimize those functions.

Constructed wetlands vary widely in their pollutant removal capabilities, but can effectively remove a number of contaminants (Bastian and Hammer, 1993; Bingham, 1994; Brix, 1993; Corbitt and Bowen, 1994; Environmental Protection Agency, 1993). Among the most important removal processes are the purely physical processes of sedimentation via reduced velocities and filtration by hydrophytic vegetation. These processes account for the strong removal rates for suspended solids, the particulate fraction of organic matter (particulate BOD), and sediment-attached nutrients and metals. Oils and greases are effectively removed through impoundment, photodegradation, and microbial action. Similarly, pathogens show good removal rates in constructed wetlands via sedimentation and filtration, natural die-off, and ultraviolet degradation. Dissolved constituents such as soluble organic matter, ammonia, and ortho-phosphorus tend to have lower removal rates. Soluble organic matter is largely degraded aerobically by bacteria in the water column, plant-attached algal and bacterial associations, and microbes at the sediment surface. Ammonia is removed largely through microbial nitrification (aerobic)–denitrification (anaerobic), plant uptake, and volatilization, while nitrate is removed largely through denitrification and plant uptake. In both cases, denitrification is typically the primary removal mechanism. The microbial degradation processes are relatively slow, particularly the anaerobic steps, and require longer residence times, a factor which contributes to the more variable performance of constructed wetland systems for these dissolved constituents. Phosphorus is removed mainly through soil sorption processes which are slow and vary based on soil composition, and through plant assimilation and subsequent burial in the litter compartment. Consequently, phosphorus removal rates are variable and typically trail behind those of nitrogen. Metals are removed largely through adsorption and complexation with organic matter. Removal rates for metals are variable, but are consistently high for lead, which often is associated with particulate matter. Constructed wetlands can be expected to achieve or exceed the pollutant removal rates as estimated for wet pond detention basins and dry detention ponds.

The use of constructed wetlands for storm water treatment is still an emerging technology, and there are no widely accepted design criteria. However, certain general design considerations do exist. It is important first to drop storm water inflow velocities and provide opportunity for initial sediment deposition with facilities which can be periodically maintained and which avoid the likelihood of entraining deposited sediment in subsequent inflows. It is important to maximize the nominal hydraulic residence time and to maximize the distribution of inflows over the treatment area, avoiding designs which may allow for hydraulic short-circuiting. Emergent macrophytic vegetation plays a key role, intimately linked with that of the sediment biota, by providing attachment sites for periphyton, by physically filtering flows, as a major storage vector for carbon and nutrients, as an energy source for sediment microbial metabolism, and as a gas exchange vector between sediments and air. Thus, it is important to design for a substantial native emergent vegetative component. Anaerobic sediment conditions should be ensured to allow for long-term burial of organic matter and phosphorus. A controlled rate of discharge is the last major physical design feature. While an adjustable outfall may seem desirable for finetuning system performance, regulatory agencies often require a fixed design to preclude subsequent inappropriate modifications to this key feature. The outfall should be fitted with some form of skimmer or other means to retain oil and grease. Plants must be chosen to withstand the pollutant loading and the frequent fluctuation in water depth associated with the design treatment volume. It is advisable to consult a wetlands botanist to choose the proper vegetation.

Location of constructed wetlands in the landscape can be an important factor in their effectiveness. Mitsch (1993) observed in a comparison of experimental systems using phosphorus as an example that retention as a function of nutrient loading will generally be less efficient in downstream wetlands than in smaller upstream wetlands. He also cautioned that the downstream wetlands could retain more mass of nutrients, and that a placement tradeoff might be optimal. Mitsch observed that creation of in-stream wetlands is a reasonable alternative only in lower-order streams, that such wetlands are susceptible to reintroduction of accumulated pollutants in large flow events as well as being unpredictable in terms of stability. Such systems would likely require higher maintenance and management costs.

Constructed wetlands are most effective as part of a BMP system which includes minimization of initial runoff volumes through the positioning of pervious landscaping features; routing of runoff to maximize infiltration; use of pervious pavement, grass swales, swale checks, or other measures; pretreatment of collected runoff to minimize sediment and associated pollutant loads; and off-line attenuation of larger storm event runoff to optimize wetland performance and minimize downstream erosion-related water quality impacts.

Properly constructed and maintained wetlands can provide very high removal of pollutants from storm water. Constructed wetlands can be used to reduce storm water runoff peak discharges as well as provide water quality benefits. Constructed wetlands can serve a dual role in controlling storm water pollution and providing a pleasing natural area. Wetlands are highly valued by residents; therefore they can be given high visibility, they can serve as attractive centerpieces to developments and recreation areas, and they typically increase property values (Schueler, 1987; Shaver, 1992). Constructed wetland systems can provide groundwater recharge in the area, thus lessening the impact of impervious surfaces. This recharge can also provide a groundwater subsidy to the surficial aquifer, which can benefit local vegetation and decrease irrigation needs.

Constructed wetlands may contribute to thermal pollution and cause downstream warming. This may preclude their use in areas where sensitive aquatic species live. They are not a competitive option compared to other treatment methods where space is a major constraint. The ponded water may be a safety hazard to children.

Constructed wetlands have an establishment period during which they require regular inspection to monitor hydrologic conditions and ensure vegetative establishment. Vegetation establishment monitoring and long-term operation and maintenance, including maintenance of structures, monitoring of vegetation, and periodic removal of accumulated sediments, must be provided for to ensure continued function (Wetzel, 1993; Bingham, 1994). Maintenance costs vary depending on the degree to which the wetlands are intended to serve as popular amenities. Frequent initial maintenance to remove opportunistic species is typically required if a particular diverse, hydrophytic regime is desired. Operators of wetlands may need to control nuisance insects, odors, and algae.

Natural and restored wetlands. The many water quality improvement functions and values of wetlands are now widely recognized. At the same time, concern has grown over the possible harmful effects of toxic pollutant accumulation and the potential for long-term degradation of wetlands from altered nutrient and hydraulic loading that can occur with the use of wetlands for water treatment. Because of these concerns, the use of natural wetlands as treatment systems is restricted by federal law (Fields, 1993). Most natural wetlands are considered to be waters of the United States and are entitled under the CWA to protection from degradation by non–point souce (NPS) pollution. Natural wetlands do function within the watershed to improve water quality, and protection or restoration of wetlands to maintain or enhance water quality is acceptable practice. However, NPS pollutants should not be intentionally diverted to wetlands for primary treatment. Wetlands must be part of an integrated landscape approach to NPS control, and cannot be expected to compensate for insufficient use of BMPs within the upgradient contributing area. Restored wetlands are subject to the same restrictions as unmodified natural wetlands. Wetlands created from upland habitat for the purpose of mitigating the loss of other wetlands as required by regulatory agencies are generally also subject to the same restrictions as natural wetlands. Constructed wetlands, which have been defined as a subset of created wetlands that are designed and developed specifically for water treatment (Fields, 1993), clearly are not intended for the same protections as natural wetlands, and can serve as valuable treatment BMPs.

Wet retention basins. Wet retention ponds, also called wet detention basins, or wet basins or ponds, maintain a permanent pool of water in addition to temporarily detaining storm water. The permanent pool of water enhances the removal of many pollutants. These ponds fill with storm water and release most of it over a few days, slowly returning to their normal depth of water. Several mechanisms in wet ponds remove pollutants, including settling of suspended particulates; biological uptake, or consumption of pollutants by plants, algae, and bacteria in the water; and decomposition of some pollutants. Wet ponds have some capacity to remove dissolved plant nutrients, an important characteristic that protects lakes, rivers, and estuaries from eutrophication.

Wet ponds can be used in most locations where there is enough space to locate the pond. Because of the permanent pool of water, wet ponds can remove moderate to high amounts of most pollutants and are more effective in removing plant nutrients than most other devices. Also, the large volume of storage in the pond helps to reduce peak storm water discharges which, in turn, helps control downstream flooding and reduces scouring and erosion of stream banks.

Construction costs for wet ponds can be somewhat high because the ponds must be large enough to hold the required volume of runoff and to contain the permanent pool of water. Maintenance costs run about 3% to 5% of the construction cost per year (Schueler, 1987).

Wet ponds should be designed to displace the older storm water with the newer storm water, which ensures the proper amount of holding time. If the design is improper, short-circuiting can occur where the newer storm water flows directly to the outlet, bypassing the main part of the wet pond. Short-circuiting causes the new storm water to be released too soon, preventing pollutant removal and settling of sediment. Basic considerations for the installation of wet retention ponds are the location, inflow runoff volume, hydraulic residence time, permanent pool size, and maintenance. Volumes of storm water runoff and normal discharge available for the permanent pool must be calculated by trained hydrologists before a wet pond is constructed. Long, narrow ponds or wedge-shaped ponds are preferred shapes to minimize short-circuiting of storm flows. These shapes also will lessen the effects of wind, which can stir up sediment and sediment-bound pollutants. Pond shape, pond depth, and surrounding fringe areas must be considered to maximize the effectiveness of the basin. Marsh plants around the pond help remove pollutants, provide habitat, and hide debris.

Because people find these ponds to be aesthetically pleasing, wet ponds can be sited in both low- and high-visibility areas. Quite often, residents feel that the permanent pool of water enhances property values as well as the aesthetic value of the area. The outlet must be sized to provide adequate time for pollutant removal, yet discharge the storm water before the next storm occurs. Wet retention ponds have been used to provide wildlife habitat, and they may be a focal point for a recreation area. Wet ponds are one of the most effective and reliable devices for removing pollutants from storm water.

One disadvantage of wet retention basins is that they may contribute to thermal pollution and cause downstream warming. This may preclude their use in areas where sensitive aquatic species live. Wet ponds are not well suited to very small developments because of their large size. Wet ponds may flood prime wildlife habitat; and sometimes there are problems with nuisance odors, algae blooms, and rotting debris when the ponds are not properly maintained. Wetland plants may need to be harvested or removed periodically to prevent releasing of plant nutrients into the water when the plants die. The pool of water presents an attractive play area to children; hence, there may be safety problems.

The maintenance costs of wet ponds are estimated at 3% to 5% of construction cost per year. Wet ponds require regular inspection, removal of sediment according to a regular schedule of maintenance, regular mowing, and regular cleaning and repair of inlets and outlets. Operators of wet ponds must control nuisance insects, weeds, odors, and algae; inspect and repair pond bottoms; and harvest deciduous vegetation prior to the onset of fall, as necessary. Mosquitoes can be controlled in wet ponds with fish of the Gambusia family which eat the mosquito larvae. The Gambusia can survive the winters in North Carolina if the permanent pool is at least 3 ft deep. Another control method which does not use insecticides is monthly application of briquettes containing bacteria which cause a disease in mosquitoes. The application needs to be done only in the warmer months. The bacteria can be purchased at hardware and garden stores.

Pollution prevention and good housekeeping for municipal operations

The final minimum control measure required for meeting the statutory MEP requirement involves the operations of the municipal entity itself. These operations should include an effective operation and maintenance program, and adequate training for municipal employees and contractors, to prevent or reduce pollutant runoff from municipal operations. The plan should include at least the following elements:

1. Maintenance activities, maintenance schedules, and long-term inspection procedures for structural and other storm water controls to reduce floatables and other pollutants discharged from the separate storm sewers.[3]
2. Controls for reducing or eliminating the discharge of pollutants from streets, roads, highways, municipal parking lots, maintenance and storage yards, and waste transfer stations—including programs that promote recycling. Controls for discharges from maintenance and storage yards should include controls for discharges from salt and sand storage locations and snow disposal areas operated by the municipality.
3. Programs to promote the minimal use of pesticides. All city personnel and commercial entities engaged in the application of pesticides should be licensed according to the applicable pesticide application law.
4. Procedures for the proper disposal of waste removed from the separate storm sewer systems and areas listed above in (2), including dredge spoil, accumulated sediments, floatables, and other debris.
5. Ways to ensure that new flood management projects assess the impacts on water quality and examine existing projects for incorporation of additional water quality protection devices or practices. The network of open channels in the community should receive regularly scheduled, routine maintenance to prevent sediment buildup and overgrowth of vegetation.

The program should include local government employee training addressing these prevention measures in government operations (such as park, golf course, and open-space maintenance; fleet maintenance; planning, building oversight, and storm water system maintenance). These practices or programs address non–storm water problems but also have storm water pollution prevention benefits.

Street cleaning can reduce pollutants in runoff if it is performed regularly. Another benefit of street cleaning is that pipes and outlets in detention structures and ponds are less likely to become clogged. Typical street-sweeping requirements for a municipal storm water management plan might include

---

[3]As noted previously in this chapter, maintenance of structural storm water pollution controls has been a major problem that has gone largely unresolved. The EPA is preparing guidance on this topic [63 FR 01735 (Jan. 9, 1998).

- Sweeping of arterial streets 8 times per year, with emphasis on sweeping after deicing and sanding applications
- Sweeping of residential streets 4 times per year
- Removal of deicing and sanding materials from bridges as soon as feasible

Disposal of street-sweeping wastes may pose a problem because of possible high levels of lead, copper, zinc, and other wastes from automobile traffic. Testing of street sweepings may be appropriate to determine appropriate disposal or reuse alternatives. Some municipalities and industries have found that street sweepings can be used as cover in sanitary landfills.

In areas where salt is used, reduced application or alternative agents, consistent with the need for safety, will reduce pollution of area water bodies. Sand is an alternative that is less harmful to vegetation and aquatic life. Storage facilities can be constructed or modified to prevent salt exposure to rainfall. As noted above, additional street sweeping should be scheduled after periods of freezing weather in order to remove road salt and sand from bridges and roadways as soon as is feasible.

Permit Requirements

The phase I MS4s are required to have a comprehensive storm water management program to effectively prohibit the discharge of nonstorm water (e.g., process waste water, sanitary sewage, wash waters, dumped oil, and household hazardous wastes), to the MS4, and to reduce pollutants in discharges from the MS4 to the maximum extent practicable.

Industrial and High-Risk Runoff

A program to identify and control pollutants in storm water discharges to the MS4 from industrial and high-risk facilities such as municipal landfills; other treatment, storage, or disposal facilities for municipal waste (e.g., transfer stations, incinerators); hazardous waste treatment, storage, disposal, and recovery facilities and facilities that are subject to Superfund Reauthorization Act (SARA) Title III, Section 313; and any other industrial or commercial discharges the permittee determines are contributing a substantial pollutant loading to the MS4 shall be implemented. The program shall include inspections and a monitoring program, and a list of industrial storm water sources discharging to the MS4 shall be maintained and updated as necessary.

Inspections at closed landfills and publicly owned treatment works will be accomplished during visual monitoring. The MS4 operator will conduct an inspection program for open landfills, EPCRA, hazardous TSDs, and other high-risk industries.

Monitoring Program

The MS4 operator is required to perform monitoring to estimate the quantity of pollutants discharged from the entire MS4, which will be used to demonstrate improvement in storm water discharge quality. The MS4 operator is required to monitor in wet weather a specified number of times per year and to screen the system in dry weather to locate illicit discharges. Three different types of monitoring programs may be used:

- Dry-weather screening program
- Wet-weather screening program
- Industrial and high-risk runoff monitoring program

The permittees are required [40 CFR 122.26(d)((2)(iii)(C) and (D)] to monitor the MS4 to provide data necessary to assess the effectiveness and adequacy of storm water management plan (SWMP) control measures; estimate annual cumulative pollutant loadings from the MS4; estimate event mean concentrations and seasonal pollutants in discharges from major outfalls; identify and prioritize portions of the MS4 requiring additional controls; and identify water quality improvements or degradation. The permittees are responsible for conducting any additional monitoring necessary to accurately characterize the quality and quantity of pollutants discharged from the MS4.

Due to the variability of storm water discharges, the cost of the monitoring program needs to be balanced with the monitoring objectives and the more important goal of actually implementing controls that will directly affect the quality of the storm water discharged. However, the municipalities must realize that the EPA will have to make future permitting decisions based on the monitoring data collected during the permit term. The public will also be looking for evidence of pollutant reductions. Where the required permit term monitoring proves insufficient to show pollutant reductions, the EPA may be forced to resort to limitations in the next permit. Two types of monitoring are required by the permit: storm event representative monitoring and floatables monitoring.

Representative monitoring. The monitoring of the discharge of representative outfalls during actual storm events will provide information on the quality of runoff from the MS4, a basis for estimating annual pollutant loads, and a mechanism to evaluate reductions in pollutants discharged from the MS4. Results from the monitoring program will be submitted annually on discharge monitoring reports.

(1) Requirements. The permittees are required to monitor for the parameters listed in the permit throughout the permit term. Monitoring will be conducted at the monitoring locations indicated in the permit.

- *Parameters:* The EPA established permit parameter monitoring requirements based on the information available regarding storm water discharges and potential impacts of these discharges.
- *Frequency:* The frequency of annual monitoring is based on monitoring at least one representative storm event per season. The permittees are to monitor once per season. Monitoring frequency is based on the permit year, not a calendar year. The first complete calendar year monitoring could be less than the stated frequency.

(2) Representative monitoring—rapid bioassessment option. Biological monitoring techniques offer the ability to indirectly assess the quality of storm water discharges from the municipal separate storm sewer system by assessing the "health" of the receiving water. Rapid bioassessment protocols evaluate the number, diversity, and relative "pollution tolerance" of aquatic species in the receiving water bodies (e.g., streams, rivers, lakes, estuaries). Either fish or benthic organisms (bottom-dwelling insects, etc., that serve as food supply for higher organisms) can be studied. Comparing the types and numbers of organisms collected from water bodies receiving discharges from the MS4 to those collected from a "reference site" relatively unimpacted by urban runoff provides an indication of how degraded the water body is. For example, a healthy stream would typically have greater species diversification and a higher number of species that require clean water to survive and reproduce. A degraded stream would have relatively fewer species and a larger proportion of species that are tolerant of pollution.

While rapid bioassessments do not directly measure the quality of storm water discharges, they can be an important (and cost-effective) tool in tracking trends in water quality. The permittees have the option of replacing a portion of the parameter representative monitoring required by the permit with a rapid bioassessment monitoring program. Upon approval by the EPA, the permittees may replace the representative monitoring for years 2, 3, and 5 with rapid bioassessment of at least two receiving waters plus a reference site. Representative monitoring of actual storm water discharges will still be required during years 1 and 4.

Floatables monitoring. Floatable surveys shall be made to investigate trends in water quality issues related to artificial debris and floatables. The comparison of yearly survey results should allow the permittees and the EPA to assess the impact of the SWMP elements as they relate to the reduction and elimination of floatables discharge from the MS4.

Storm Water Reporting Requirements

Municipal separate storm sewer systems are required by 40 CFR 122.42(c)(1) to submit an annual status report that includes

- Status of implementing the components of the storm water management program that are established as permit conditions
- Proposed changes to the storm water management program
- Revisions, if any, to the assessment of controls and the fiscal analysis originally reported in the permit application
- Summary of data, including monitoring data, accumulated throughout the reporting year
- Annual expenditures and budget for the year following each annual report
- A summary describing the number and nature of enforcement actions, inspections, and public education programs
- Identification of water quality improvements or degradation

The reports will be used by the permitting authority to aid in evaluating compliance with permit conditions and, where necessary, to modify the permit to address changed conditions. The permittees are required to do annual evaluations on the effectiveness of the SWMP and to institute or propose modifications necessary to meet the overall permit standard of reducing the discharge of pollutants to the maximum extent practicable. Copies of these reports will be made available to the public.

References

Bastian, R. K., and D. A. Hammer, 1993. "The Use of Constructed Wetlands for Wastewater Treatment and Recycling," pp. 59–68. In G. A. Moshiri (ed.), *Constructed Wetlands for Water Quality Improvement,* CRC Press, Boca Raton, FL.

Bingham, D. R., 1994. "Wetlands for Stormwater Treatment," pp. 243–262. In D. M. Kent (ed.), *Applied Wetlands Science and Technology,* Lewis Publishers, CRC Press, Boca Raton, FL.

Birch, P. B., and H. E. Pressley (eds.), 1992. Stormwater Management Manual for the Puget Sound Basin, review draft, Dept. of Ecology Publication No. 90-73.

Brix, H., 1993. "Wastewater Treatment in Constructed Wetlands: System Design, Removal Processes, and Treatment Performance," pp. 9–22. In G. A. Moshiri (ed.), *Constructed Wetlands for Water Quality Improvement,* CRC Press, Boca Raton, FL.

Corbitt, R. A., and P. T. Bowen, 1994. "Constructed Wetlands for Wastewater Treatment," pp. 221–241. In D. M. Kent (ed.), *Applied Wetlands Science and Technology,* Lewis Publishers, CRC Press, Boca Raton, FL.

Environmental Protection Agency, 1988. *Design Manual: Constructed Wetlands and Aquatic Plant Systems for Municipal Wastewater Treatment,* EPA/625/1-88/022, Office Of Research and Development, Washington.

Environmental Protection Agency, 1990. "Urban Targeting and BMP Selection," *Information and Guidance Manual for State Nonpoint Source Program Staff Engineers and Managers,* The Terrene Institute, EPA No. 68-C8-0034.

Environmental Protection Agency, 1992. *Storm Water Management for Industrial Activities: Developing Pollution Prevention Plans and Best Management Practices,* EPA 832-R-92-006, Office of Water, Washington.

Environmental Protection Agency, 1993. *Investigation of Inappropriate Pollutant Entries into Storm Drainage Systems—A User's Guide,* Office of Research and Development, EPA 600/R-92/238, Washington.

Environmental Protection Agency, 1922. "How to Set Up a Local Program to Recycle Used Oil," EPA/530-SW-89-039A.

Environmental Protection Agency Region 6: *Fact Sheet for Draft National Pollutant Discharge Elimination System (NPDES) Permit No. TXS000601, for the City of Corpus Christi Municipal Separate Storm Sewer System, Dallas, TX, date unknown.*

——, August 3, 1994. Tulsa Municipal Separate Storm Sewer System Permit Fact Sheet, Dallas, TX.

——, December 22, 1994. *Fact Sheet for Draft National Pollutant Discharge Elimination System (NPDES) Permit No. OKS000101, for the City of Oklahoma City Municipal Separate Storm Sewer System, Dallas, TX.*

Fields, S., 1993. "Regulations and Policies Relating to the Use of Wetlands for Nonpoint Source Pollution Control," pp. 151–158. In R. K. Olson (ed.), *Created and Natural Wetlands for Controlling Nonpoint Source Pollution,* C.K. Smoley, CRC Press, Boca Raton, FL.

Government Printing Office: *Federal Register,* Office of the Federal Register, National Archives and Records Administration, Washington.

——, 1988. *United States Code,* Washington.

——, July 1, 1992. *Code of Federal Regulations,* Office of the Federal Register, National Archives and Records Administration, Washington.

Hammer, D. A., 1994. "Guidelines for Design, Construction and Operation of Constructed Wetlands for Livestock Wastewater Treatment," pp. 155–181. In P. J. DuBowy and R. P. Reaves (eds.), *Constructed Wetlands for Animal Waste Management: Proceedings of Workshop,* Department of Forestry and Natural Resources, Purdue University, West Lafayette, IN.

King County Solid Waste Division, 1990. "Local Hazardous Waste Management Plan for Seattle-King County: Final Plan and Environmental Impact Statement for the Management of Small Quantities of Hazardous Waste in the Seattle-King County Region."

Mitsch, W. J., 1993. "Landscape Design and the Role of Created, Restored, and Natural Riparian Wetlands in Controlling Nonpoint Source Pollution," pp. 43–70. In R. K. Olson (ed.), *Created and Natural Wetlands for Controlling Nonpoint Source Pollution,* C.K. Smoley, CRC Press, Boca Raton, FL.

Osmond, D. L., et al., 1995. WATERSHEDSS: Water, Soil and Hydro-Environmental/ Decision Support System (can be found at the website: http://h20sparc.wq.ncsu.edu).

Phillips, N., 1992. *Decisionmaker's Stormwater Handbook: A Primer,* The Terrene Institute, Washington.

Schueler, T. R., 1987. *Controlling Urban Runoff: A Practical Manual for Planning and Designing Urban BMPs,* Publication no. 87703, Metropolitan Washington Council of Governments, Washington.

——, P. A. Kumble, and M. A. Heraty, 1992. *A Current Assessment of Urban Best Management Practices: Techniques for Reducing Non-Point Source Pollution in the Coastal Zone,* Publication no. 92705, Metropolitan Washington Council of Governments, Washington.

Shaver, E., 1992. "Sand Filter Design for Water Quality Treatment." In *Stormwater Management: Urban Runoff Management Workshop,* Book 2, Environmental Protection Agency, Washington.

Urbonas, Ben, and Peter Stahre, 1993. *Storm Water: Best Management Practices and Detention for Water Quality, Drainage, and CSO (Combined Sewer Overflow) Management,* Prentice-Hall, Englewood Cliffs, NJ.

Wetzel, R. G., 1993. "Constructed Wetlands: Scientific Foundations Are Critical," pp. 3–8. In G. A. Moshiri (ed.), *Constructed Wetlands for Water Quality Improvement,* CRC Press, Boca Raton, FL.

Chapter

5

Storm Water Associated with Industrial Activity

This is the first of five chapters which deal with the specific requirements for storm water discharges from *industrial* (nonconstruction) facilities. This chapter describes what types of storm water discharges must have permits.

Chapter 6 describes the permit application process and the types of applications available. Chapters 7 and 8 describe how to prepare a *storm water pollution prevention plan* (*SWPPP*) for an industrial facility. Chapter 9 describes the requirements and procedures for storm water sampling. There is additional information on requirements for specific states in App. A.

Regulated Discharges

As noted in Chap. 3, only point source discharges of storm water into waters of the United States are regulated by National Pollutant Discharge Elimination System (NPDES) permits. Chapter 3 defined the terms *storm water, discharge, point source,* and *waters of the United States.*

For industrial storm water discharge permits, only storm water discharges *associated with industrial activity* are required to have permits. The definition of this term is lengthy and detailed. It includes "the discharge from any conveyance which is used for collecting and conveying storm water and which is directly related to manufacturing, processing or raw materials storage areas at an industrial plant. The term does not include discharges from facilities or activities excluded from the NPDES program" [40 CFR 122.26(b)(14)].

The EPA definition also provides several examples of industrial activity:

> ...the term includes, but is not limited to, storm water discharges from industrial plant yards; immediate access roads and rail lines used or traveled by carriers of raw materials, manufactured products, waste material, or by-products used or created by the facility; material handling sites; refuse sites; sites used for the application or disposal of process waste waters; sites used for the storage and maintenance of material handling equipment; sites used for residual treatment, storage, or disposal; shipping and receiving areas; manufacturing buildings; storage areas (including tank farms) for raw materials, and intermediate and finished products; and areas where industrial activity has taken place in the past and significant materials remain and are exposed to storm water [40 CFR 122.26(b)(14)].

The reference to *immediate access roads and rail lines* is limited to those roads which are exclusively or primarily dedicated for use by the industrial facility. The EPA does not intend that these regulations extend to public roads such as county, state, or federal roads and highways. In addition, the materials hauled on these roads must be part of an actual industrial operation, not simply part of a preliminary reconnaissance or sampling program.

The EPA provides examples of material handling activities covered by the storm water regulations: "...material handling activities include the storage, loading and unloading, transportation, or conveyance of any raw material, intermediate product, finished product, by-product or waste product" [40 CFR 122.26(b)(14)].

Pilot plants and research facilities are required to obtain storm water discharge permits, provided that the manufacturing operations of the full-scale facility would require a permit (Environmental Protection Agency, 1992).

One important aspect of the EPA definition is the exclusion of storm water discharges from those portions of the industrial facility which are not actively involved in industrial activities: The term excludes areas located on plant lands separate from the plant's industrial activities, such as office buildings and accompanying parking lots, as long as the drainage from the excluded areas is not mixed with storm water drained from the above-described areas [40 CFR 122.26(b)(14)]. In addition, off-site stockpiles of final product from an industrial facility do not require permit coverage, because they are not located at the site of the industrial facility (Environmental Protection Agency, 1992).

Areas associated with industrial activity do not include commercial or retail facilities. This is an important distinction; in some cases (such as construction), the EPA has chosen to regulate only those activities which are significant enough to be inherently *industrial* in nature.

There is no exclusion for federal or state-owned facilities. Many military facilities have operations that fit within the categories which must receive permit coverage. Some municipally owned facilities are excluded under the Transportation Act of 1991, as described in Chap. 3.

Specific Areas Associated with Industrial Activity

The EPA requires storm water discharge permits only for specific types of industrial activities. The activities requiring permits are defined in two ways: by a narrative description or by a *Standard Industrial Classification* (*SIC*) code. SIC codes are standard numeric codes assigned to each type of industrial process in the United States by the President's Office of Management and Budget (Office of Management and Budget, 1987).

There is an important distinction between these two types of categories: industrial sites identified by SIC code are required to obtain permit coverage only if the *primary site activity* is within the SIC codes listed. If the listed activity is not the primary site activity, it is considered an *auxiliary activity* which does not require permit coverage. For categories defined by a narrative description, however, a permit is required if *any of the described activity* occurs on site. Therefore, the narrative categories are more inclusive.

There are seven categories of industrial activity defined by narrative description:

1. Subchapter N industries
2. Hazardous waste treatment, storage, or disposal facilities
3. Landfills
4. Power generation facilities
5. Sewage treatment plants
6. Construction activities
7. Water quality violators or significant polluters

These categories are described in detail later in this chapter. Construction activities are also described in Chap. 10.

There are five categories of industrial activity defined by SIC codes:

1. Heavy manufacturing
2. Light manufacturing
3. Mining

4. Recyclers
5. Industrial transportation

These categories are also described in detail later in this chapter.

Narrative category: Subchapter N industries

Subchapter N of Title 40 of the *Code of Federal Regulation* (CFR) includes all the effluent guidelines and standards for various types of industrial facilities. Subchapter N contains 40 CFR Sections 401 through 471. Facilities subject to any of the following types of limitations or guidelines under 40 CFR, subchapter N (except facilities which are exempt under the light-industry exclusion) must obtain NPDES permits to discharge storm water:

- Storm water effluent limitation guidelines
- New source performance standards (NSPSs)
- Toxic pollutant effluent standards

According to the Environmental Protection Agency, the industries in these categories have generally been identified by EPA as the most significant dischargers of process wastewaters in the country. As such, these facilities are likely to have storm water discharges associated with industrial activity for which permit applications should be required [55 FR 47989 (Nov. 16, 1990)].

Because these industries are described as a narrative category, a permit is required if *any* of the described activity occurs on site.

The following sections list the industry groups covered by storm water effluent limitations guidelines, new source performance standards, or toxic pollutant effluent standards. Many of these industry groups overlap. Some also overlap with industry groups required to obtain storm water discharge permits under other provisions of the EPA rules.

Storm water effluent limitation guidelines. The following types of industries are required to obtain NPDES storm water discharge permits because they are covered by storm water effluent limitations or guidelines (Environmental Protection Agency, 1992):

- Cement manufacturing (40 CFR 411)
- Feedlots (40 CFR 412)
- Fertilizer manufacturing (40 CFR 418)
- Petroleum refining (40 CFR 419)
- Phosphate manufacturing (40 CFR 422)

- Steam electric power generating (40 CFR 423)
- Coal mining (40 CFR 434)
- Mineral mining and dressing (40 CFR 436)
- Ore mining and dressing (40 CFR 440)
- Asphalt emulsion (40 CFR 443)

Industries subject to storm water effluent limitations are not eligible for the *light-industry* exemption from storm water discharge permitting described later in this chapter.

New source performance standards. Section 306 of the Clean Water Act requires the EPA to develop performance standards for all new sources described in that section. These standards apply to all facilities which go into operation after the date the standards are promulgated. Section 511(c) of the Clean Water Act requires the EPA to comply with the National Environmental Policy Act prior to issuance of a permit under the authority of Section 402 of the CWA to facilities defined as a new source under Section 306.

Facilities which are subject to the performance standards for new sources must provide the EPA with an environmental information document pursuant to 40 CFR 6.101 prior to seeking coverage under this permit. This information is be used by the EPA to evaluate the facility under the requirements of the National Environmental Policy Act (NEPA) in an environmental review. The permittee must obtain a copy of EPA's final finding prior to the submittal of a discharge permit application. The EPA's finding may include a requirement for mitigation. If so, this mitigation must be implemented, or the permit coverage may be terminated.

A *new source* is defined as any source discharging a toxic pollutant, the construction of which is commenced after proposal of an effluent standard or prohibition applicable to such source if such effluent standard or prohibition is thereafter promulgated in accordance with Section 307. NSPSs apply only to discharges from those facilities or installations that were constructed after the promulgation of NSPSs.

Most effluent guidelines listed in 40 CFR Subchapter N include new source performance standards, and facilities subject to these standards are required to submit a storm water permit application.

Industries subject to new source performance standards are not eligible for the *light-industry* exemption from storm water discharge permitting described in this chapter.

Toxic pollutant effluent standards. For purposes of determining storm water discharge permit requirements, *toxic pollutant effluent stan-*

dards include only the standards established pursuant to CWA Section 307(a)(2) and codified at 40 CFR Part 129. Part 129 applies only to manufacturers of certain pesticide products which are defined as toxic pollutants (Environmental Protection Agency, 1993). These pesticide products include aldrin/dieldrin, DDT, endrin, benzidine, and polychlorinated biphenyls (PCBs).

Industries subject to toxic pollutant effluent standards may be eligible for the light-industry exemption from storm water discharge permitting described earlier in this chapter, provided that they meet all other qualifications for this exemption.

Narrative category: Hazardous waste treatment, storage, or disposal facilities

Hazardous waste treatment, storage, or disposal facilities, including those that are operating under interim status or a permit under subtitle C of the Resource Conservation and Recovery Act (RCRA), must obtain NPDES storm water discharge permits. A facility that stores hazardous waste for less than 90 days is not considered to be a treatment, storage, or disposal facility and therefore is not required to submit a storm water discharge permit application (Environmental Protection Agency, 1992). Because these industries are described as a narrative category, a permit is required if *any* of the described activity occurs on site.

Land disposal units and incinerators as well as *boilers and industrial furnaces* (*BIFs*) that burn hazardous waste may receive a diverse range of industrial wastes. Waste receiving, handling, storage, and processing, in addition to actual waste disposal, can be a significant source of pollutants at waste disposal facilities. The EPA has summarized case studies documenting surface water impacts and groundwater contamination of land disposal units. Evaluation of 163 case studies revealed surface water impacts at 73 facilities. Elevated levels of organic chemicals, including pesticides, and metals have been found in groundwater and/or surface water at many sites [55 FR 47989 (Nov. 16, 1990)].

Incinerators and BIFs burn hazardous materials such as spent solvents, contaminated fuels, and so on. The primary purpose of an incinerator is waste disposal, but the primary purpose of a BIF is to provide heat or steam or to generate electricity for use in manufacturing processes.

Incinerators and BIFs will typically manage the same types of wastes as landfills, and therefore present similar risks with respect to waste transportation, handling, and storage. In addition, a wide range of toxic pollutants potentially present in fuel stocks, material accepted for disposal, air emission particulate, and ash at these facilities have the potential to contaminate storm water runoff.

Narrative category: Industrial waste landfills

Landfills, land application sites, and open dumps that receive industrial wastes must obtain NPDES storm water discharge permits. *Industrial waste* is waste received from the manufacturing portions of facilities under any of the other industrial categories under this program. It does include construction debris.

The EPA considers the construction of new cells at a landfill to be routine landfill operations that are covered by the landfill's industrial storm water permit. Therefore, no separate construction permit is required to construct the cell, even if the area disturbed by construction of the cell is greater than 5 acres. However, where a new landfill is being constructed and 5 acres or more of land is disturbed, the construction activity should be covered by a construction permit. Some authorized states may address these situations differently (Environmental Protection Agency, July 1993).

Because these industries are described as a narrative category, a permit is required if *any* of the described activity occurs on site.

Landfills which are capped and closed must be judged on a case-by-case basis. A permit application should be filed for such facilities. The EPA has excluded from coverage under the general permit those storm water discharges from inactive landfills on federal lands where an operator cannot be identified. The EPA is addressing these discharges in conjunction with distinct permitting efforts addressing storm water discharges from inactive mining operations and inactive oil and gas operations on federal lands.

As described in Chap. 3, under Section 1068(c) of the Transportation Act of 1991, the EPA shall not require any municipality with a population of less than 100,000 to apply for or obtain a permit for any storm water discharge associated with most industrial activities. However, these municipalities are required to obtain storm water discharge permits for uncontrolled sanitary landfills which they own or operate. Section 1068(d) of the Transportation Act defines *uncontrolled sanitary landfill* to mean a landfill or open dump, whether open or closed, that does not meet the requirements for run-on and runoff controls established pursuant to subtitle D of the Solid Waste Disposal Act (RCRA). Even landfills that are in compliance with subtitle D requirements and that are owned by small municipalities may be required to obtain NPDES permits for storm water discharges if they are significant contributors of pollutants to waters of the United States or if they contribute to a violation of a water quality standard.

The EPA has published criteria for solid waste disposal facilities, including municipal solid waste landfills (MSWLFs), pursuant to subtitle D of the Solid Waste Disposal Act [56 FR 50978 (Oct. 9, 1991)].

Several provisions of these regulations specifically address run-on and runoff from the active portions of regulated units. Owners or operators of all MSWLF units are required to design, construct, and maintain a run-on control system to prevent flow onto the active portion of the MSWLF unit during the peak discharge from a 25-year storm (40 CFR 258.25). In addition, all MSWLF units are required to design, construct, and maintain a runoff control system from the active portion of the landfill to collect and control at least the water volume resulting from a 24-h, 25-year storm.

The EPA rules issued pursuant to subtitle D of the Solid Waste Disposal Act do not require that the collected runoff be sampled or treated. However, the EPA intended that this runoff be subjected to NPDES permit requirements, which may require sampling and/or treatment. Runoff from the active portion of the unit must be handled so that all MSWLF units are operated in compliance with NPDES requirements [40 CFR 258.27(a)]. Any discharges of a nonpoint source of pollution from an MSWLF unit into waters of the United States must also be in conformance with any established water quality management plan developed under the CWA.

Older landfills are of greatest concern to the EPA because they may have received large volumes of hazardous waste and, in general, their use of design controls was very limited. States have reported to the EPA that of the 1100 municipal solid waste landfills which monitored discharges to surface water, 660 were cited for surface water impacts. The EPA believes that newer and future solid waste landfills may present lower risks because subtitle C regulations keep most hazardous waste out of solid waste landfills. In addition, design controls for solid waste landfills have improved, and are expected to continue to improve with the implementation of subtitle D requirements [56 FR 50981 (Oct. 9, 1991)].

Narrative category: Steam electric power generation

Steam electric power generating facilities, including coal handling sites, must obtain NPDES storm water discharge permits. This would include single-user facilities, such as a steam electric power generating facility for a university campus. However, steam production for heating and cooling is not covered by permit requirements. Cogeneration facilities are regulated if they are based on the use of dual fuels. However, cogeneration facilities based on heat capture only are not regulated.

The EPA originally required storm water discharges from transformer storage sites to receive storm water discharge permits. However, after further investigation, the EPA determined that the Toxic Substances Control Act (TSCA) addresses pollutants associated with transformers that may enter receiving water through storm water discharges. Under TSCA, transformers are required to be stored in a manner that prevents rainwater from reaching the stored PCBs or PCB items [40 CFR 761.65(b)(1)(i)]. Therefore, the EPA does not require storm water discharge permits for transformer storage areas. Storm water discharges from electric substations are also not regulated (Environmental Protection Agency, 1992).

The EPA acknowledges that certain discharges are regulated under the Atomic Energy Act and are therefore exempt from EPA regulation. However, the EPA may require permits from other storm water discharges from nuclear power facilities.

Coal handling activities at coal-fired steam electric plants can be a significant source of pollutants in storm water discharges. Runoff from coal handling areas can have high levels of total suspended solids, sulfate, iron, aluminum, mercury, copper, arsenic, selenium, and manganese as well as an acidic pH (Environmental Protection Agency, 1982). However, coal piles which are located off-site (not at the site of the steam electric power generation) are not required to be permitted, because they are not located on the site of a facility which is engaged in industrial activity (Environmental Protection Agency, 1992).

Spills and leaks from fuel handling sites, including loading and unloading areas and storage tanks, at oil-fired steam electric power generating facilities are potential significant sources of pollutants to storm water runoff. Given the large amounts of oil managed at these facilities, many of the pollutant sources associated with oil handling and storage are expected to be similar to those at petroleum refineries (Environmental Protection Agency, 1979b).

Because these industries are described as a narrative category, a permit is required if *any* of the described activity occurs on site.

As described in Chap. 3, Section 1068(c) of the Transportation Act of 1991 provides that the EPA shall not require any municipality with a population of less than 100,000 to apply for or obtain a permit for most industrial activities. However, power plants owned or operated by such municipalities are still required to obtain permits to discharge storm water.

Narrative category: Sewage treatment

Sewage treatment plants have been required to obtain NPDES permits to discharge treated sewage effluent since the passage of the Clean Water Act. The Water Quality Act of 1987, however, now requires permit coverage for storm water discharges from such facilities.

Storm water discharge permits must be obtained for treatment works treating domestic sewage or any other sewage sludge or wastewater treatment device or system, used in the storage, treatment, recycling, and reclamation of municipal or domestic sewage, including land dedicated to the disposal of the sewage sludge that is located within the confines of the facility. Off-site pumping stations are not required to obtain storm water discharge permits. Sewage treatment plants typically include sludge composting and storage of chemicals such as ferric chloride, alum, polymers, and chlorine. The plants may experience spills and bubble-overs which contribute to storm water pollution. (See Fig. 5.1.)

Only sewage facilities with a design flow of 1.0 million gallons per day (gal/day) or more, or which are required to have an approved pretreatment program under 40 CFR Part 403, are included. Farmlands, domestic gardens, and lands used for sludge management where sludge is beneficially reused and which are not physically located within the confines of the sewage treatment facility, or areas that are in compliance with Section 405 of the Clean Water Act, are not included. (Section 405 of the Clean Water Act regulates the disposal of sewage sludge.) If the facility collects all storm water from the plant site and treats it as part of the normal inflow that is processed through the treatment plant, no storm water discharge permit is required.

As described in Chap. 3, the Transportation Act of 1991 exempted all publicly owned treatment works (POTWs) owned by small municipalities (less than 100,000), regardless of whether they meet the other cri-

Figure 5.1 Chemical handling area at sewage treatment plant.

teria for regulation described above. However, permit coverage may still be required if the POTW is a significant contributor of pollutants to waters of the United States or if it contributes to a violation of a water quality standard.

Because these industries are described as a narrative category, a permit is required if *any* of the described activity occurs on site.

Narrative category: Construction activity

Storm water discharges from construction activities involving at least 5 acres of disturbed land must be covered by an NPDES permit, as discussed in detail in Chap. 10.

Narrative category: Water quality violators or significant polluters

The EPA or an authorized state NPDES program administrator may determine that a particular discharge contributes to a violation of a water quality standard or is a significant contributor of pollutants to waters of the United States. This determination, which is made on a case-by-case basis, will result in the requirement that the facility obtain a storm water discharge permit. To make such a determination, the regulatory agency must follow the specific administrative procedures of the Clean Water Act and other applicable regulations. The determination must be based on fact and cannot be arbitrary.

SIC code category: Heavy manufacturing

Heavy manufacturing includes several industrial activities and processes that are generally conducted outdoors and exposed to storm water:

- SIC code 24: lumber and wood products (except 2434: wood kitchen cabinets). These facilities are engaged in operating sawmills, planing mills, and other mills engaged in producing lumber and wood basic materials.
- SIC code 26: paper and allied products (except 265: paperboard containers and boxes, and 267: converted paper and paperboard products).
- SIC code 28: chemicals and allied products (except 283: drugs, and 285: paints, varnishes, lacquers, enamels, etc.).
- SIC code 29: petroleum refining and related activities.
- SIC code 311: leather tanning and finishing. Such processes use chemicals such as sulfuric acid and sodium dichromate; detergents; and a variety of raw and intermediate materials.

- SIC code 32: stone, clay, glass, and concrete products (except 323: glass products made of purchased glass). These facilities manufacture glass, clay, stone, and concrete products from raw materials in the form of quarried and mined stone, clay, and sand.
- SIC code 33: primary metal industries, including facilities that smelt and refine ferrous and nonferrous metals from ore, pig, or scrap, and those that manufacture related products.
- SIC code 3441: structural metal fabricating.
- SIC code 373: ship and boat building and repair.

When taken as a group, these industries are expected to have one or many of the following activities or processes occurring on-site: storage of raw materials, intermediate products, final products, by-products, waste products, or chemicals outside; smelting; refining; production of significant emissions from stacks or air exhaust systems; loading or unloading of chemical or hazardous substances; use of unhoused manufacturing and heavy industrial equipment; and generation of significant dust or particulates. Accordingly, these are classes of facilities which can be viewed as generating storm water discharges associated with industrial activity requiring a permit.

Because these industrial activities are identified by SIC code, the site is required to obtain permit coverage only if the *primary site activity* is within the SIC codes listed.

Primary metal industries. Primary metal facilities (SIC code 33) are engaged in the manufacturing of ferrous metals and metal products and the primary and secondary smelting and refining of nonferrous metals. In addition, facilities engaged in the molding, casting, or forming of ferrous or nonferrous metals are included in this group.

Due to the nature of processes and activities commonly occurring at these facilities, a number of sources can potentially contribute significant amounts of pollutants to storm water. Sources of pollutants include outdoor storage and material handling activities, particulate- and dust-generating processes, and slag quench processes. Open-air storage and handling of raw materials, products, and wastes are common practices at many of these facilities. In addition, dust- and particulate-generating processes, particularly at smelting and refining facilities, are considered potential sources of pollutants in storm water discharges. Many of these types of facilities also use a high volume of water for operations such as spray quenching, heat treating, and die cooling, which when coupled with the old age of many primary metals facilities, can create the potential for nonstorm water to be discharged to the storm water collection systems.

Ship and boat building and repair. A number of industrial activities at shipbuilding and ship repairing facilities can be significant sources of pollutants to storm water discharges, including improper controls on activities such as ship bottom cleaning, bilge water disposal, loading and unloading of fuels, metal fabrication and cleaning operations, and surface preparation and painting (Environmental Protection Agency, 1979a, 1993a).

SIC code category: Light manufacturing

Light manufacturing includes a large number of industrial activities and processes that may be conducted indoors so that exposure to storm water is minimal:

- SIC code 20: food and kindred products, including process foods such as meats, dairy food, fruit, and flour.
- SIC code 21: tobacco products, including cigarettes, cigars, chewing tobacco, and related products.
- SIC code 22: textile mill products, producing yarn and so on, and/or dye and finish fabrics.
- SIC code 23: apparel and other textile products, which produce clothing by cutting and sewing purchased woven or knitted textile products.
- SIC code 2434: wood kitchen cabinets.
- SIC code 25: furniture and fixtures.
- SIC code 265: paperboard containers and boxes.
- SIC code 267: converted paper and paperboard products (except containers and boxes).
- SIC code 27: printing and publishing, including bookbinding and plate making.
- SIC code 283: drugs (pharmaceuticals).
- SIC code 285: paints, varnishes, lacquers, enamels, and allied products.
- SIC code 30: rubber and miscellaneous plastic products.
- SIC code 31: leather and leather products (except 311: leather tanning and finishing).
- SIC code 323: glass products made of purchased glass.
- SIC code 34: fabricated-metal products (except 3441: structural metal fabricating).

- SIC code 35: industrial and commercial machinery and computer equipment.
- SIC code 36: electronic and other electric equipment and components.
- SIC code 37: transportation equipment (except 373: ship and boat building and repair).
- SIC code 38: instruments and related products, including measuring, analyzing, and controlling instruments; photographic, medical, and optical goods; and watches and clocks.
- SIC code 39: miscellaneous manufacturing industries, including jewelry, silverware, plated ware, musical instruments, dolls, toys, games, sporting and athletic goods, pens, pencils, artists' materials, novelties, buttons, notions, brooms, brushes, signs, burial caskets, and hard surface floor coverings.
- SIC code 4221: farm products warehousing and storage.
- SIC code 4222: refrigerated warehousing and storage.
- SIC code 4225: general warehousing and storage.

Because these industrial activities are identified by SIC codes, the site is required to obtain permit coverage only if the *primary site activity* is within the SIC codes listed.

Under current EPA regulations, these types of industrial activities can be conducted without obtaining a storm water discharge permit, if no material handling equipment or activities, raw materials, intermediate products, final products, waste materials, by-products, or industrial machinery is exposed to storm water [40 CFR 122.26(b)(14)].[1]

When considered as a class, most of the activity at these facilities is undertaken in buildings; emissions from stacks will be minimal or nonexistent; the use of unhoused manufacturing and heavy industrial equipment will be minimal; outside material storage, disposal, or han-

[1]After the EPA published the regulations concerning phase I of the National Storm Water Program, the Natural Resources Defense Council (NRDC) filed suit against the EPA on December 10, 1990. The NRDC argued that (1) Congress did not allow an exemption from permit requirements for industries without storm water exposure, and (2) these industries should therefore be required to obtain permit coverage in the same manner as the heavy industries discussed above. The ninth circuit court of appeals remanded this issue to the EPA on June 4, 1992. The EPA published a notice in the *Federal Register* on January 9, 1998, proposing to extend the waiver of permit requirements for facilities that have no exposure to storm water. If this proposal is finalized, only those industrial facilities that have exposure to storm water will be required to obtain storm water discharge permits. This will affect not only the "light industries," but all other industries as well.

dling generally will not be a part of the manufacturing process; and generating significant dust or particulates would be atypical. As such, these industries are more akin or comparable to retail, commercial, or service industries, and storm water discharges from these facilities are not "associated with industrial activity."

The simple fact that a manufacturing building, rail spur, or access road is exposed to storm water does not necessarily constitute exposure of materials. In addition, a covered dumpster does not constitute exposure of materials, provided that the container is completely covered and nothing can drain out of holes in the bottom or be lost during loading onto a garbage truck. An air vent may constitute exposure of materials if particulate matter has accumulated around the air vent and is subject to washoff by storm water. Therefore, air vents must be considered on a case-by-case basis.

Some of these types of facilities handle oil drums or other contained materials which are exposed during loading and unloading operations. If there is a reasonable potential for leaks or spills from these containers which could be exposed to storm water, discharges from the exposed area would be subject to storm water permitting requirements (Environmental Protection Agency, July 1993).

If material handling equipment or activities, raw materials, intermediate products, final products, waste materials, by-products, or industrial machinery is stored outside in a structure with a roof but with no sides, and if wind-blown rain, snow, or runoff comes into contact with the equipment, material, or activities, then discharges from the area will be subject to storm water permitting requirements (Environmental Protection Agency, 1993).

Because of the potential savings available by avoiding permit responsibilities for these types of facilities, it is generally worthwhile to make some changes in the facility to eliminate storm water exposure. For example, changing from uncovered to covered dumpsters, eliminating or covering outside materials storage areas, and cleaning up exposed waste products or scrap machinery may all be cost-effective and advisable, if these steps eliminate storm water exposure and therefore eliminate storm water discharge permit requirements.

SIC code category: Mining and oil and gas extraction

According to the EPA,

> oil, gas, and mining facilities are among those industrial sites that are likely to discharge storm water runoff that is contaminated by process wastes, toxic pollutants, hazardous substances, or oil and grease. Such contamina-

> tion can include disturbed soils and process wastes containing heavy metals or suspended or dissolved solids, salts, surfactants, or solvents used or produced in oil and gas operations [55 FR 47989 (Nov. 16, 1990)]

However, the 1987 Water Quality Act recognized that the storm water in some facilities is "channeled around plants and operations through a series of ditches and other structural devices in order to prevent pollution of the storm water by harmful contaminants." Therefore, there are significant exemptions for facilities which practice good storm water management.

Several categories of active or inactive mining operations and oil and gas exploration, production, processing, or treatment operations, or transmission facilities are required to obtain storm water permit coverage, including the following:

- SIC code 10: metal mining
- SIC code 11: anthracite mining
- SIC code 12: coal mining
- SIC code 13: oil and gas extraction
- SIC code 14: nonmetallic minerals, except fuels

Because these industrial activities are identified by SIC code, the site is required to obtain permit coverage only if the *primary site activity* is within the SIC codes listed.

Active mining facilities. Several major exemptions will allow many mining facilities to operate without permit coverage. Because of these exemptions, only contaminated mine discharges are required to be permitted. Any mining area where storm water discharges do not contact any overburden, raw material, intermediate products, finished products, by-products, or waste products is exempt from permit requirements.

Roads for mining operations will not be required to obtain storm water discharge permit coverage unless storm water runoff from such roads mixes with storm water that is contaminated by contact with overburden, raw materials, intermediate products, finished products, by-products, or waste products. When roads are constructed from materials such as overburden or by-products, an application for an NPDES storm water discharge permit is required (Environmental Protection Agency, 1992).

Mining claims. Sites where mining claims are being maintained prior to disturbances associated with the extraction, beneficiation, or processing of mined materials are not included in the permit requirements. Similarly, no discharge permit is required for sites where minimal activities required for the sole purpose of maintaining the mining claim are undertaken.

Inactive mines. Inactive mining areas which have an identifiable owner-operator must be permitted under the same conditions as an active mine, because the EPA believes that some of these mining sites represent a significant source of contaminated storm water runoff. An inactive mining site is one where there has been past extraction, beneficiation, or processing of mining materials, but with no current active mining. However, in such cases the exclusion discussed above for uncontaminated discharges will still apply. The EPA is issuing a separate general permit for inactive mining areas on federal lands which have no identifiable owner or operator.

Reclaimed mines. Inactive mines which have undergone a complete reclamation do not require storm water discharge permits. These include the following:

- *Reclaimed coal mines:* areas of coal mining operations which no longer meet the definition of a reclamation area under 40 CFR 434.11(1) because the performance bond issued to the facility by the appropriate SMCRA authority has been released.
- *Other reclaimed mines:* areas of noncoal mining operations which have been released from applicable state or federal reclamation requirements after December 17, 1990. The EPA decided to require storm water discharge permits for reclaimed mining areas released prior to this date because "EPA does not have sufficient evidence to suggest that each state's previous reclamation rules and/or Federal requirements, if applicable, were necessarily effective in controlling future storm water contamination" [55 FR 47989 (Nov. 16, 1990)].

Oil and gas extraction. Most oil and gas operations are exempt from storm water discharge permit requirements. However, if a facility meets one of the following criteria, a discharge permit must be obtained:

- The facility has had a reportable quantity discharge in storm water at any time since November 16, 1987, pursuant to 40 CFR 117.21, 40 CFR 302.6, or 40 CFR 110.6.
- The facility contributes to a violation of a water quality standard.

A *reportable quantity* of oil is the amount that violates applicable water quality standards or causes a film or sheen or a discoloration of the water surface or adjoining shorelines or causes a sludge or emulsion to be deposited beneath the surface of the water or upon adjoining shorelines (40 CFR 110.6). Reportable quantities for other substances are listed in terms of pounds released during any 24-h period, in 40 CFR 117.3 and 40 CFR 302.4.

Information from sources such as non–point source assessments developed pursuant to Section 319(a) of the CWA indicate that significant water quality impacts can be caused by wet-weather failure of on-site waste disposal systems at oil and gas exploration and production operations (such as storm-induced overflows of reserve pits used to hold spent drilling muds and cuttings).

The American Petroleum Institute (API) estimates that there are about 850,000 active oil and gas wells, 219,000 tank batteries, and 150,000 injection wells in the United States. The API also estimates that *spill prevention control and countermeasure* (*SPCC*) plans have been developed for about 130,000 of these facilities. Many of these sites include multiple components (e.g., active wells, a tank battery, and injection wells).

Oil and gas facilities that discharge a reportable quantity of oil or a hazardous substance in storm water after October 1, 1992, must submit a Notice of Intent (NOI) to be covered by the EPA general permit within 14 days of their first knowledge of the release, and they must prepare and begin complying with a storm water discharge permit within 60 days.[2]

SIC code category: Recyclers

Some types of industrial activities which are concerned with the salvage and reclamation of materials are required to obtain storm water discharge permit coverage. These include metal scrap yards, battery reclaimers, salvage yards, and automobile junkyards, including but not limited to those classified as

- SIC code 5015: motor vehicle parts, used—wholesale or retail
- SIC code 5093: scrap and waste materials, including the following wholesale businesses: automotive wrecking for scrap, bag reclaiming, waste bottles, waste boxes, fur cuttings and scraps, iron and steel scrap, general junk and scrap, metal and waste scrap, nonfer-

[2] The procedure for filing a Notice of Intent for general permit coverage is described in Chap. 6, and the requirements for a storm water pollution prevention plan are described in Chaps. 7 and 8.

rous metals scrap, waste oil, plastics scrap, rags, rubber scrap, scavenging, scrap and waste materials, textile waste, wastepaper (including paper recycling), and wiping rags (including washing and reconditioning).

Figure 5.2 illustrates a small automobile parts recycling operation. Only *industrial* recyclers are included in the storm water regulations. *Commercial* recycling operations, such as gas stations or automotive repair shops that collect tires or batteries, or municipal waste collection sites that collect bottles, cans, and newspapers, are not regulated.

According to EPA, automotive fluids and greases from automobile drivelines are a significant potential source of pollutants to storm water discharges from automobile junkyards. Drivelines include the engine, transmission, differential or transaxle, fuel, brake, and coolant (radiator) systems. Automotive fluids and greases from these areas typically include engine oil, fuel, transmission fluid or oil, rear-end oil, suspension joint and bearing greases, antifreeze, brake fluid, power steering fluid, and the oil and grease leaking from and covering various components (e.g., oil and grease on the exterior of an engine). The procedures used for fluids capture during the dismantling process will affect the potential to contribute pollutants to storm water.

Figure 5.2 Small auto parts recycling operation.

Figure 5.3 Small battery recycling area.

Based on an evaluation of the battery reclamation industry, the EPA also identified handling, storage, and processing of lead acid batteries, as well as by-product and waste handling at reclamation facilities, as having a significant potential for pollutants in storm water discharges. (See Fig. 5.3.)

Because these industrial activities are identified by SIC code, the site is required to obtain permit coverage only if the *primary site activity* is within the SIC codes listed. If this is an auxiliary activity of the site, permit coverage is not required.

SIC code category: Industrial transportation

Several types of industrial transportation activities are included in the storm water discharge permit requirements:

- SIC code 40: railroad transportation
- SIC code 41: local and interurban highway passenger transit
- SIC code 42: trucking and warehousing (except 4221: farm products warehousing and storage, 4222: refrigerated warehousing trucking and warehousing and storage, and 4225: general warehousing and storage)
- SIC code 43: U.S. Postal Service

- SIC code 44: water transportation
- SIC code 45: transportation by air
- SIC code 5171: petroleum bulk stations and terminals

Discharges must be permitted only for facilities that provide vehicle maintenance shops, equipment-cleaning operations, or airport deicing operations. Only those portions of the facility that are involved in vehicle maintenance (including vehicle rehabilitation, mechanical repairs, painting, fueling, and lubrication), equipment cleaning operations, or airport deicing operations are covered by the permit requirements. Parking lots used to store vehicles prior to maintenance are considered to be a component of the vehicle maintenance activity and should therefore be covered by a storm water discharge permit.

The presence of a vehicle maintenance or equipment-cleaning operation on an industrial site does not trigger a storm water discharge permit requirement unless industrial transportation (as defined using the SIC codes listed above) is the *primary* industrial activity of the site. (See Fig. 5.4.)

Figure 5.4 Industrial transportation maintenance facility.

An off-site vehicle maintenance facility supporting one company would not be required to apply for a storm water discharge permit if that company were not primarily engaged in providing transportation services and therefore would not be classified as SIC code 42. The maintenance facility would be considered an auxiliary operation to the manufacturing facility. The EPA has determined that off-site vehicle maintenance facilities which primarily service trucks used for local transportation of goods or for local services are generally considered supporting establishments that do not assume a transportation SIC code. Such facilities are classified according to the SIC code of the facility they support (Environmental Protection Agency, 1993b).

If the maintenance facility is located on the same site as the manufacturing operation, it is included in the areas associated with industrial activity and must therefore be covered by a storm water discharge permit (Environmental Protection Agency, 1992).

The operation of fire trucks and police cars is classified under public order and safety (SIC code 92). Therefore, the operator of a facility primarily engaged in servicing these vehicles is not required to apply for a storm water discharge permit (Environmental Protection Agency, 1992).

Gas stations and commercial automotive repair facilities are not included in the definition of *industrial transportation* because these facilities are commercial or retail in nature.

Figure 5.5 presents a flowchart illustrating the decisions that must be made in order to determine whether a particular transportation facility would be required to obtain a storm water discharge permit.

Transportation facilities owned by municipal entities. As noted in Chap. 3, by Section 1068(c) of the Transportation Act of 1991, the EPA shall not require any municipality with a population of less than 100,000 to apply for or obtain a permit for most industrial activities. Therefore, vehicle maintenance facilities and other transportation-related facilities owned and operated by small municipalities may operate without a storm water discharge permit. For this reason, school bus maintenance facilities owned and/or operated by municipal entities (including school boards, school districts, or other municipal entities) may not be required to apply for NPDES storm water discharge permits. However, private-contract school bus companies are required to apply (Environmental Protection Agency, 1992).

Municipalities are required to obtain storm water discharge permits for airports which they own or operate. Permit coverage may also be required for other industrial transportation facilities if the facility is a significant contributor of pollutants to waters of the United States or if it contributes to a violation of a water quality standard.

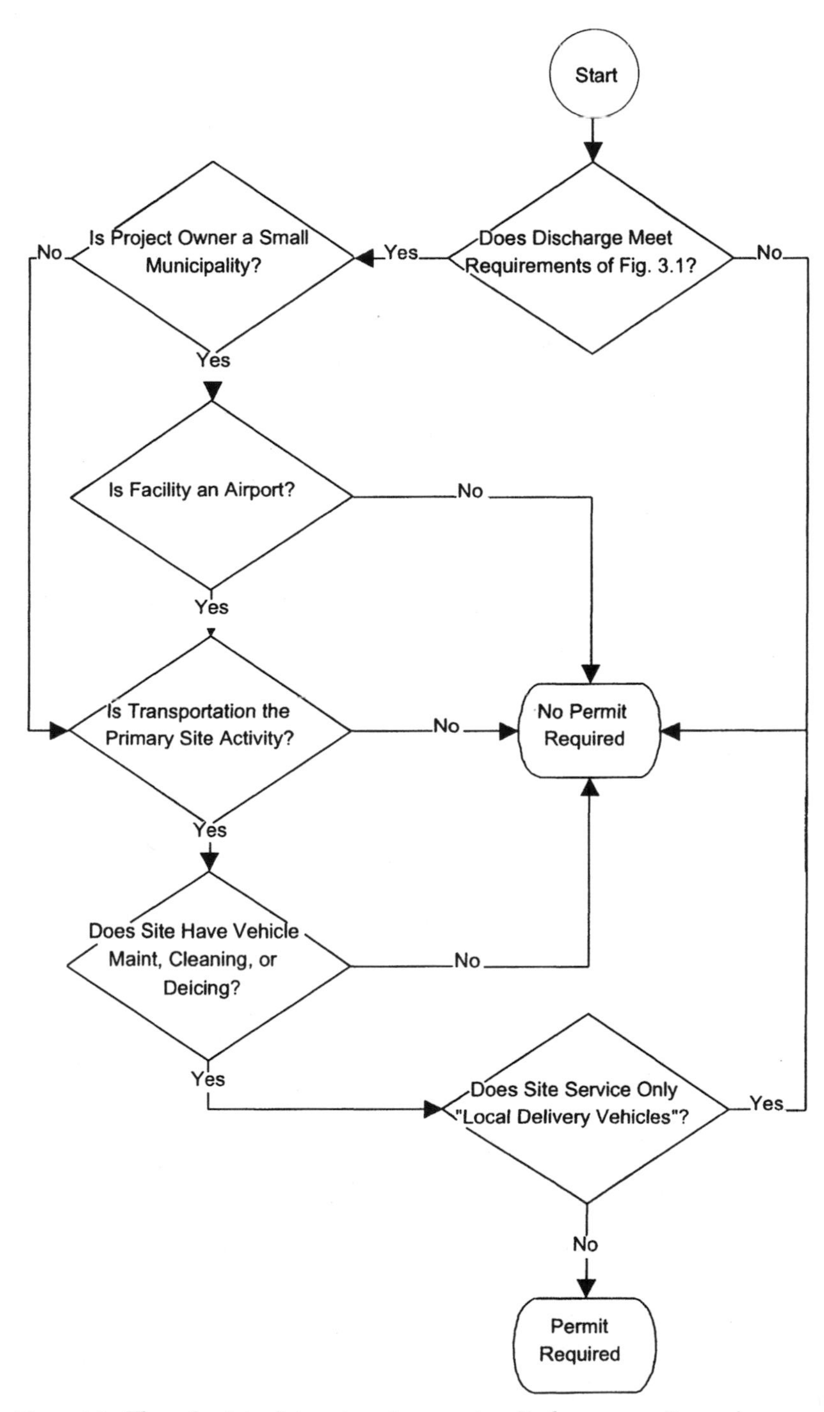

Figure 5.5 Flow chart to determine storm water discharge permit requirements for industrial transportation facilities.

Airport requirements. Airports or airline companies must apply for a storm water discharge permit for locations where deicing chemicals are applied. This includes, but is not limited to, runways, taxiways, ramps, and areas used for the deicing of airplanes. The operator of the airport should apply for the storm water discharge permit, while the individual airline companies should be included as coapplicants. The EPA has the discretion to issue individual permits to each discharger or to issue an individual permit to the airport operator and have other dischargers to the same system act as copermittees. Facilities primarily engaged in performing services that incidentally use airplanes (such as crop dusting or aerial photography) are classified according to the service performed (Environmental Protection Agency, 1992).

Deicing activities at airports can be a significant source of pollutants to storm water discharges. The amount of deicing fluids used depends on the temperature and the amount and type of precipitation. For example, freezing rain may require more deicing fluids than many snowfalls. Ethylene glycol, urea, and ammonium nitrate are the primary ingredients of other deicing compounds used at airports. These chemicals can have a significant oxygen demand in water. When deicing operations are performed, large volumes of ethylene glycol are sprayed on aircraft and runways. Data from Stapleton International Airport in Denver, Colorado, indicate that storm water discharges contained levels of up to 5050 milligrams per liter (mg/L) ethylene glycol during a monitoring period from December 1986 to January 1987. Deicing fluids have been implicated in several fish kills across the nation.

References

Environmental Protection Agency, 1979a. *Development Document for Proposed Effluent Limitations Guidelines and Standards for the Shipbuilding and Repair Point Source Category,* EPA 440/1/-79/076-b, December.

——, 1979b. *Development Document for Effluent Limitations Guidelines and Standards for Pretreatment Standards for the Petroleum Refineries Point Source Category,* EPA 440/1/-79/014b.

——, 1982. *Final Development Document for Effluent Limitations Guidelines and Standards and Pretreatment Standards for the Steam Electric Point Source Category,* EPA No. 440/1-82/029.

——, 1992. *NPDES Storm Water Program Question and Answer Document,* vol. 1, EPA 833-F-93-002, March 16.

——, 1993a. *Proposed Guidance Specifying Management Measures for Sources of Nonpoint Pollution in Coastal Waters,* EPA 340-B-92-002, Office of Water, January.

——, 1993b. *NPDES Storm Water Program Question and Answer Document,* vol. 2, EPA 833-F-93-002B, July.

Government Printing Office: *Federal Register,* Office of the Federal Register, National Archives and Records Administration, Washington.

——, 1992. *Code of Federal Regulations,* Office of the Federal Register, National Archives and Records Administration, Washington, July 1.

Office of Management and Budget, 1987. *Standard Industrial Classification Manual,* Executive Office of the President, Washington.

Chapter

6

Industrial Storm Water Discharge Permit Applications

This is the second of five chapters (Chaps. 5 to 9) which deal with the specific requirements for storm water discharges from industrial (non-construction) facilities. This chapter describes the permit application process and the types of applications available.

EPA Storm Water Permitting Strategy

In establishing phase I of the National Storm Water Program (NSWP), EPA set up a "risk-based" permitting strategy which allows for regulation of storm water discharges in at least four tiers [57 FR 11393 (Apr. 2, 1992)]:

- *Tier I: baseline permitting*—permits that establish a minimum set of requirements which must be met by all permittees, regardless of location or type of discharge.
- *Tier II: watershed permitting*—permits that establish a set of more stringent, water-quality-based requirements for targeted permittees discharging into a specific watershed.
- *Tier III: industry-specific permitting*—permits that establish a set of requirements which must be met by permittees within a certain industrial classification, regardless of the location of the discharge.
- *Tier IV: facility-specific permitting*—permits that establish a set of requirements which must be met by a specific permittee for a spe-

cific discharge. Some facilities which are not eligible for general permits or group permit applications must submit individual permit applications. EPA or state regulators may require an individual permit application from any regulated discharger.

It is important not to misinterpret the list of tiers as a list of four separate "phases" through which the NSWP will pass. In fact, it is likely that all four tiers of permitting will coexist, each used as appropriate for local circumstances. The EPA and various states are already using each of the four tiers of regulation established according to this strategy.

The primary purpose of baseline permitting, through the baseline general permits, is to provide an expedient means of offering permit coverage to most regulated industries. During the 5-year duration of the initial baseline general permits (1992 to 1997), the EPA gathered data to assess the effectiveness of the general permits. As the initial baseline general permits expired, they have been replaced by industry sector permits, facility-specific permits, or revised baseline general permits, as needed.

Types of NPDES Storm Water Permit Applications

Initially, three types of permit applications were available for complying with the National Pollutant Discharge Elimination System (NPDES) storm water discharge permit requirements:

1. *General permits.* Most authorized states have issued general permits for many types of industrial storm water discharges. For nonauthorized states, the EPA published two general permits in September 1992: one industry-specific permit covering construction activities only, and one baseline general permit covering all other types of industrial activities. The 1992 EPA baseline general permit for industrial discharges expired in 1997, and it has now been replaced by the EPA multisector general permit for industrial activities (discussed under item 2 below). Another separate general permit is planned for inactive mining, landfills, and oil and gas operations on federal lands. Coverage under a general permit is usually obtained by filing a Notice of Intent (NOI). Most authorized states have issued NOI forms for general permit coverage.

2. *Group applications.* These applications were received in two parts in 1991 and 1992. Each application represented a group of similar industrial facilities. About 1200 such applications, including a total of about 60,000 facilities, were processed by EPA. As a result of the work done in processing the group applications, the EPA issued a *multisec-*

tor general permit providing coverage for 29 different industry sectors in 1995. When EPA issued the multisector general permit, it was offered as a permit option not only for facilities that participated in the original group application process, but also for any facility eligible for general permit coverage within any of the sectors included in the permit. When the 1992 baseline general permit expired in 1997, EPA expanded the 1995 multisector general permit to provide a complete replacement for the baseline general permit. This was done by expanding the coverage of some of the original 29 industry sectors, and by adding a new industry sector to cover those industries not already included in the original 29.

3. *Individual permit applications.* These are still required for certain facilities that are not eligible for coverage under a baseline or multisector general permit.

The EPA and most state regulatory agencies prefer that as many discharges as possible be covered under a general permit, because these agencies do not have the resources to process a large number of individual permit applications. However, individual permit applications are required for the following discharges:

- *Discharges subject to more stringent storm water effluent limitation guidelines.* The EPA multisector general permit offers coverage to storm water discharges subject to certain effluent limitation guidelines, such as runoff from phosphate fertilizer manufacturing, asphalt emulsion manufacturing, storage piles at cement kilns, dewatering discharges at sand and gravel mines, and coal piles at steam electric facilities. Before the multisector permit was available, such discharges had to be covered under an individual permit. However, the multisector general permit does not authorize all storm water discharges subject to effluent guidelines. For example, storm water discharges subject to the effluent guidelines at 40 CFR Part 436, the effluent guidelines for mine drainage under 40 CFR Part 440, and the effluent guidelines for acid or alkaline mine drainage under 40 CFR Part 20434 are not covered under the multisector general permit. These discharges must be covered by an individual permit.
- *Discharges covered by existing permit.* Storm water discharges which are already covered by an individual NPDES permit are not eligible for general permit coverage. However, facilities with an existing NPDES permit for process wastewaters and/or other non–storm water discharges are allowed to obtain general permit coverage for their storm water discharges. The discharges covered by the individual permit and the discharges covered by the general

permit may be mixed prior to discharge. However, the individual permit should address the monitoring requirements and compliance point for numeric limitations. Storm water discharges that were subject to an NPDES permit that was terminated by the permitting authority are not eligible for coverage under the multisector permit. However, storm water discharges that were subject to a permit that was terminated as a result of the permittee's request are eligible for coverage.

- *Discharges excluded by EPA on the basis of water quality concerns, and discharges that would adversely affect a listed endangered or threatened species.* The multisector general permit prohibits coverage for facilities that have an adverse impact on endangered species. Under these circumstances an individual storm water discharge permit would be issued and consultation undertaken. The multisector permit requires the permittee to certify that there will be no adverse impact, as described later in this chapter.
- *Discharges mixed with nonstorm water.* As noted later in this chapter, the EPA general permit allows certain types of nonstorm discharges to be mixed with storm water discharges, without requiring additional permit coverage. The EPA recognizes that discharging some classes of nonstorm water via separate storm sewers or otherwise mixed with storm water discharges is largely unavoidable and/or poses little, if any, environmental risk. However, where a storm water discharge is mixed with *unallowable* types of nonstorm water, the discharger should gain permit coverage of the non–storm water portion of the discharge. Some states do not provide coverage for any nonstorm discharges under the general permits for storm water discharges.

State Permitting

As described in App. A, states which administer their own NPDES permitting programs deal with general permits, individual permits, and group permit applications according to their own laws and regulations. As noted in Chap. 1, most states have the authority to issue individual and general permits for storm water discharges. Within these states, the general permits for industrial and construction discharges may differ from the EPA general permits. However, there are also many similarities. Some states have chosen to issue general permits for additional industrial categories.

All state permit programs must meet the minimum requirements established by EPA under the Clean Water Act, as noted in Chap. 1. However, it is important to recognize that the permits issued by EPA and by the individual states are merely *tools* to implement the program regulations. This means, e.g., that a permit issued by EPA for use in nonauthorized states is not binding on an authorized state. The authorized states customarily review and consider the permits issued by EPA. However, each state ultimately decides upon the appropriate permit requirements for itself. As long as these permit requirements satisfy the requirements of the regulations established by EPA, they may differ from the permits issued by EPA for use in nonauthorized states. In this way, it is possible that the requirements of a particular state permit may be less stringent than the corresponding requirements of an EPA permit. In all cases, however, the state permit requirements must meet the minimum standards of the Clean Water Act, as explained in Chap. 1.

Municipal Notification and Requirements

All industrial facilities which discharge storm water into a large or medium-size municipal separate storm sewer system (MS4) must submit a notification to the operators of the storm sewer systems before the first discharge, including a copy of the Notice of Intent.

As described in Chap. 1, a large MS4 is one which serves an urbanized population of at least 250,000. A medium-size MS4 is one which serves an urbanized population of 100,000 to 250,000. An MS4 includes not only enclosed storm sewer systems but also "roads with drainage systems, municipal streets, catch basins, curbs, gutters, ditches, manmade channels, or storm drains." The system may by operated by an incorporated city, county, special district, or any other public entity created under state law.

During the early portion of the development of regulations to implement the National Storm Water Program, the EPA considered the possibility of regulating industrial storm water discharges by municipal permit requirements only. In other words, industrial facilities discharging into a large or medium-size municipal storm sewer system would not have been required to obtain EPA or state permits directly. Instead, these dischargers would have been indirectly regulated by the requirements imposed on the municipal separate storm sewer systems receiving discharges from these industrial facilities.

Although the EPA eventually decided to require discharge permits from all industrial dischargers, many industries are still likely to be regulated by municipal operators. Some of the requirements of these municipal operators may exceed the requirements of the facility's NPDES permit for storm water discharges.

The application requirements for large and medium-size MS4s impose special conditions with respect to certain types of industrial facilities, including

- Landfills
- Hazardous waste treatment, disposal, and recovery facilities
- Facilities subject to SARA Title III, Section 313 regulations
- Facilities contributing a substantial pollutant load to the municipal separate storm sewer system

For these types of facilities, the storm water management program developed by the municipal operator must identify priorities and procedures for inspections and for establishing and implementing control measures for these discharges. The municipal operator must also implement a monitoring program for discharges from these facilities. The municipal operator is required to effectively prohibit non–storm water discharges.

If an industrial facility discharges into a nonmunicipal storm sewer system (such as a federal government system), then the facility must still obtain an NPDES storm water discharge permit. The industrial facility may also choose to operate as a copermittee with the operator of the nonmunicipal storm sewer system.

In the preliminary rules for phase II of the National Storm Water Program, EPA has proposed a major expansion in the number of municipal entities required to obtain storm water discharge permit coverage [63 FR 01735 (Jan. 9, 1998)]. This expansion could mean that many more industrial facilities will be required to notify the operators of their municipal separate storm sewer system when they apply for permit coverage for their storm water discharges. Chapter 4 provides further details on the municipal permit requirements, including the proposed phase II requirements.

Individual Permit Application

Dischargers not eligible for coverage under a general permit must apply for coverage under an individual NPDES permit for each discharge. As noted previously, EPA or state regulators may also direct any regulated discharger to submit an individual permit application.

Within nonauthorized states, the requirements for an individual permit application are reflected in EPA Form 1 and Form 2F. There are simplified application requirements for construction activities. For other industrial activities, however, these forms require the development and submittal of relatively detailed site-specific information, including

- A site map showing topography and/or drainage areas and site characteristics

- An estimate of impervious surface areas and total area drained by each outfall
- A narrative description of significant materials and materials management practices
- A certification that outfalls have been tested or evaluated for nonstorm water discharges
- Information on significant leaks and spills of toxic or hazardous pollutants that occurred at the facility in the last 3 years
- Sampling data from one representative storm event

This information is intended to be used to develop the site-specific conditions generally associated with individual permits. The following sections describe each portion of the permit application in greater detail.

Site maps and site description

The site map, including topography and site drainage, should have arrows indicating the site drainage and entering and leaving points. Existing permit application regulations at 40 CFR 122.21(f)(7) require all permit applicants to submit as part of EPA Form 1 a topographic map extending 1 mi beyond the property boundaries of the source and depicting

- The facility and each intake and discharge structure
- Each hazardous waste treatment, storage, or disposal facility
- Each well where fluids from the facility are injected underground
- Those wells, springs, other surface water bodies, and drinking water wells listed in the map area in public records or otherwise known to the applicant within 0.25 mi of the facility property boundary [47 FR 15304, Apr. 8, 1982]

A narrative description must also be submitted to accompany the drainage map. The narrative will provide a description of on-site features including existing structures (buildings which cover materials and other material covers; dikes; diversion ditches, etc.) and nonstructural controls (employee training, visual inspections, preventive maintenance, and housekeeping measures) that are used to prevent or minimize the potential for release of toxic and hazardous pollutants; a description of significant materials that are currently being treated or in the past have been treated, stored, or disposed of outside; and the method of treatment, storage, or disposal used. The narrative will include a description of activities at material loading and unloading areas; the location, manner, and frequency with which pesticides, herbicides, soil conditioners, and fertilizers are applied; a description of the soil; and a description of the areas which are predominately responsible for first flush runoff (defined in Chap. 9).

Description of significant materials and management practices

The application must include a description of each past or present area used for outdoor storage or disposal of significant materials, including hazardous substances, fertilizers, pesticides, and raw materials used in the production or processing of food. The list must include any chemical the facility is required to report pursuant to Section 313 of Title III of SARA. The EPA does not require information on past practices occurring prior to 3 years before the date on which the application is submitted.

Certification of nonstorm discharges

Permit applicants for storm water discharges associated with industrial activity must also certify that all the outfalls covered in the permit application have been tested or evaluated for non–storm water discharges which are not covered by an NPDES permit. If this determination can be made by inspection of schematics or piping diagrams, then no testing is required. Chapter 7 provides more details on conducting evaluations for non–storm discharges.

This requirement is included to assist operators of municipal separate storm sewer systems in meeting their requirement to effectively prohibit non–storm water discharges to the storm sewer system. The certification requirement would not apply to outfalls where storm water is intentionally mixed with process wastewater streams which are already identified in and covered by a permit.

Sampling data

The following sampling data must be provided for one representative storm event:

1. Laboratory analysis results for any pollutants limited in an effluent guideline to which the facility is subject
2. Laboratory analysis results for any pollutant listed in an NPDES permit for process wastewater
3. Laboratory analysis results for oil and grease, pH, biochemical oxygen demand (BOD), chemical oxygen demand (COD), total suspend-

ed solids (TSS), total phosphorus, and nitrate plus nitrite and total kjeldahl nitrogen

4. Laboratory analysis results for any pollutant known to be in the discharge
5. Flow measurements or estimates of flow during the storm event
6. Date and duration of storm event

Chapter 9 provides more information on storm water sampling.

General Permit Coverage

As noted in Chap. 1, a series of court rulings have established the authority of the EPA to utilize innovative concepts such as general permits to regulate discharges under the Clean Water Act. The EPA has used general permits as a tool to accommodate the large number of dischargers included in the National Storm Water Program. Most traditional NPDES permits have been individual permits, with requirements specific to the facility named in the permit. In addition, the traditional NPDES permit gives numeric effluent limitations on various pollutants and specifies a minimum discharge sampling interval.

In contrast to the traditional NPDES permit requirements, general permits for storm water discharges have requirements which are broad enough to be applied to a wide range of industrial facilities in many different areas of the United States. General permits do not require treatment, but instead emphasize *best management practices,* many of which are nonstructural.

EPA is now in its second generation of general permits. The baseline general permit, which was available from 1992 through 1997, was almost totally devoid of numeric effluent limitations. The EPA multisector general permit, issued in 1995, is more like the traditional NPDES permit, because it includes effluent limitation guidelines for some types of discharges. However, it is still a general permit, with a strong emphasis on best management practices.

Because of the fundamental differences between the EPA general permit for storm water discharges and previous NPDES permits, some dischargers are confused about exactly how to meet the new permit requirements. The new EPA general permit is so flexible that it is almost like being able to "write your own permit." Some people, even representatives of dischargers, are uncomfortable with having this much flexibility. In the EPA's view, however, this flexibility provides dischargers with the opportunity to develop best management practices which are the most appropriate and effective for each facility.

The EPA multisector general permit requires the submittal of a Notice of Intent (NOI), which states the permittee's intention to discharge according to the terms and provisions of the general permit. Compliance with the provisions of the general permit involves the preparation and maintenance of a storm water pollution prevention plan (as described in Chaps. 7 and 8). In addition, some types of facilities are required to perform storm water discharge sampling (as described in Chap. 9).

The EPA does not currently assess permit fees for general permit coverage, but several states do. The EPA has worked on a federal permit fee system which would impose fees on permit holders in states without NPDES permit authority. This fee system was developed under the 1990 Budget Reconciliation Act, which required the EPA to develop a system of user fees to pay for services rendered. It is reasonable to expect the EPA to implement a fee system for storm water discharge permit applications.

Industry sector approach

To develop the multisector storm water general permit, EPA relied on data collected through the group application process in 1991 and 1992. The EPA divided the industrial activities represented by the group application members into different industry sectors. EPA reviewed the descriptions of industrial activities, materials exposed to storm water, and best management practices on a sector-by-sector basis. Based upon this review, EPA identified pollutants of concern in each industry sector, sources of these pollutants, and best management practices used by industry to control the pollutants of concern. The multisector general permit requires the implementation of a pollution prevention plan as the basic storm water control strategy for each industry sector.

To analyze the storm water sampling data submitted by the group applicants, EPA further subdivided many of the industry sectors into subsectors. This subdivision grouped together those members of group applications with very similar industrial activities [typically at the three-digit Standard Industrial Classification (SIC) code level]. Based upon this subsector sampling data analysis and using additional information where necessary, the EPA determined monitoring requirements for certain subsectors. Chapter 9 describes these monitoring requirements in detail.

Types of discharges covered by the multisector storm water permit

Coverage under the multisector storm water general permit is potentially available to all industrial activities described in the following list. In this list, the range of SIC codes is provided for most industry sectors.

In the SIC code numbers, the letter X could be replaced by any digit from 0 to 9, making a complete SIC code.

Sector A: Timber products

Subsector A1 (SIC code 2421)—general sawmills and planing mills

Subsector A2 (SIC code 2491)—wood preserving

Subsector A3 (SIC code 2411)—log storage and handling

Subsector A4

- SIC code 2426—hardwood dimension and flooring mills
- SIC code 2429—special product sawmills, not elsewhere classified
- SIC code 243X (except 2434)—millwork, veneer, plywood, and structural wood (except wood kitchen cabinet manufacturers)
- SIC code 244X—wood containers
- SIC code 245X—wood buildings and mobile homes
- SIC code 2493—reconstituted wood products
- SIC code 2499—wood products not elsewhere classified

The multisector general permit includes the effluent limitation guideline in 40 CFR Part 429 Subpart I for discharges resulting from spraydown of lumber and wood products in storage yards (wet decking) [62 FR 37448 (July 11, 1997)].

Sector B: Paper and allied products manufacturing

Subsector B1 (SIC code 261X)—pulp mills

Subsector B2 (SIC code 262X)—paper mills

Subsector B3 (SIC code 263X)—paperboard mills

Subsector B4 (SIC code 265X)—paperboard containers and boxes

Subsector B5 (SIC code 267X)—converted paper and paperboard products, except containers and boxes

Sector C: Chemical and allied products manufacturing

Subsector C1 (SIC code 281X)—industrial inorganic chemicals

Subsector C2 (SIC code 282X)—plastics materials and synthetic resins, synthetic rubber, cellulosic, and other artificial fibers except glass

Subsector C4 (SIC code 284X)—soaps, detergents, and cleaning preparations; perfumes, cosmetics, and other toilet preparations

Subsector C5 (SIC code 285X)—paints, varnishes, lacquers, enamels, and allied products

Subsector C6 (SIC code 286X)—industrial organic chemicals

Subsector C7 (SIC code 287X)—nitrogenous and phosphatic basic fertilizers, mixed fertilizers, pesticides, and other agricultural chemicals

Subsector C8 (SIC code 289X)—miscellaneous chemical products

Subsector C9 (SIC code 3952, limited to list)—inks and paints, including china painting enamels, india inks, drawing ink, platinum paints for burnt wood or leather work, paints for china painting, artist's paints and artist's watercolors

Subsector Ci (SIC codes 2833 to 2836)—Medicinal chemical and botanical products; pharmaceutical preparations; in vitro and in vivo diagnostic substances; biological products, except diagnostic substances [62 FR 37454 (July 11, 1997)]

Sector D: Asphalt paving and roofing materials manufacturers and lubricant manufacturers

Subsector D1 (SIC code 295X)—asphalt paving and roofing materials

Subsector D2 (SIC code 299X)—miscellaneous products of petroleum and coal

Sector E: Glass, clay, cement, concrete, and gypsum product manufacturing

Subsector E1

- SIC code 321X—flat glass
- SIC code 322X—glass and glassware, pressed or blown
- SIC code 323X—glass products made of purchased glass
- SIC code 3281—cut stone and stone products, benches, blackboards, tabletops, pedestals, etc. [62 FR 37454 (July 11, 1997)]
- SIC code 3291—abrasive products [62 FR 37454 (July 11, 1997)]
- SIC code 3292—asbestos products, tiles, building materials, except paper, insulating pipe coverings [62 FR 37454 (July 11, 1997)]
- SIC code 3296—mineral wool, insulation [62 FR 37454 (July 11, 1997)]
- SIC code 3299—nonmetallic mineral products, not elsewhere classified, plaster of Paris and papier-mâché, etc. [62 FR 37454 (July 11, 1997)]

Subsector E2: SIC code 3241—hydraulic cement

Subsector E3

- SIC code 325X—structural clay products
- SIC code 326X—pottery and related products
- SIC code 3297—nonclay refractories

Subsector E4

- SIC code 327X—concrete, gypsum, and plaster products
- SIC code 3295—minerals and earths, ground, or otherwise treated

Sector F: Primary metals

Subsector F1 (SIC code 331X)—steel works, blast furnaces, and rolling and finishing mills

Subsector F2 (SIC code 332X)—iron and steel foundries

Subsector F3 (SIC code 333X)—primary smelting and refining of nonferrous metals

Subsector F4 (SIC code 334X)—secondary smelting and refining of nonferrous metals

Subsector F5 (SIC code 335X)—rolling, drawing, and extruding of nonferrous metals

Subsector F6 (SIC code 336X)—nonferrous foundries (castings)

Subsector F7 (SIC code 339X)—miscellaneous primary metal products

Sector G: Metal mining (ore mining and dressing). This sector does not include inactive metal mining activities occurring on federal lands where an operator cannot be identified. These discharges are to be covered under a separate general permit. Discharges subject to effluent guidelines for mine drainage [40 CFR 440] are not eligible for coverage.

Subsector G1 (SIC code 101X)—iron ores

Subsector G2 (SIC code 102X)—copper ores

Subsector G3 (SIC code 103X)—lead and zinc ores

Subsector G4 (SIC code 104X)—gold and silver ores

Subsector G5 (SIC code 106X)—ferroalloy ores, except vanadium

Subsector G6 (SIC code 108X)—metal mining services

Subsector G7 (SIC code 109X)—miscellaneous metal ores

Sector H: Coal mines and coal mining–related facilities

SIC code 12XX—coal mines and coal mining–related facilities

Sector I: Oil and gas extraction. Permit coverage is required only for oil and gas facilities that have had a reportable quantity release, as described in Chap. 5.

Subsector I1 (SIC code 131X)—crude petroleum and natural gas

Subsector I2 (SIC code 132X)—natural gas liquids

Subsector I3 (SIC code 138X)—oil and gas field services

SIC code 2911—petroleum refining [62 FR 37454 (July 11, 1997)]

Sector J: Mineral mining and dressing. This sector does not include inactive mineral mining activities occurring on federal lands where an operator cannot be identified. These discharges are to be covered under a separate general permit.

Subsector J1

- SIC code 141X—dimension stone
- SIC code 142X—crushed and broken stone, including riprap
- SIC code 148X—nonmetallic minerals, except fuels

Subsector J2 (SIC code 144X)—sand and gravel

Subsector J3 (SIC code 145X)—clay, ceramic, and refractory materials

Subsector J4

- SIC code 147X—chemical and fertilizer mineral mining
- SIC code 149X—miscellaneous nonmetallic minerals, except fuels

The multisector general permit authorizes mine dewatering discharges from construction sand and gravel and industrial sand and crushed stone mines in the areas of EPA regions II and X, where EPA is the permitting authority [62 FR 37448 (July 11, 1997)].

Sector K: Hazardous waste treatment storage or disposal facilities. This sector includes hazardous waste treatment, storage, or disposal facilities, including those that are operating under interim status or a permit under Subtitle C of RCRA [40 CFR 122.26(b)(14)(iv)]. No SIC code is available for these facilities.

Sector L: Landfills and land application sites. This sector includes landfills, land application sites, and open dumps that receive or have received any industrial wastes, including those that are subject to regulation under Subtitle D of RCRA [40 CFR 122.26(b)(14)(v)]. No SIC code is available for these facilities.

Inactive landfills or land application sites occurring on federal lands where an operator cannot be identified are not included. These facilities should be covered under a separate EPA general permit specifically written for these facilities, and for abandoned mines under the same conditions.

Sector M: Automobile salvage yards

SIC code 5015—facilities engaged in dismantling or wrecking used motor vehicles for parts recycling or resale and for scrap

Sector N: Scrap recycling facilities

SIC code 5093—processing, reclaiming, and wholesale distribution of scrap and waste materials

Sector O: Steam electric generating facilities. This sector includes steam electric power generating facilities, including coal handling sites [40 CFR 122.26(b)(14)(vii)]. No SIC code is available for these facilities.

Sector P: Land transportation facilities. This sector includes only those land transportation facilities that have vehicle and equipment maintenance shops and/or equipment-cleaning activities.

Subsector P1 (SIC code 40XX)—railroad transportation

Subsector P2 (SIC code 41XX)—local and highway passenger transportation

Subsector P3 (SIC code 42XX)—motor freight transportation and warehousing

Subsector P4 (SIC code 43XX)—U.S. Postal Service

Subsector P5 (SIC code 5171)—petroleum bulk stations and terminals

Sector Q: Water transportation facilities. This sector includes only those water transportation facilities that have vehicle (vessel) and equipment maintenance shops and/or equipment-cleaning operations.

SIC code 44XX—water transportation

Sector R: Ship and boat building or repair yards

SIC code 373X—ship and boat building or repair yards

Sector S: Air transportation facilities

SIC code 45XX—air transportation facilities that have vehicle maintenance shops, material handling facilities, equipment-cleaning operations, or airport and/or aircraft deicing/anti-icing operations

Sector T: Treatment works. Treatment works treating domestic sewage or any other sewage sludge or wastewater treatment device or system used in the storage, treatment, recycling, or reclamation of municipal or domestic sewage with a design flow of 1.0 Mgal/day or more or required to have an approved pretreatment program [40 CFR 122.26(b)(ix)]. No SIC code is available for these facilities.

Sector U: Food and kindred products

Subsector U1 (SIC code 201X)—meat products

Subsector U2 (SIC code 202X)—dairy products

Subsector U3 (SIC code 203X)—canned, frozen, and preserved fruits, vegetables, and food specialties

Subsector U4 (SIC code 204X)—grain mill products

Subsector U5 (SIC code 205X)—bakery products

Subsector U6 (SIC code 206X)—sugar and confectionery products

Subsector U7 (SIC code 207X)—fats and oils

Subsector U8 (SIC code 208X)—beverages

Subsector U9

- SIC code 209X—miscellaneous food preparations and kindred products
- SIC code 21XX—tobacco products

Sector V: Textile mills, apparel, and other fabric product manufacturing

Subsector V1 (SIC code 22XX)—textile mill products

Subsector V2 (SIC code 23XX)—apparel and other finished products made from fabrics and similar materials

SIC code 3131—boot and shoe cut stock and findings (leather soles, inner soles, other boot and finished wood heels) [62 FR 37454 (July 11, 1997)]

SIC codes 3142 to 3144—house slippers; men's dress, street, and work shoes; women's dress, street, and work shoes

SIC code 3149—footwear, except rubber, including athletic shoes

SIC code 3151—leather gloves and mittens

SIC code 3161—luggage and cases

SIC code 3171—women's handbags and purses, leather

SIC code 3172—personal leather goods, e.g., billfolds, key cases, coin purses, checkbooks

SIC code 3199—leather goods, not elsewhere classified, e.g., saddlery, belts, holsters, leather aprons

Sector W: Furniture and fixtures

SIC code 25XX—furniture and fixtures

SIC code 2434—wood kitchen cabinets

Sector X: Printing and publishing

SIC code 2732—book printing

SIC code 2752—commercial printing, lithographic

SIC code 2754—commercial printing, gravure

SIC code 2759—commercial printing, not elsewhere classified

SIC code 2796—platemaking and related services

Sector X includes all facilities in major SIC group 27 [62 FR 37448 (July 11, 1997)]

Sector Y: Rubber, miscellaneous plastic products, and miscellaneous manufacturing industries

Subsector Y1

- SIC code 301X—tires and inner tubes
- SIC code 302X—rubber and plastics footwear
- SIC code 305X—gaskets, packing, and sealing devices and rubber and plastics hose and belting
- SIC code 306X—fabricated rubber products, not elsewhere classified

Subsector Y2

- SIC code 308X—miscellaneous plastics products
- SIC code 393X—musical instruments
- SIC code 394X—dolls, toys, games, and sporting and athletic goods
- SIC code 395X—pens, pencils, and other artists' materials
- SIC code 396X—costume jewelry, costume novelties, buttons, and miscellaneous notions, except precious metal
- SIC code 399X—miscellaneous manufacturing industries

Sector Z: Leather tanning and finishing

SIC code 311X—leather tanning and finishing

This sector includes facilities that make fertilizer solely from leather scraps and leather dust.

Sector AA: Fabricated metal products

Subsector AA1

- SIC code 3429—cutlery, hand tools, and general hardware
- SIC code 3441—fabricated structural metal
- SIC code 3442—metal doors, sashes, frames, molding, and trim
- SIC code 3443—fabricated plate work (boiler shops)
- SIC code 3444—sheet-metal work
- SIC code 3451—screw machine products
- SIC code 3452—bolts, nuts, screws, rivets, and washers
- SIC code 3462—metal forgings and stampings
- SIC code 3471—electroplating, plating, polishing, anodizing, and coloring
- SIC code 3494—valves and pipe fittings not elsewhere classified
- SIC code 3496—miscellaneous fabricated wire products
- SIC code 3499—miscellaneous fabricated metal products not elsewhere classified
- SIC code 391X—jewelry, silverware, and plated ware

Subsector AA2 (SIC code 3479)—coating, engraving, and allied services

Sector AA includes all facilities in major SIC group 34 [62 FR 37448 (July 11, 1997)].

Sector AB: Transportation equipment, industrial or commercial machinery

SIC code 35XX (except 357)—industrial and commercial machinery (except computer and office equipment)

SIC code 37XX (except 373)—transportation equipment (except ship and boat building and repair)

Sector AC: Electronic, electrical, photographic, and optical goods

SIC code 36XX—electronic and other electrical equipment and components, except computer equipment

SIC code 38XX—measuring, analyzing, and controlling instruments; photographic, medical, and optical goods; watches and clocks

SIC code 357—computer and office equipment

Sector AD: Other facilities. These are facilities which may not clearly fall into one of the sectors of the modified multisector general permit and coverage of certain other storm water discharges which are designated for permitting.

EPA multisector general permit NOI requirements

To obtain coverage under the multisector general permit, a Notice of Intent must be submitted at least 2 days prior to the commencement of the industrial activity at the facility, using a form provided by EPA (or a copy of the EPA form). One NOI is generally sufficient for an entire industrial site, including several separate outfalls. However, separate NOIs must be submitted for each separately located industrial facility, even if the facilities are under common ownership. If there are multiple operators at a site, each operator must submit an NOI. The EPA will confirm the receipt of the NOI and will provide the applicant with a permit number [Environmental Protection Agency, 1993].

The NOI form requires that facility operator information be provided, including the legal name of the person, firm, public organization, or any other entity that operates the facility or site. Colloquial names are not acceptable. The name of the operator may or may not be the same as the name of the facility. The responsible party is the legal entity that controls the facility's operation, rather than the plant or site manager. The complete address and telephone number of the operator must also be provided, along with the legal status of the operator (i.e., as a federal, state, or other government entity, or as a private entity).

Information about the location of the facility or site must also be provided. This must include the facility's or site's official or legal name and complete street address, including city, state, and ZIP code. The street address cannot be a post office box. Either the latitude and longitude of the facility to the nearest 15 seconds or the quarter, section, township, and range (to the nearest quarter section) of the approximate center of the site must also be provided. The NOI form must indicate whether the facility is located on Indian lands.

The NOI form must list up to two four-digit SIC codes that best describe the principal products or services provided at the facility or site. The most important SIC code must be listed first. This information is intended to give the EPA an indication of the nature of the industrial activity at the facility. Some industrial activities do not have SIC codes that accurately describe the principal products produced or services provided. For these activities, use the following two-character codes:

- HZ = hazardous waste treatment, storage, or disposal facilities
- LF = landfills, land application sites, and open dumps
- SE = steam electric power generating facilities
- TW = treatment works

The Notice of Intent must include the permit number of additional NPDES permits for any discharges from the site (including non–storm water discharges) that are currently authorized by an NPDES permit. If the NOI is being submitted to add a copermittee to an existing storm water permit, then the existing permit number should be entered.

The monitoring status of the facility must be entered, using one of the following options:

1. Not subject to monitoring requirements under the conditions of the permit
2. Subject to monitoring requirements and required to submit data
3. Subject to monitoring requirements but not required to submit data
4. Subject to monitoring requirements but submitting certification for monitoring exclusion

Monitoring requirements are described in Chap. 9.

If the storm water discharges to a municipal separate storm sewer system, the name of the municipality or county that operates the MS4 must be provided on the NOI. The name of the receiving water of the discharge from the MS4 should also be listed. If the facility discharges storm water directly to receiving water(s), enter the name of the receiving water(s).

The NOI form has specific certification language concerning compliance with the Endangered Species Act (ESA) and the National Historic Preservation Act (NHPA):

> To the best of my knowledge, the discharges covered under this permit, and construction of BMPs to control storm water runoff, are not likely to and will not likely adversely affect any species identified in Addendum H of the multisector storm water general permit or are otherwise eligible for coverage due to previous authorization under the Endangered Species Act.
>
> To the best of my knowledge, I further certify that such discharges, and construction of BMPs to control storm water runoff, do not have an effect on properties listed or eligible for listing on the National Register of Historic Places under the National Historic Preservation Act, or are otherwise eligible for coverage due to a previous agreement under the National Historic Preservation Act.

The same ESA and NHPA requirements apply to the construction general permit as well. The information provided in Chap. 11 concerning ESA and NHPA compliance for construction storm water discharges applies to storm water discharges covered under the multisector general permit for industrial discharges.

The person signing the NOI must have a sufficient level of authority and responsibility within the organization to help ensure compliance with the terms and conditions of the permit. For a sole proprietorship, the proprietor must sign the NOI. For a partnership, a general partner must sign.

For a corporation, the person signing the NOI must be a either (1) a responsible corporate officer, which includes only the president, secretary, treasurer, or vice president of the corporation in charge of a principal business function, or any other person who performs similar policy- or decision-making functions for the corporation; or (2) the manager of one or more manufacturing, production, or operating facilities employing more than 250 persons or having gross annual sales or expenditures exceeding $25 million (in second-quarter 1980 dollars), if authority to sign documents has been assigned or delegated to the manager in accordance with corporate procedures.

For a government agency, the person signing the NOI must be a principal executive officer or ranking elected official. A principal executive officer of a federal agency includes the chief executive officer of the agency or a senior executive officer having responsibility for the overall operations of a principal geographic unit of the agency (e.g., regional administrators of the EPA). Within the military, base commanders represent the appropriate level of authority.

In addition to the ESA and NHPA certifications described previously, the person signing the NOI is making the following certification:

> I certify under penalty of law that this document and all attachments were prepared under my direction or supervision in accordance with a system designed to assure that qualified personnel properly gather and evaluate the information submitted. Based on my inquiry of the person or persons who manage the system, or those persons directly responsible for gathering the information, the information submitted is, to the best of my knowledge and belief, true, accurate, and complete. I am aware that there are significant penalties for submitting false information, including the possibility of fine and imprisonment for knowing violations.

Notice of termination

The EPA general permit allows the discharger to submit a Notice of Termination (NOT) form (or photocopy thereof) when the storm water discharges associated with industrial activity from a facility have been eliminated. This assists the EPA in tracking the status of the discharger. The NOT form requires the following information:

- *Permit information,* including the NPDES storm water general permit number that is being terminated. The reason for the termination must also be provided. Only two reasons are acceptable: Either the permittee is no longer the operator of the facility, or the storm water discharge has been terminated.

- *Facility operator identification*—the name, address, and telephone number of the operator addressed by the Notice of Termination.
- *Facility and site location information*—the name, mailing address, and location of the facility for which the notification is submitted. Where a street address for the site is not available, the location of the approximate center of the site must be described in terms of the latitude and longitude to the nearest 15 seconds, or the section, township, and range to the nearest quarter.
- *Certification*—the following certification is required:

> I certify under penalty of law that all storm water discharges associated with industrial activity from the identified facility that are authorized by a NPDES general permit have been eliminated or that I am no longer the operator of the facility or construction site. I understand that by submitting this Notice of Termination, I am no longer authorized to discharge storm water associated with industrial activity under this general permit, and that discharging pollutants in storm water associated with industrial activity to waters of the United States is unlawful under the Clean Water Act where the discharge is not authorized by a NPDES permit. I also understand that the submittal of this notice of termination does not release an operator from liability for any violations of this permit or the Clean Water Act.

The NOT must be signed in accordance with the signatory requirements of 40 CFR 122.22, which were summarized earlier in this chapter with regard to the NOI form.

Storm water pollution prevention plan

Each facility covered by the multisector permit must develop a storm water pollution prevention plan (SWPPP) tailored to the facility's site-specific conditions. The pollution prevention plan is the most important part of the permit and is to be designed with the goal of eliminating, minimizing, or reducing the amount of pollution in storm water discharged from the site. Most authorized states have similar requirements, although some states use different terminology. The SWPPP must be completed and ready to implement at the time the facility begins industrial operations.

Under EPA multisector general permit requirements, the SWPPP is not submitted with the NOI. Instead, the SWPPP is retained on-site by the discharger. Federal, state, and local regulatory agencies have the authority to review the SWPPP at any time. If the reviewing agency finds that the SWPPP is not in compliance with general permit requirements, the discharger has 30 days to revise the SWPPP to achieve compliance. Chapters 7 and 8 provide further details on the SWPPP.

References

Environmental Protection Agency, 1993. *NPDES Storm Water Program Question and Answer Document,* vol. 2, EPA 833-F-93-002BRAVO, July.
Government Printing Office, *Federal Register*, Office of the Federal Register, National Archives and Records Administration, Washington.
——, 1992. *Code of Federal Regulations,* Office of the Federal Register, National Archives and Records Administration, Washington, July 1.

Chapter

7

Basic Elements of Industrial Storm Water Pollution Prevention Plans

This is the third of five chapters which deal with the specific requirements for storm water discharges from industrial (nonconstruction) facilities. This chapter describes how to prepare the basic elements of a *storm water pollution prevention plan* (*SWPPP*) for an industrial facility.

Requirements for Storm Water Pollution Prevention Plans

The EPA multisector general permit for storm water discharges from industrial activities requires that a storm water pollution prevention plan be completed at the time that the facility operator submits the Notice of Intent (NOI) for permit coverage. The facility should begin to implement the SWPPP immediately. The SWPPP should have the following basic sections, which are described in this chapter:

- *Pollution prevention team.* This section identifies the specific personnel responsible for developing and implementing the plan, including the specific responsibilities of each individual.
- *Potential pollutant sources.* This section, along with a site map, describes and illustrates the existing drainage conditions of the site, indicates the industrial activities of the site, presents an inventory of significant materials, and discusses the management practices for these materials. This section also lists any recent spills and leaks,

describes the procedures used to evaluate and eliminate nonstorm discharges, and includes any available storm water sampling data for the site. It concludes with a discussion of the risks of storm water pollution at the site.

- *Measures and controls.* This section describes the operations and mechanisms that will be used to eliminate or minimize storm water pollution at the facility. These will include a housekeeping and maintenance plan, spill prevention and response procedures, storm water management controls, inspections and record keeping, employee training, and sediment and erosion controls. A specific schedule with interim milestones for implementing each measure and control must also be included.
- *Monitoring requirements.* This section describes the required annual site compliance evaluation and certification, and any storm water sampling required for the facility.
- *Other regulatory requirements.* This section describes the applicable state and local requirements for the facility, and the methods used to ensure compliance with these requirements.
- *Certification and signature.* The completed SWPPP must be signed by an authorized individual (see Chap. 6 for details) and include the following certification language:

> I certify under penalty of law that this document and all attachments were prepared under my direction or supervision in accordance with a system designed to assure that qualified personnel properly gathered and evaluated the information submitted. Based on my inquiry of the person or persons who manage the system, or those persons directly responsible for gathering the information, the information submitted is, to the best of my knowledge and belief, true, accurate, and complete. I am aware that there are significant penalties for submitting false information, including the possibility of fine and imprisonment for knowing violations.

Industry-Specific BMP Requirements

In addition to the general measures and controls discussed in this chapter, the EPA multisector general permit includes certain pollution prevention plan requirements which are specific to a given industry sector. Some industries are also required to consider pollutant-specific best management practices. Chapter 8 describes these specific requirements.

In the case where a facility has industrial activities occurring on-site that meet the eligibility requirements of more than one sector, those industrial activities are considered to be "colocated" activities. Storm water discharges from colocated industrial activities are authorized by the EPA multisector general permit, provided that the permittee complies with the additive pollution prevention plan and monitoring requirements applicable to the colocated industrial activity.

Philosophy of Storm Water Pollution Prevention Plans

Storm water pollution prevention plans should be dynamic documents which are revised as appropriate to reflect changes in the facility's operations. The EPA multisector general permit requires that the permittee amend the plan whenever there is a change in design, construction, operation, or maintenance which has a significant effect on the potential for the discharge of pollutants to waters of the United States. The SWPPP should also be amended if the plan proves to be ineffective in eliminating or significantly minimizing pollutants. In addition, permittees are required to inspect their sites, evaluate the accuracy and effectiveness of their plans, and modify the plan as necessary.

The use of storm water pollution prevention plans as a means to control pollutants in storm water discharges is a much different approach from the numeric effluent limitations used in traditional National Pollutant Discharge Elimination System (NPDES) permits. It is an example of the flexible and innovative approaches available to the EPA in carrying out the NPDES program. As discussed in Chap. 1, court rulings have established that the EPA has considerable discretionary authority to use pollution prevention requirements when quantitative limitations are infeasible or unwieldy. The pollution prevention requirements operate as limitations on effluent discharges that reflect the application of best available technology and best conventional technology, as required by the Clean Water Act.

The EPA believes that the pollution prevention approach is the most environmentally sound and cost-effective way to control the discharge of pollutants in storm water runoff from industrial facilities. This position is supported by the results of a comprehensive technical survey that the EPA completed in 1979. The survey found that there are two classes of management practices used in industry to control the non-routine discharge of pollutants from sources such as storm water runoff, drainage from raw material storage and waste disposal areas, and discharges from places where spills or leaks have occurred. The first class of management practices includes those that are low in cost, are applicable to a broad class of industries and substances, and generally are considered essential to a good pollution control program.

Some examples of practices in this class are good housekeeping, employee training, and spill prevention procedures. The second class includes management practices that provide a second line of defense against the release of pollutants, including containment, mitigation, cleanup, and treatment. The multisector general permit requirements for development and implementation of a storm water pollution prevention plan include both classes of management practices.

Pollution Prevention Team

As a first step in the process of developing and implementing the SWPPP, permittees must identify a qualified individual or team of individuals to be responsible for developing the plan and assisting the facility or plant manager in its implementation. The team should include the expertise of all relevant departments within the plant to ensure that all aspects of plant operations are considered when the plan is developed. Pollution prevention teams may consist of one individual where appropriate (e.g., in certain small businesses with limited storm water pollution potential).

The pollution prevention team members should be specifically identified. It is not sufficient simply to say, "The team shall include representatives of the plant safety office, the operations office...." Instead, specific individuals must be identified, if not by name, then at least by position. For example, "The team shall include the director of plant safety, the assistant manager of operations...." The team leader must also be identified, and must have overall responsibility for preparation and implementation of the SWPPP.

The plan must clearly describe the responsibilities of each team member as they relate to specific components of the plan. In addition to enhancing the quality of communication between team members and other personnel, clear delineation of responsibilities will ensure that every aspect of the plan is addressed by a specified individual or group of individuals.

Potential Pollutant Sources

Storm water pollution prevention plans must be based on an accurate understanding of the pollution potential of the site. The first part of the plan requires an evaluation of the sources of pollution at a specific industrial site. The permit proposes that the source identification portion of the plan identify all activities and significant materials which may potentially be significant pollutant sources.

The source identification portion of the plan includes the following sections:

1. *Drainage:* a drainage site map with a narrative description
2. *Inventory of exposed materials:* a narrative description of significant materials, materials management practices, pollutant control measures, and storm water treatment
3. *Significant spills and leaks:* a list of significant spills and leaks of toxic or hazardous pollutants that occurred at the facility after the date that the NOI is submitted
4. *Non–storm water discharges:* a certification that the facility has been tested or evaluated for non–storm water discharges, and that these have been eliminated (except for allowable discharges)
5. *Sampling data:* a summary of existing sampling data describing pollutants in storm water discharges
6. *Risk identification:* an evaluation of activities likely to be significant sources of pollutants to storm water discharges

Each of these items is discussed in the following sections.

Drainage and site activities

The SWPPP must contain a map of the site that shows the pattern of storm water drainage on the site, including the location of all storm water outfalls covered by all NPDES permits for the site. The EPA requires only that the portions of the drainage area within the facility's boundaries be identified.

The EPA does not require that the map include site topography, although such information would clearly be pertinent to site drainage. The map should indicate an outline of the portions of the drainage area of each storm water outfall, each existing structural control measure to reduce pollutants in storm water runoff, the location of all surface water bodies (including wetlands), places where significant materials are exposed to rainfall and runoff, and locations of major spills and leaks that occurred in the 3 years prior to the date that the NOI is submitted. Nonstructural features such as grass swales and vegetative buffer strips also should be shown.

The map must show areas where the following industrial activities take place: fueling, vehicle and equipment maintenance and/or cleaning, loading and unloading, material storage (including tanks or other vessels used for liquid or waste storage), material processing, and waste disposal. For areas of the facility that generate storm water discharges with a reasonable potential to contain significant amounts of pollutants, the map must indicate the probable direction of storm water flow and the pollutants likely to be in the discharge. Flows with a significant potential to cause soil erosion also must be identified. To increase the readability of the map, the inventory of the types of discharges contained in each outfall may be kept as an attachment to the site map.

Inventory of exposed materials

Facility operators are required to carefully conduct an inspection of the site and related records to identify significant materials that are or may be exposed to storm water. The inventory must address materials that have been handled, stored, processed, treated, or disposed of in a manner allowing exposure to storm water within 3 years prior to the date that the NOI is submitted. Findings of the inventory must be documented in detail in the pollution prevention plan.

At a minimum, the plan must describe the methods and location of on-site storage or disposal; management practices used to minimize contact of materials with rainfall and runoff; existing structural and nonstructural controls that reduce pollutants in runoff; and any treatment the runoff receives before it is discharged to surface waters or a separate storm sewer system. The description must be updated whenever there is a significant change in the types or amounts of materials, or material management practices, that may affect the exposure of materials to storm water.

The inventory of exposed materials is intended to address materials that potentially may be exposed to precipitation, including chemicals used and by-products formed at the site. Significant materials include, but are not limited to, the following: raw materials; fuels; solvents, detergents, and plastic pellets; finished materials, such as metallic products; raw materials used in food processing or production; hazardous substances designated under Section 101(14) of CERCLA; any chemical the facility is required to report pursuant to EPCRA Section 313; chemicals or compounds listed in effluent limitation guidelines to which the facility is subject; chemicals or compounds specifically controlled or limited in any other NPDES permit for the facility; fertilizers; pesticides; and waste products, such as ashes, slag, and sludge that have the potential to be released with storm water discharges. [See 40 CFR 122.26(b)(8).]

Significant spills and leaks

The storm water pollution prevention plan must include a list of any significant spills and leaks of toxic or hazardous pollutants that occurred in the 3 years prior to the date that the NOI is submitted. Significant spills include, but are not limited to, releases of oil or hazardous substances in excess of quantities that are reportable under Section 311 of CWA [see 40 CFR 110.10 and 117.21] or Section 102 of CERCLA [see 40 CFR 302.4].

Significant spills may also include chronic releases of oil or hazardous substances that are not in excess of reporting requirements. Instances of chronically repeated smaller spills can constitute significant spills if such spills, taken together, add significant amounts of pollutants to storm water discharges.

Significant spills may also include releases of materials that are not classified as oil or a hazardous substance, but that could potentially add significant amounts of pollutants. These discharges can also cause water quality impacts.

The listing should include a description of the causes of each spill or leak, the actions taken to respond to each release, and the actions taken to prevent similar such spills or leaks in the future. This effort will aid the facility operator in examining existing spill prevention and response procedures and in developing any additional procedures necessary to comply with the permit.

Some spills, such as those that occur inside buildings which drain to a sanitary sewer, are not potential sources of pollution to storm water discharges and thus do not need to be identified in the SWPPP. However, spills to sumps or secondary containment areas that receive storm water discharges should generally be identified in the SWPPP because such devices can overflow during large or repeated storm events, or storm water may be drained and discharged from such devices. It is also important to identify spills that occur on impervious surfaces exposed to precipitation or that otherwise drain to a storm drain even when the spill is cleaned up before any of it enters a storm drain. Listing such events provides an indication of potential pollutant sources that may occur in the future and helps direct priorities for developing and implementing spill response measures.

Storm water pollution prevention plans are to be updated to address significant spills and leaks that occur during the term of the permit. This information is necessary to ensure that major potential sources of pollution to storm water discharges are identified.

Non–storm water discharges

Two types of non–storm water discharges must be addressed in the SWPPP: allowable non–storm water discharges and illicit non–storm water discharges.

Allowable non–storm water discharges. The provisions of the EPA multisector general permit prohibit most non–storm water discharges, except the following:

- Discharges from fire-fighting activities
- Fire hydrant flushing

- Potable water sources, including water line flushing and drinking fountain sources
- Landscape irrigation drainage
- Lawn watering
- Routine external building wash water (without detergents or other contaminants)
- Pavement wash waters, where spills or leaks of toxic or hazardous materials have not occurred (unless all spilled material has been removed) and where detergents are not used
- Air-conditioning condensate (but not including cooling water from cooling towers, heat exchangers, or other sources)
- Springs
- Uncontaminated groundwater
- Foundation or footing drains where flows are not contaminated with process materials such as solvents

To be authorized under the EPA multisector general permit, these sources of nonstorm water (except flows from fire-fighting activities) must be identified in the SWPPP prepared for the facility. Where such discharges occur, the plan must also identify and ensure the implementation of appropriate pollution prevention measures for the non–storm water component(s) of the discharge.

Some state storm water discharge permits prohibit non–storm water discharges altogether; separate permits are required for all nonstorm discharges. Under the EPA multisector general permit, however, the sources of nonstorm water listed above are allowed, but they must be identified in the SWPPP prepared for the facility. Where such discharges occur, the plan must also identify and ensure the implementation of appropriate pollution prevention measures for the non–storm water component(s) of the discharge. For example, to reduce pollutants in irrigation drainage, a plan could identify low-maintenance lawn areas that do not require the use of fertilizers or herbicides; for higher-maintenance lawn areas, a plan could identify measures such as limiting fertilizer use based on seasonal and agronomic considerations, decreasing herbicide use with an integrated pest management program, introducing natural vegetation or hardier species, and reducing water use (thereby reducing the volume of irrigation drainage).

Pollution prevention measures are not required to be identified and implemented for non–storm water flows from fire-fighting activities because these flows will generally be unplanned emergency situations where it is necessary to take immediate action to protect the public. The EPA does not provide specific guidance on fire-fighting activities undertaken as part of routine training of fire-fighting personnel, or routine checking of fire-fighting equipment. It is reasonable to assume that since these discharges are not emergencies, they would not be eligible for unrestricted discharge as would emergency fire-fighting flows. However, they may still be eligible for discharge as a potable water source (under the appropriate circumstances). If such discharges are anticipated, the storm water pollution prevention plan should identify the frequency and amount of such discharges, all flammable substances used for the training or test activity, and the pollution prevention measures used to reduce or eliminate the discharge of pollutants produced by these activities.

Illicit non–storm water discharges. Many facilities have cross-connections between the storm and sanitary sewer systems. In addition, floor drains commonly are connected to the storm sewer system in many industrial plants. These cross-connections and floor drain connections to the storm sewer network are considered to be illicit connections and represent potential non–storm water discharges. Technically, these connections violate the Clean Water Act, unless the facility has already obtained a separate NPDES permit for the connection.

Rinse waters used to clean or cool objects often discharge to floor drains that may be connected to separate storm sewers. Large amounts of rinse waters may originate from industries that use regular washdown procedures for cleaning and/or cooling. For example, bottling plants use rinse waters for removing waste products, debris, and labels; and rinse water is sometimes sprayed over the final products of a metal plating facility to cool them.

Condensate return lines of heat exchangers often discharge to floor drains. Heat exchangers, particularly those used under stressed conditions such as in the metal finishing and electroplating industry, typically develop pinhole leaks, which may result in contamination of condensate by process wastes.

These and other non–storm water discharges to a storm sewer may be intentional, based on the belief that the discharge does not contain pollutants; or they may be inadvertent, for the operator may be unaware that a floor drain is connected to the storm sewer. The connection may have been legal at the time of installation. In many cases, operators of industrial facilities may be unaware of illicit discharges or leakage from underground storage tanks or other nonvisible systems.

Certification of testing for non–storm water discharges. Each pollution prevention plan must include a certification, signed by an authorized individual, that discharges from the site have been tested or evaluated for the presence of non–storm water discharges. The certification must describe possible significant sources of nonstorm water, the results of any test and/or evaluation conducted to detect such discharges, the test method or evaluation criteria used, the dates on which tests or evaluations were performed, and the on-site drainage points directly observed during the test or evaluation. The certification must be completed within 180 days after submittal of an NOI to be covered by the EPA multisector general permit. The required language of the certification form is as follows:

> I certify under penalty of law that this document and all attachments were prepared under my direction or supervision in accordance with a system designed to assure that qualified personnel properly gathered and evaluated the information submitted. Based on my inquiry of the person or persons who manage the system, or those persons directly responsible for gathering the information, the information submitted is, to the best of my knowledge and belief, true, accurate, and complete. I am aware that there are significant penalties for submitting false information, including the possibility of fine and imprisonment for knowing violations.

Certification may not be feasible where there is no access to an outfall, manhole, or other point of access to the conduit that ultimately receives the discharge. Where a certification is not feasible, it is not required. However, the source identification section of the SWPPP must indicate why the certification is not feasible and must identify potential significant sources of nonstorm water at the site. The discharger must also notify the EPA that the certification is not feasible. Since the EPA allows a very broad range of methods to be used to perform the testing or evaluation (as described below), there should not be many circumstances in which it is absolutely infeasible to perform any kind of test or evaluation. It would not be unreasonable to expect EPA or state regulatory agencies to be skeptical of a notice that the certification was not feasible, especially if the discharger makes it a practice to submit such notices instead of actually performing the tests and evaluations necessary to complete the certifications.

Methods of identifying non–storm water discharges. A comprehensive evaluation of the storm sewers at a facility may draw on several methods, including the following (The Cadmus Group, Inc., 1990):

1. *Schematics*. Where they exist, accurate piping schematics can be inspected as a first step in evaluating the integrity of the separate storm sewer system. The use of schematics is limited because schematics usually reflect the design of the piping system and may not reflect the actual configuration constructed. Schematics should be updated or corrected based on additional information found during inspections.

2. *Evaluation of drainage map and inspections*. Drainage maps should identify the key features of the drainage system—each of the inlet and discharge structures, the drainage area of each inlet structure, and units such as storage or disposal units or material loading

areas, which may be the source of an illicit discharge or improper dumping. In addition, floor drains and other water disposal inlets that are thought to be connected to the sanitary sewer can be identified. A site inspection can be used to augment and verify map development.

3. *End-of-pipe screening.* Discharge points or other access points such as manhole covers can be inspected for the presence of dry-weather discharges and other signs of non–storm water discharges. Dry-weather flows can be screened by a variety of methods. Inexpensive on-site tests include measuring pH; observing for oil sheens, scums, and discoloration of pipes and other structures; as well as colormetric detection tests for chlorine, detergents, metals, and other parameters. In some cases, it may be appropriate to collect samples for more expensive analysis in a laboratory for fecal coliform, fecal streptococcus, conventional pollutants, volatile organic carbon, or other appropriate parameters.

4. *Water balance.* Many industrial facilities measure the volume of effluent discharged to the sanitary sewer system and the volume of water supplied to the facility. A significantly higher volume of water supplied to the facility relative to that discharged to the sanitary sewer and other consumptive uses may be an indication of illicit connections. This method is limited by the accuracy of the flowmeters used.

5. *Dry-weather testing.* Where storm sewers do not discharge during dry-weather conditions, water can be introduced into floor drains, toilets, and other points where non–storm water discharges are collected. Storm drain outlets are then observed for possible discharges.

6. *Dye testing.* Dry-weather discharges from storm sewers can occur for a number of legitimate reasons including groundwater infiltration or the presence of a continuous discharge subject to an NPDES permit. Where storm sewers do have a discharge during dry-weather conditions, dye testing for illicit connections can be done. Dye testing involves introducing fluorometric or other types of dyes into floor drains, toilets, and other points where non–storm water discharges are collected. Storm drain outlets are then observed for possible discharges.

7. *Manhole and internal television inspection.* Inspection of manholes and internal inspection of storm sewers either physically or by television are used to identify potential entry points for illicit connections. Dry-weather flows, material deposits, and stains are often indicators of illicit connections. TV inspections are relatively expensive, and generally should be used only after a storm sewer has been identified as having illicit connections.

The EPA has specifically not listed smoke tests because of the potential to misapply such tests in evaluating the presence of non–storm water discharges to storm sewers. Smoke tests (blowing smoke from a downstream point in a pipe up through the pipe) are often ineffective at finding non–storm water discharges to separate storm sewers, because line traps which are intended to block sewer gas (and will prevent the passage of smoke) commonly are used on non–storm water drain systems. However, because line traps are less frequently used on storm drains, smoke testing can be useful for detecting cross-connections from storm drains to sanitary sewers. Note that in some industrial facilities which handle explosive vapors, such as refineries, storm drains may have line traps for safety reasons.

Sampling data

Any existing data on the quality or quantity of storm water discharges from the facility must be described in the SWPPP. These data may be useful for locating areas that have contributed pollutants to storm water. The description should include a discussion of the methods used to collect and analyze the data. Sample collection points should be identified in the plan and shown on the site map.

Risk identification

The description of potential pollution sources culminates in a narrative discussion of the risk potential that sources of pollution pose to storm water quality. A formal risk assessment is not required, but the discussion should clearly point to activities, materials, and physical features of the facility that have a reasonable potential to contribute significant amounts of pollutants to storm water. Any such activities, materials, or features must be addressed by the measures and controls subsequently described in the plan. Factors to consider in evaluating the reasonable pollution potential of runoff from various portions of an industrial plant include these:

- Loading and unloading of dry bulk materials or liquids
- Outdoor storage of raw materials, intermediate products, by-products, or finished products
- Outdoor manufacturing or processing activities
- Significant dust- or particulate-generating processes
- Illicit connections or management practices
- On-site waste disposal practices
- The manner and frequency in which pesticides, herbicides, fertilizers, or soil enhancers are applied at the site

Other factors that need to be considered include the toxicity of chemicals; quantity of chemicals used, produced, or discharged; likelihood of these materials coming into contact with storm water; and the history of significant leaks or spills of toxic or hazardous pollutants. The assessment must identify the pollutant parameter or parameters (i.e., biochemical oxygen demand, suspended solids, and so on) associated with each source.

Outside drum storage areas, waste dumping sites, and bulk material piles are all obvious sources of potential storm water pollution, but some sources may be less obvious. Sites previously used for other industrial purposes may have residual matter from former occupants which represent a pollutant source. Facilities located downwind from another industrial facility may contain particulate airborne emissions. A comprehensive soil testing and storm water monitoring program may be required to identify these unexpected pollutant sources.

The whole purpose of risk identification is to produce a prioritized list of potential pollutant sources, considering the possibility of storm water exposure, the amount of exposed material, and the consequences of the pollutant. This prioritized list is then used to provide guidance for the selection and implementation of pollution prevention measures.

Measures and Controls

Following completion of the source identification and assessment phase, the permittee must evaluate, select, and describe the pollution prevention measures, best management practices (BMPs), and other controls that will be implemented at the facility. BMPs include processes, procedures, schedules of activities, prohibitions on practices, and other management practices that prevent or reduce the discharge of pollutants in storm water runoff. These BMPs should result in an improvement over the baseline levels of pollutants identified in storm water discharges with due consideration to economic feasibility.

The EPA emphasizes the implementation of pollution prevention measures and BMPs that reduce possible pollutant discharges at the source. *Source reduction measures* include, among others, preventive maintenance, chemical substitution, spill prevention, good housekeeping, training, and proper materials management. Where such practices are not appropriate to a particular source or do not effectively reduce pollutant discharges, the EPA supports the use of *source control measures* and BMPs such as material segregation or covering, water diversion, and dust control. Like source reduction measures, source control measures and BMPs are intended to keep pollutants out of storm water. The remaining classes of BMPs, which involve recycling or treatment of storm water, allow the reuse of storm water or attempt to lower pollutant concentrations prior to discharge.

The pollution prevention plan must discuss the reasons why each selected control or practice is appropriate for the facility, and how each will address one or more of the potential pollution sources identified in the first part of the plan. The plan must include a schedule specifying the time or times during which each control or practice will be implemented. Not all measures must be implemented immediately; the implementation may be phased over the life of the permit.

In addition, the plan should discuss ways in which the controls and practices relate to one another and, when taken as a whole, produce an integrated and consistent approach for preventing or controlling potential storm water contamination problems.

The EPA recognizes that some facilities will have adequate measures and controls that have been successful in preventing pollutant discharges in storm water. Under the EPA general permit, these facilities are only required to document such practices in a pollution prevention plan and continue them.

The following five categories describe options for reducing pollutants in storm water discharges from industrial plants:

1. Implement best management practices to prevent pollution.
2. Use traditional storm water management practices.
3. Eliminate pollution sources.
4. Divert storm water discharge to municipal sewage treatment plants.
5. Provide end-of-pipe treatment.

A comprehensive storm water management program for a given plant may include controls from each of these categories. Development of comprehensive control strategies should be based on a consideration of plant characteristics.

Best management practices

The term *best management practices* can describe a wide range of management procedures, schedules of activities, prohibitions on practices, and other management practices to prevent or reduce the pollution of waters of the United States. BMPs also include operating procedures, treatment requirements, and practices to control plant site runoff, drainage from raw materials storage, spills, or leaks. Many BMPs involve planning, reporting, training, preventive maintenance, and good housekeeping.

Many industrial facilities currently employ BMPs as part of normal plant operation. For example, preventive maintenance and good housekeeping are routinely used in the chemical and related industries to reduce equipment downtime and to promote a safe work environment for employees. Good-housekeeping BMPs generally are aimed at preventing spills and similar environmental incidents by stressing the importance of proper management and employee awareness. Experience has shown that many spills of hazardous chemicals can be attributed, in one way or another, to human error. Improper procedures, lack of training, and poor engineering are among the major causes of spills [58 FR 61146 (November 19, 1993)].

In preparing an industrial SWPPP, the following categories of BMPs should be considered:

- Good housekeeping
- Preventive maintenance
- Spill prevention and response procedures
- Inspections
- Employee training
- Record keeping and internal reporting procedures

These BMP categories are described in the following sections.

Good housekeeping. Good housekeeping involves using common sense to identify ways to maintain a clean and orderly facility and keep contaminants out of separate storm sewers. It includes establishing protocols to reduce the possibility of mishandling chemicals or equipment and training employees in good housekeeping techniques. These protocols must be described in the plan and communicated to appropriate plant personnel. These measures also ensure that discharges of wash waters to separate storm sewers are avoided.

Where indoor activities are not a potential source of pollutants, good housekeeping measures do not have to be addressed for such areas.

Preventive maintenance. Preventive maintenance procedures should be addressed in the SWPPP. Permittees must develop a preventive maintenance program that involves regular inspection and maintenance of storm water management devices and other plant equipment and systems. The program description should identify the devices, equipment, and systems that will be inspected; provide a schedule for inspections and tests; address appropriate adjustment, cleaning, repair, or replacement of devices, equipment, and systems; and maintain complete records on the equipment and systems.

For storm water management devices such as oil-water separators and catch basins, the preventive maintenance program should provide for periodic removal of debris to ensure that the devices are operating efficiently. Maintenance is important because the control measures may be of little or no use if the devices have not been properly maintained.

For other plant equipment and systems, the program should reveal and enable the correction of conditions that could cause breakdowns or failures which could result in the release of pollutants.

Spill prevention and response procedures. The SWPPP should reflect requirements for spill prevention control and countermeasures (SPCC) plans required under Section 311 of the CWA, and may incorporate any part of the SPCC plan into the SWPPP by reference. The SWPPP should also ensure that solid and hazardous waste is managed in accordance with requirements established under the Resource Conservation and Recovery Act (RCRA). Management practices required under RCRA should be expressly incorporated into the SWPPP.

Based on an assessment of possible spill scenarios, the SWPPP must specify appropriate material handling procedures, storage requirements, containment or diversion equipment, and spill cleanup procedures that will minimize the potential for spills and, in the event of a spill, enable proper and timely response. Areas and activities that typically pose a high risk for spills include loading and unloading areas, storage areas, process activities, and waste disposal activities.

Areas where potential spills can occur and their accompanying drainage points should be identified and described clearly in the SWPPP. Where appropriate, specification of material handling procedures and storage requirements in the plan should be considered. Procedures for cleaning up spills should be identified in the plan and made available to the appropriate personnel. The necessary equipment to implement a cleanup should be available to personnel.

Most facilities already have spill prevention and response plans in place. For storm water pollution prevention, spill response protocols should stress containment and neutralization of the spill, rather than utilization of wash-off procedures. The facility should develop its spill response plan in conjunction with the local fire department to ensure that wash-off procedures will be used only as a last resort in response to a spill incident.

If a release of a hazardous substance or oil occurs above a certain *reportable quantity* (*RQ*) threshold, the discharger must notify EPA's National Response Center (NRC) at (800) 424-8802 [or (202) 426-2675 in the Washington, DC, area] within the first 24 h after the discharger learns of the release [40 CFR Parts 110, 117, and 302].

Within 14 days of knowledge of the RQ release, a written description of the release (including the type and estimate of the amount of material released), an account of the circumstances leading to the release, and the date of the release must be incorporated into the SWPPP; and the SWPPP must be reviewed and revised if necessary to identify measures to prevent and respond to such releases. This information also must be submitted to the appropriate EPA regional office within the first 14 days.

If more than one RQ release occurs during a calendar year (or for the first year of the permit, after submittal of an NOI), then each additional release should be reported to the National Response Center. The SWPPP should be revised to include a written description of the dates on which such releases occurred, the type and estimate of the amount of material released, and the circumstances leading to the release. However, these additional releases need not be reported to the EPA regional office.

Any point-source discharge of pollutants to waters of the United States without a permit is prohibited under Section 301 of the CWA. Therefore, the requirement that a discharge of oil or a hazardous substance be reported only if the amount discharged exceeds the reportable quantity does not imply that the EPA general permit freely allows the discharge of smaller quantities of these substances. The EPA general permit requires dischargers to develop and implement BMPs and pollution prevention measures to reduce and/or control pollutants in the discharge, even in cases where the discharge does not contain hazardous substances or contains hazardous substances at levels significantly lower than reportable quantities.

Facilities subject to SARA Title III Section 313 requirements are subject to additional spill prevention requirements, as described in Chap. 8.

Inspections. *Inspection* is the process by which a discharger can evaluate whether the pollution prevention measures which have already been installed or applied are still effective. In most cases, inspection of pollution prevention measures requires that an inspector look at all the disturbed areas and material storage areas of the site. Typical inspections should include examination of pipes, pumps, tanks, supports, foundations, dikes, and drainage ditches. Material handling areas should be inspected for evidence of, or the potential for, pollutants entering the drainage system.

The EPA multisector general permit refers to two separate classes of inspections: an annual comprehensive site compliance evaluation and more frequent (but less comprehensive) periodic site inspections. The annual comprehensive site compliance evaluation is considered to be a monitoring activity, as described in detail later in this chapter. It is required for all industry sectors. For the more frequent periodic inspections, qualified facility personnel must be identified to inspect designated equipment and areas of the facility at appropriate intervals specified in the plan. A set of tracking or follow-up procedures must be used to ensure that appropriate actions are taken in response to the inspections. Periodic inspection requirements vary among the industry sectors included in the multisector general permit, as described in Chap. 8.

There are primarily three things an inspector should look for when inspecting a pollution prevention measure:

1. Was the measure installed or performed correctly?
2. Has there been damage to the measure since it was installed or performed?
3. What should be done to correct any problems with the measure?

The inspector must prepare a report documenting the findings of the inspection. The inspector should also request any required maintenance or repair. If the SWPPP should be changed to allow for unexpected conditions, then the inspector should make the changes or notify the appropriate person to make the changes.

There are no formal requirements for inspectors. Any person authorized and considered qualified by the operator may complete inspections and sign the inspection reports. The inspector(s) will generally be key members of the pollution prevention team and will be identified as such in the SWPPP. The EPA recognizes that experience is the best way to develop an understanding of pollution prevention measures. Qualified personnel must have sufficient technical abilities to conduct the inspection or evaluation and should have knowledge of the operations at the facility. The inspector should have detailed knowledge about the site's storm water pollution prevention plan, particularly the following portions:

- The location and type of control measures
- The construction requirements for the control measures
- Maintenance procedures for each of the control measures
- Spill prevention and cleanup measures
- Inspection and maintenance record-keeping requirements

Employee and contractor training. The SWPPP must describe a program for informing personnel at all levels of responsibility of the components and goals of the plan. The training program should address topics such as good housekeeping, materials management, and spill response procedures. Where appropriate, contractor personnel must be trained in relevant aspects of storm water pollution prevention.

A schedule for conducting training must be provided in the plan. Several industry sectors specify a minimum frequency for training of once per year, as noted in Chap. 8. This is the recommended minimum frequency. However, more frequent training may be necessary at facilities with high turnover of employees or where employee participation is essential to the storm water pollution prevention plan.

Probably the most important aspect of employee training is motivation. At most industrial facilities, it is not too difficult to understand how to improve storm water quality; the most effective steps are probably simple, day-to-day commonsense activities that do not involve highly technical processes. However, these simple steps often require some additional work, and some changes in the established routine. Employees often provide very useful ideas and suggestions, or they may simply begin doing useful things that were not even mentioned in the formal training program, if they are motivated to do so.

Very few people really want to pollute the environment. Once they realize that simple actions such as site cleanliness can make a difference in storm water quality, many people are willing to modify their behavior.

Several years ago, an internal site inspection at a quarry in a beautiful area of the Rocky Mountains revealed that the heavy equipment operators at the quarry had made it a practice for years to simply drive their equipment to a corner of the site and drain the engine oil and other fluids onto the ground when the equipment required clean oil. Obviously, this created the potential for storm water pollution. When confronted with the problem, the employees were embarrassed about the situation; they knew that this was not the proper disposal method for waste oil. However, the practice had been initiated by previous workers (probably by a single worker). The current employees felt that they might as well continue this practice, since the area was already heavily soaked with oil. However, the inspection provided the motivation for them to remove the contaminated soil, establish appropriate maintenance procedures, and dispose of waste oil properly.

Record-keeping and internal reporting procedures. The pollution prevention plan must describe procedures for developing and retaining records on the status and effectiveness of plan implementation. A record-keeping system ensures adequate implementation of the SWPPP. The record-keeping requirements begin with the preparation of the SWPPP and NOI (or permit application) and continue through the Notice of Termination (NOT) and afterward. The following records should be maintained on-site:

1. *SWPPP.* The storm water pollution prevention plan must be filed on-site for reference by employees and contractors, and for review by

regulatory authorities or the public. For convenience, it should be organized into a binder with major sections corresponding to the plan requirements outlined in this chapter.

2. *NOI.* A copy of the Notice(s) of Intent should be filed with the SWPPP. Copies of any additional NOIs prepared as a result of additional operators becoming involved in the project should also be filed as they are generated.

3. *Inspection reports.* All inspection reports should be filed with the SWPPP as they are generated. Blank inspection and maintenance forms should be prepared ahead of time. The inspection forms should be specific to the facility. The forms should list each of the measures to be inspected on the site. The form should include blanks for the inspector to fill in: inspector's name, date of inspection, condition of the measure or area inspected, maintenance or repair performed, and any changes which should be made to the SWPPP to control or eliminate unforeseen pollution of storm water.

4. *Maintenance reports.* Reports should be made on regular and special maintenance activities performed on the site, including waste disposal activities.

5. *Industrial activity reports.* In addition to the inspection and maintenance reports, the operator should keep records of the industrial activity on the site.

6. *Spill reports.* Records of releases of a hazardous substance in excess of reportable quantities established at 40 CFR 117.3 or 40 CFR 302.4 describing each release that has occurred at any time after the date of 3 years prior to the issuance of this permit, measures taken in response to the release, and measures taken to prevent recurrence must be included in plans.

7. *Other materials.* Other materials relevant to the permit, including correspondence with regulatory authorities, photographs, and so on, should be included.

8. *NOT.* A copy of the Notice of Termination for the NPDES permit is included.

The entire file must be kept until at least 1 year after the permit coverage ends. In addition, all sampling reports must be retained for at least 6 years after the date of sampling or until the termination of permit coverage (whichever is longer).

All SWPPPs required under the permit are considered reports that shall be available to the public under Section 308(b) of the CWA. However, if the storm water pollution plan contains trade secrets or other confidential information, the permittee may claim any portion of the plan as confidential, in accordance with 40 CFR Part 2.

The amount of record keeping required will vary based upon the industrial activities occurring on-site and the number of inspections required. For example, automobile salvage yard facilities must consider maintaining an organized inventory of materials used at the facility that may contaminate rainfall or storm water runoff. See Chap. 8 for more details.

Traditional storm water management practices

"Traditional" storm water management practices are measures which reduce pollutant discharges by reducing the volume of storm water discharges by diverting, infiltrating, reusing, or otherwise managing storm water runoff so as to reduce the discharge of pollutants.

Based on an assessment of the potential of various sources at the plant to contribute pollutants to storm water discharges associated with industrial activity, the SWPPP must provide that traditional storm water management measures determined to be reasonable and appropriate be implemented and maintained.

Traditional storm water management practices such as grass swales, catch basins, infiltration devices, inlet controls, unlined retention or detention basins, and oil and grit separators can sometimes be applied to an industrial setting. Traditional storm water management practices can include water reuse activities, such as the collection of storm water for later uses such as irrigation or dust control. Appropriate snow removal activities may be considered, such as selecting a site for removed snow and selecting and using deicing chemicals. In addition, other types of controls such as spill prevention measures can be considered to prevent catastrophic events that can lead to surface or groundwater contamination.

The plan must identify practices that are reasonable and appropriate for the facility. The plan should describe the particular pollutant source area or activity to be controlled by each storm water management practice. Reasonable and appropriate practices must be implemented and maintained according to the provisions prescribed in the plan.

In selecting storm water management measures, it is important to consider the potential effects of each method on other water resources, such as groundwater. Although storm water pollution prevention plans primarily focus on storm water management, facilities must also consider potential groundwater pollution problems and take appropriate steps to avoid adversely impacting groundwater quality. For example, if the water table is unusually high in an area, an infiltration pond may contaminate a groundwater source unless special preventive measures are taken. In some cases, it is appropriate to limit traditional storm water management practices to those areas of the drainage system that generate storm water with relatively low levels of pollutants (e.g., many rooftops, parking lots, and so on). Under EPA's July 1991 groundwater protection strategy, states are encouraged to develop comprehensive state groundwater protection programs (CSGWPPs). Efforts to control storm water should be compatible with state groundwater objectives as reflected in CSGWPPs.

In some instances, facilities may have to develop an elaborate set of structural controls, such as detention and retention ponds, to provide treatment to storm water runoff prior to its discharge off-site. For most industries, simple modifications to existing facility material handling practices will suffice. Material storage areas can often be enclosed or covered to prevent exposure of the material to storm water runoff. Diversion structures can prevent storm water from entering material storage areas.

The pollution prevention plan must identify areas that, due to topography, activities, soils, cover materials, or other factors, have a high potential for significant soil erosion. The plan must identify measures, such as the placement of vegetation, that will be implemented to limit erosion in these areas and provide filtering of storm water runoff.

Elimination of pollution sources

In some cases, elimination of pollution sources may be the most cost-effective way to control pollutants in storm water discharges associated with industrial activity. Options for eliminating pollution sources include reducing on-site air emissions affecting runoff quality, changing chemicals used at the facility, and modifying material management practices such as moving storage areas into buildings.

Diversion of discharge to sewage treatment plant

Where storm water discharges contain significant amounts of pollutants that can be removed by a sewage treatment plant, the storm water discharge can be discharged to the sanitary sewage system. Such diversions must be coordinated with the operators of the sewage treatment plant and the collection system to avoid exacerbating problems with *combined sewer overflows* (*CSOs*), basement flooding, or wet-weather operation of the treatment plant. Where CSO discharges, flooding, or plant operation problems can result, on-site storage followed by a controlled release during dry-weather conditions may be considered.

End-of-pipe treatment

At many types of industrial facilities, it may be appropriate to collect and treat the runoff from targeted areas of the facility. This approach was taken with nine industrial categories with national effluent guideline limitations for storm water discharges. To meet the numeric effluent limitation, most, if not all, facilities must collect and temporarily store on-site the runoff from targeted areas of the plant.

The effluent guideline limitations do not apply to discharges whenever a particular rainfall event or a sequence of rainfall events causes an overflow of storage devices designed, constructed, and operated to contain a design storm. The 10-year, 24-h storm or the 25-year, 24-h storm commonly is used as the design storm in the effluent guideline limitations.

Monitoring Plan

Many facilities are required to perform periodic sampling of their storm water runoff as a condition of the EPA's general permit. Chapter 9 provides details on storm water sampling and monitoring. Sampling data collected during the term of the permit must be summarized in the SWPPP.

Comprehensive site compliance evaluation

Under the EPA multisector general permit, the minimum monitoring requirement for all facilities consists of an annual site compliance evaluation. The SWPPP must describe the scope and content of comprehensive site inspections that qualified personnel will conduct to

1. Confirm the accuracy of the description of potential pollution sources contained in the plan
2. Determine the effectiveness of the plan
3. Assess compliance with the terms and conditions of the permit

As noted earlier, the comprehensive site evaluations, are not the same as the periodic or other inspections described for certain industries in Chap. 8. However, it is acceptable to combine a periodic inspection with a comprehensive site compliance evaluation if the two inspections are scheduled to be performed on approximately the same date. Of course, all the requirements for both types of inspections must be met.

The SWPPP must indicate the frequency of such evaluations, which in most cases must be at least once per year. The individual(s) who will conduct the inspections must be identified in the plan and should be member(s) of the pollution prevention team.

Material handling and storage areas and other potential sources of pollution must be visually inspected for evidence of actual or potential pollutant discharges to the drainage system. Inspectors also must observe erosion controls and structural storm water management devices to ensure that each is operating correctly. Equipment needed to implement the pollution prevention plan, such as that used during spill response activities, must be inspected to confirm that it is in proper working order. The following steps will be completed in connection with the annual inspection:

- *SWPPP review.* Review the SWPPP, and draw up a list of those items which are part of material handling, storage, and transfer areas covered by the plan.
- *Materials review.* Verify the list of equipment and materials in these areas covered in the plan.
- *Operations review.* Review facility operations for the past year to determine if any more areas should be included in the original plan or any existing areas were modified so as to require plan modification. Change the plan as appropriate.
- *Drainage area inspection.* Inspect storm water drainage areas for evidence of pollutants entering the drainage system.
- *Source reduction measures.* Evaluate the effectiveness of measures to reduce pollutant loadings, and decide whether additional measures are needed.
- *Source control measures.* Observe structural measures, sediment controls, and other storm water BMPs to ensure proper operation.
- *Equipment review.* Inspect any equipment needed to implement the plan, such as spill response equipment.
- *Plan revision.* Revise the plan as needed within 2 weeks of inspection (potential pollutant source description and description of measures and controls).
- *Plan implementation.* Implement any necessary changes in a timely manner, but within 12 weeks of the inspection.
- *Inspection report.* Prepare a report that summarizes inspection results and follow-up actions, the date of inspection, and list of personnel who conducted the inspection.
- *Certification.* Document all incidents of noncompliance in the inspection report. Where there are no incidents of noncompliance, the inspection report must contain a certification that the facility is in compliance with the plan. Figure 7.1 contains a certification form used for this purpose.
- *Record keeping.* Sign the report and keep it with the plan.

ANNUAL CERTIFICATION OF COMPLIANCE

This certification is completed by an authorized signatory after each annual site compliance evaluation. It indicates that the Storm Water Pollution Prevention Plan was evaluated as part of an inspection, is adequate for control of facility storm water discharges, and that the facility is in compliance with the Plan. If changes to the Plan or the site are necessary as a result of the inspection, these changes should be performed before this certification is completed.

I certify under penalty of law that this document and all attachments were prepared under my direction or supervision in accordance with a system designed to assure that qualified personnel properly gathered and evaluated the information submitted. Based on my inquiry of the person or persons who manage the system, or those persons directly responsible for gathering the information, the information submitted is, to the best of my knowledge and belief, true, accurate, and complete. I am aware that there are significant penalties for submitting false information, including the possibility of fine and imprisonment for knowing violations.

Signed: ______________________________

Name: ______________________________

Title: ______________________________

Company: ______________________________

Address: ______________________________

Telephone: ______________________________

Date: ______________________________

Figure 7.1 Certification of comprehensive site compliance evaluation.

The results of each site inspection must be documented in a report signed by an authorized company official. The report must describe the scope of the inspection, the personnel making the inspection, the date(s) of the inspection, and any major observations relating to implementation of the SWPPP. Comprehensive site evaluation reports must be retained for at least 3 years after the date of the evaluation.

Based on the results of each inspection, the description of potential pollution sources and measures and controls included in the SWPPP must be revised as appropriate within 2 weeks after each inspection. Changes in the procedural operations must be implemented on the site in a timely manner—and never more than 12 weeks after completion of the inspection. Procedural changes that require construction of structural measures and controls are allowed up to 3 years for implementation. In both instances, an extension may be requested from the EPA or state agency that has authority over NPDES permit enforcement.

The pollution prevention plan for the facility must be revised where necessary to address the findings and reflect the recommendations of the inspection. Additionally, an annual certification must be prepared, indicating that the SWPPP was evaluated as part of an inspection, that the plan is adequate for control of facility storm water discharges, and that the facility is in compliance with the plan.

Where annual site inspections are shown in the plan to be impracticable for inactive mining sites due to remote location and inaccessibility of the site, comprehensive site compliance evaluations are to be conducted at least once every 3 years. The first site inspection must take place 2 years after the site becomes inactive.

A copy of the Annual Comprehensive Site Compliance Evaluation worksheet will be completed and filed on-site. The worksheet will be signed by an authorized signatory authority. The completed worksheet should include the following certification language:

> I certify under penalty of law that this document and all attachments were prepared under my direction or supervision in accordance with a system designed to assure that qualified personnel properly gathered and evaluated the information submitted. Based on my inquiry of the person or persons who manage the system, or those persons directly responsible for gathering the information, the information submitted is, to the best of my knowledge and belief, true, accurate, and complete. I am aware that there are significant penalties for submitting false information, including the possibility of fine and imprisonment for knowing violations.

Monitoring requirements

For the multisector permit, monitoring is designed to assess the effectiveness of the SWPPP and to provide an incentive to reduce pollution.

The EPA multisector permit requires most specific industry sectors to monitor discharges quarterly during years 2 and 4 of permit coverage. Facilities with pollutant concentrations lower than benchmarks in year 2 are exempt from monitoring in year 4. The EPA multisector permit has a list of parameters to be monitored for each industry based upon review of information from group applications, including sampling data and descriptions of industrial activity, and significant materials. Facilities may exempt themselves from monitoring *on a pollutant-by-pollutant basis* if they can certify that there are no sources of a pollutant present. All facilities that monitor must submit the sampling data. Facilities are encouraged to review monitoring results and revise the SWPPP where pollutants are above benchmark concentrations. Routine visual monitoring is required to assess problems in the EPA multisector permit. Chapter 9 provides a detailed description of the sampling and monitoring requirements for each industry sector.

Numeric effluent limitations in the EPA multisector permit

The multisector permit provides coverage for four types of storm water discharges subject to effluent guidelines, provided that the discharges are not already subject to an existing individual NPDES storm water permit. These discharges include contaminated storm water runoff from phosphate fertilizer manufacturing facilities, runoff associated with asphalt paving or roofing emulsion production, runoff from material storage piles at cement manufacturing facilities, and runoff from coal piles at steam electric generating facilities. The multisector permit establishes technology-based effluent limitations for coal pile runoff from all facilities covered by the permit. The inclusion of numeric effluent limitations allows facilities with discharges subject to these guidelines to obtain coverage for their storm water discharges under one permit. In addition, the permit establishes numeric effluent limitations for mine dewatering discharges composed entirely of storm water or groundwater seepage from construction sand and gravel, industrial sand, and crushed stone mines located in region VI (the states of Louisiana, New Mexico, Oklahoma, and Texas).

The source of the numeric effluent limitations in the multisector permit includes the following guidelines:

- Phosphate fertilizer manufacturing runoff at 40 CFR Part 418
- Asphalt paving and roofing emulsions production runoff at 40 CFR Part 443
- Cement manufacturing materials storage piles runoff at 40 CFR Part 411

- Steam electric power generating coal pile runoff at 40 CFR Part 423
- Construction sand and gravel, industrial sand, and crushed stone mine dewatering at 40 CFR Part 436

Other Regulatory Requirements

Many states, cities, and regional water quality agencies have their own requirements for storm water runoff. The pollution prevention plan should be developed with these requirements in mind.

The EPA multisector general permit requires that the SWPPP list all applicable local and state regulatory requirements, and summarize the activities undertaken in compliance with these requirements.

Permittees which discharge storm water associated with industrial activity through large or medium-size municipal separate storm water systems are required to submit a signed copy of their NOI to the operator of the municipal separate storm sewer system.

Facilities covered by these permits must comply with applicable requirements in municipal storm water management programs developed under NPDES permits issued for the discharge of the municipal separate storm sewer system that receives the facility's discharge, provided the discharger has been notified of such conditions. In addition, permittees that discharge storm water associated with industrial activity through a municipal separate storm sewer system serving a population of 100,000 or more must make their pollution prevention plans available to the municipal operator of the system upon request by the municipal operator.

By requiring compliance with local and state regulations, the EPA is effectively putting the full force and authority of the Clean Water Act behind the applicable local and state regulations.

References

Environmental Protection Agency, 1979. *NPDES Best Management Practices Guidance Document,* EPA 600/9-79-045, December.

The Cadmus Group, 1990. *Manual of Practice: Identification of Illicit Connections,* September.

Government Printing Office: *Federal Register,* Office of the Federal Register, National Archives and Records Administration, Washington.

———, 1992. *Code of Federal Regulations,* Office of the Federal Register, National Archives and Records Administration, Washington, July 1.

Chapter

8

Special Sections of Industrial Storm Water Pollution Prevention Plans

This is the fourth of five chapters which deal with the specific requirements for storm water discharges from industrial (nonconstruction) facilities. This chapter describes how to prepare special sections of a storm water pollution prevention plan (SWPPP) for an industrial facility. Chapter 7 described how to prepare the basic sections of the SWPPP. This chapter is based on the requirements of the EPA multisector general permit for storm water discharges associated with industrial activity.

Discharges through Large and Medium-Size Municipal Separate Storm Sewer Systems

Facilities that discharge storm water associated with industrial activity through large or medium-size municipal separate storm sewer systems are required to submit notification of the discharge to the operator of the municipal separate storm sewer system. These facilities must also comply with applicable requirements of the municipal storm water management programs, provided the discharger has been notified of such conditions. The facility's SWPPP must be made available to the municipal operator of the system upon request. The facility's SWPPP should document these requirements.

Salt Piles

The EPA multisector general permit contains special requirements for storm water discharges associated with industrial activity from salt storage facilities. Storage piles of salt used for deicing or other commercial or industrial purposes must be enclosed or covered to prevent exposure to precipitation, except for exposure resulting from adding or removing materials from the pile. This requirement applies only to runoff from storage piles discharged to waters of the United States. Facilities that collect all the runoff from their salt piles and reuse it in their processes or discharge it subject to a separate National Pollutant Discharge Elimination System (NPDES) permit do not need to enclose or cover their piles.

Coal Piles

The EPA multisector general permit establishes effluent limitations of 50 milligrams per liter (mg/L) total suspended solids and a pH range of 6.0 to 9.0 for coal pile runoff. This effluent limitation is similar to the effluent guideline limitation for coal pile runoff from facilities in the steam electric power generating point source category [see 40 CFR 423.12(b)(9)].

Any untreated overflow from facilities designed, constructed, and operated to treat the volume of coal pile runoff associated with a 10-year, 24-h rainfall event is not subject to the 50 mg/L limitation for total suspended solids. Providing a limit to effluent guidelines for events that exceed a specified storm event provides operators with a basis for installing and operating a treatment system, as the design of the system, particularly the collection devices, will depend on the design storm chosen.

Steam electric generating facilities must comply with these limitations upon submittal of the NOI. All other types of facilities must comply with this requirement as expeditiously as practicable, but in no event later than 3 years from the date of permit issuance. Coal pile runoff from steam electric facilities is already subject to an effluent limitation guideline [see 40 CFR 423].

The pollutants in coal pile runoff can be classified into specific types according to chemical characteristics (Environmental Protection Agency, 1982). The type relates to pH of the coal pile drainage. The pH tends to be of an acidic nature, primarily as a result of the oxidation of iron sulfide in the presence of oxygen and water. The potential influence of pH on the behavior of toxic and heavy metals is of particular concern. Many of the metals are amphoteric with regard to their solubility behavior. The factors affecting acidity, pH, and the subsequent leaching of trace metals are

- Concentration and form of pyritic sulfur in coal
- Size of the coal pile

- Method of coal preparation and clearing prior to storage
- Climatic conditions, including rainfall and temperature
- Concentrations of $CaCO_3$ and other neutralizing substances in the coal
- Concentration and form of trace metals in the coal
- Residence time of storm water in the coal pile

Coal piles can generate runoff with low pH values, with the acid values being quite variable. The suspended solids levels can be significant, with levels of 2500 mg/L not uncommon. Metals present in the greatest concentrations are copper, iron, aluminum, nickel, and zinc. Others present in trace amounts include chromium, cadmium, mercury, arsenic, selenium, and beryllium.

The effluent limitations for coal pile runoff in the EPA multisector general permit can be achieved by these two primary methods: (1) limiting exposure to coal by use of covers or tarpaulins and storm water run-on berms; and (2) collecting and treating the runoff. In some cases, coal pile runoff may be in compliance with the effluent limitations without covering of the pile or collection or treatment of the runoff. In these cases, the operator of the discharge will not have a control cost. The use of tarpaulins and berms to prevent exposure is expected to be practical for coal piles smaller than 30,000 cubic meters (m^3). The EPA expects that the majority of industrial facilities subject to the requirement will have coal piles smaller than 30,000 m^3.

The primary technology options for treating coal pile runoff are (1) equalization, pH adjustment, and settling; and (2) equalization, chemical precipitation treatment, settling, and pH adjustment.

Metals may be removed from wastewater by raising the pH of the wastewater to precipitate them out as hydroxides. Typically, wastewater pH levels of 9 to 12 are required to achieve the desired precipitation levels. Lime is frequently used for pH adjustment. Wastewaters which have a pH greater than 9 after lime addition will require acid addition to reduce the pH before final discharge. Polymer addition may be required to enhance the settling characteristics of the metal hydroxide precipitate. Typical polymer feed concentrations in the wastewater are 1 to 4 parts per million (ppm). The metal hydroxide precipitate is separated from the wastewater in a clarifier or a gravity thickener. Unlike settling ponds, these units continually collect and remove the sludge formed. Filters are typically used for effluent polishing and can reduce suspended solids levels below 10 mg/L. Sand and coal are the most common filter media. Vacuum filtration is a common technique for dewatering sludge to produce a cake that has good handling properties and minimum volume.

The major equipment requirements for such a system include a lime feed system, mix tank polymer feed system, flocculator-clarifier, deep-bed filter, and acid feed system. For wastewaters which have a pH of less than 6, mixers and mixing tanks are made of special materials of construction (stainless steel or lined-carbon steel). For wastewaters with pH levels greater than 6, concrete tanks are typically used. The underflow from the clarifier may require additional treatment with a gravity thickener and a vacuum filter to provide sludge which can be transported economically for landfill disposal.

SARA Title III, Section 313 (EPCRA) Facilities

The Superfund Amendments and Reauthorization Act (SARA) of 1986 resulted in the enactment of Title III of SARA, the Emergency Planning and Community Right-to-Know Act (EPCRA). Section 313 of Title III of SARA requires operators of certain facilities that manufacture, import, process, or otherwise use listed toxic chemicals to report annually their releases of those chemicals to any environmental media.

The criteria for facilities that must report under Section 313 are given at 40 CFR 372.22. A facility is subject to the annual reporting provisions of Section 313 if it meets all three of the following criteria for a calendar year:

- It is included in Standard Industrial Classification (SIC) codes 20 through 39.
- It has 10 or more full-time employees.
- It manufactures (including imports), processes, or otherwise uses a chemical listed in 40 CFR 372.65 in amounts greater than the threshold quantities specified in 40 CFR 372.25. After 1989, the threshold quantity of listed chemicals that the facility must manufacture, import, or process in order to be required to submit a release report is 25,000 pounds per year (lb/yr). The threshold for uses other than manufacturing, importing, or processing of listed toxic chemicals is 10,000 lb/yr.

There are more than 300 individually listed Section 313 chemicals, as well as 20 categories of toxic release inventory (TRI) chemicals for which reporting is required. EPA has the authority to add to and delete from this list. Approximately 175 of these chemicals are classified as *Section 313 water priority chemicals.* These are defined as chemicals or chemical categories which also are listed at 40 CFR 372.65 pursuant to SARA Title III, Section 313; are manufactured, processed, or otherwise used at or above threshold levels at a facility subject to SARA Title III, Section 313 reporting requirements; and meet at least one of the following criteria:

1. Are listed in appendix D of 40 CFR Part 122 on table II (organic priority pollutants), table III (certain metals, cyanides, and phenols), or table V (certain toxic pollutants and hazardous substances)
2. Are listed as a hazardous substance pursuant to Section 311(b)(2)(A) of the CWA at 40 CFR 116.4
3. Are pollutants for which EPA has published an acute or a chronic toxicity criterion

Targeting EPCRA Section 313 facilities

The special requirements in the EPA multisector general permit for facilities that are subject to EPCRA Section 313 reporting requirements apply only to areas of the facilities where EPCRA Section 313 chemicals are managed. The other baseline requirements of the EPA multisector general permit apply to other parts of the facility that generate storm water discharges associated with industrial activity.

The special requirements for SWPPPs at facilities that are subject to EPCRA Section 313 for Section 313 water priority chemicals primarily focus on areas of the facility where equipment used for the management, storage, and processing of Section 313 water priority chemicals is exposed to precipitation or can otherwise contribute pollutants to a storm drainage system. The burdens associated with the requirements of the EPA multisector general permit are significantly reduced for facilities that manage (including loading and unloading activities) their toxic chemicals in buildings or under cover such that there is no exposure to precipitation and where the floor drainage in the building is known to be segregated from the storm water collection system.

If the discharger submits a properly signed certification that all the Section 313 water priority chemicals handled and/or stored on-site are only in gaseous or nonsoluble liquid or solid (at atmospheric pressure and temperature) forms, then none of the additional requirements in the EPA multisector general permit is applicable to that facility.

Summary of special requirements

The special requirements in the EPA multisector general permit for facilities subject to reporting requirements under EPCRA Section 313 for a water priority chemical state that SWPPPs, in addition to the baseline requirements described in Chap. 7, must contain special provisions addressing areas where Section 313 water priority chemicals are stored, processed, or otherwise handled. The permit states that appropriate containment, drainage control, and/or diversionary structures must be provided for such areas. At a minimum, one of the following preventive systems or its equivalent must be used:

- *Drainage controls:* curbing, culverting, gutters, sewers, or other forms of drainage control to prevent or minimize the potential for storm water run-on to come into contact with significant sources of pollutants
- *Protection measures:* roofs, covers, or other forms of appropriate protection to prevent storage piles from exposure to storm water and wind

The EPA allows the discharger to choose the preventive system most appropriate for a particular facility. If a leak is discovered which may result in a significant release of a Section 313 water priority chemical to waters of the United States, then the EPA multisector general permit requires that permittees either take action to stop the leak or otherwise prevent the significant release of Section 313 water priority chemicals to waters of the United States. The temporary use of drip pans, diversions to sumps, or other measures that prevent toxic chemicals from being discharged to waters of the United States until permanent repairs can be made may constitute appropriate action.

The permit also establishes requirements for priority areas of the facility, including the following:

1. *Liquid storage areas:* liquid storage areas where storm water comes into contact with any equipment, tank, container, or other vessel used for Section 313 water priority chemicals
2. *Other material storage areas:* material storage areas for Section 313 water priority chemicals other than liquids
3. *Loading and unloading areas:* truck and railcar loading and unloading areas for liquid Section 313 water priority chemicals
4. *Other transfer, process, or handling areas:* areas where Section 313 water priority chemicals are transferred, processed, or otherwise handled

Drainage from these priority areas should be restrained by valves or other positive means to prevent the discharge of a spill or other excessive leakage of Section 313 water priority chemicals. Where containment units are employed, such units may be emptied by pumps or ejectors; however, these must be manually activated. Flapper-type drain valves must not be used to drain containment areas, as these will not effectively control spills. Valves used for the drainage of containment areas should, as far as is practical, be of manual, open-and-closed design. If facility drainage does not meet these requirements, the final discharge conveyance of all in-facility storm sewers must be equipped to be equivalent to a diversion system that could, in the event of an uncontrolled spill of Section 313 water priority chemicals, return the spilled material or contaminated storm water to the facility. Records must be kept of the frequency and estimated volume (in gallons) of discharges from containment areas.

Liquid storage areas

Appropriate measures to minimize discharges of Section 313 chemicals may include secondary containment provided for at least the entire contents of the largest single tank, plus sufficient freeboard to allow for precipitation, a strong spill contingency and integrity testing plan, and/or other equivalent measures.

Where storm water comes into contact with any equipment, tank, container, or other vessel used for Section 313 water priority chemicals, the material and construction of tanks or containers used for the storage of a Section 313 water priority chemical must be compatible with the material stored and conditions of storage, such as pressure and temperature.

A strong spill contingency plan would typically contain, at a minimum, a description of response plans, personnel needs, and methods of mechanical containment (such as the use of sorbents, booms, and collection devices); steps to be taken for removal of spill chemicals or materials; and procedures to ensure access to and availability of sorbents and other equipment. The testing component of the plan would provide for conducting integrity testing of storage tanks at set intervals such as once every 5 years, and conducting integrity and leak testing of valves and piping at a minimum frequency, such as once per year. In addition, a strong plan would include a written and actual commitment of workforce, equipment, and materials required to comply with the permit and to expeditiously control and remove any quantity of spilled or leaked chemicals that could result in a toxic discharge.

Other material storage areas

Material storage areas for Section 313 water priority chemicals other than liquids that are subject to runoff, leaching, or wind must incorporate drainage or other control features to minimize the discharge of Section 313 water priority chemicals by reducing storm water contact with these chemicals.

Truck and railcar loading and unloading areas

Appropriate measures to minimize discharges of Section 313 chemicals may include the placement and maintenance of drip pans (including the proper disposal of materials collected in the drip pans) where spillage may occur (such as at hose connections, hose reels, and filler nozzles) for use when one is making and breaking hose connections; a strong spill contingency and integrity testing plan; and/or other equivalent measures.

Other transfer, process, or handling areas

Materials used in piping and equipment must be compatible with the substances handled. Drainage from process and material handling areas must minimize storm water contact with Section 313 water priority chemicals. Additional protection—such as covers or guards to prevent exposure to wind, spraying, or releases from pressure relief vents to prevent a discharge of Section 313 water priority chemicals to the drainage system, and overhangs or door skirts to enclose trailer ends at truck loading and unloading docks must be provided as appropriate. Visual inspections or leak tests—must be provided for overhead piping conveying Section 313 water priority chemicals without secondary containment.

Other areas

The EPA multisector general permit provides that site runoff from other industrial areas of the facility that may contain Section 313 water priority chemicals or spills of Section 313 water priority chemicals must incorporate the necessary drainage or other control features to prevent the discharge of spilled or improperly disposed material and to ensure the mitigation of pollutants in runoff or leachate. The permit also establishes special requirements for preventive maintenance and good housekeeping, facility security, and employee training.

Sampling requirements for storm water discharges from EPCRA Section 313 are discussed in Chap. 9. Facilities should review monitoring data and evaluate pollution prevention measures suitable for reducing pollutants in discharges.

Sector A: Timber Products

Those facilities that have conducted activities associated with wood preserving and wood surface protection with pentachlorophenol formulations, creosote formulations, or arsenic/chromium formulations in the past must identify: areas where soils are contaminated, treatment equipment, and/or stored materials which remain as a result of these operations. The management practices being employed to minimize the contact of these materials with storm water runoff must also be identified.

Where facilities have used chlorophenolic, creosote, or chromium-copper-arsenic formulations for wood surface protection or preserving activities on-site in the past, and information is available, the facility must inventory the areas where soils are contaminated, treatment equipment, and treated materials that remain. Once these areas are identified, measures to minimize their exposure to storm water or to limit discharge of pollutants into storm water must be implemented.

All facilities must consider the implementation of BMPs in the following areas of the site: log, lumber, and other wood product storage areas; residue storage areas; loading and unloading areas; material handling areas; chemical storage areas; and equipment/vehicle maintenance, storage, and repair areas. Facilities that surface-protect and/or preserve wood products must address specific BMPs for wood surface protection and preserving activities. The SWPPP must include a discussion of how and why certain BMPs were chosen for the facility.

In addition to typical good housekeeping measures that require the maintenance of areas which may contribute pollutants to storm water in a clean and orderly manner, the SWPPP must specifically address good housekeeping measures and the specific frequency of performance of these measures which are designed to (1) limit the discharge of wood debris, (2) minimize the leachate generated from decaying wood materials, and (3) minimize the generation of dust.

The periodic removal of debris from ditches, swales, diversion, containment basins, and infiltration measures must be required.

Schedules must be developed and presented in the SWPPP for response procedures to limit the tracking of spilled materials to other areas of the site. Leaks or spills of wood surface protection or preservation chemicals must be cleaned up immediately. The SWPPP must also limit the tracking of significant materials that have been leaked or spilled on the site from containers, facility equipment, or on-site vehicles, particularly outside the area where storm water controls are in place. This may occur, e.g., during the filling of storage tanks. Vehicles or equipment used to transfer materials may come into contact with any materials spilled during the filling or emptying of tanks. As the vehicles move to other locations at the site, such material may be tracked and eventually lead to contamination of storm water discharges.

Discharges of boiler blowdown, water treatment, wastewaters, noncontact cooling waters, contact cooling waters, washdown waters from treatment equipment, and storm water that have come in contact with site areas where hand spraying of surface protection chemicals is performed are not authorized. Authorized non–storm water discharges include discharges from spraydown of lumber and wood product storage yards where no chemical additives are used in the spray water and no chemicals are applied to the wood during storage.

Facility operators must conduct visual inspections of BMPs on a quarterly basis. Inspections must be performed quarterly at processing areas, transport areas, and treated wood storage areas of facilities performing wood surface protection and preservation activities. In addition, all timber products facilities must conduct daily inspections of material handling activities and unloading and loading areas whenever activities are occurring in those areas (if activities are not occurring in those areas, no inspection is required).

Records must be maintained showing that these inspections have been performed at the required frequencies. In addition, a set of tracking or follow-up procedures must be implemented to ensure that appropriate actions are taken based on the findings of the inspections. These records should be developed on a case-by-case basis depending upon the facility's needs.

The following areas of the plant must be considered for sediment and erosion controls: loading and unloading areas, access roads, material handling areas, storage areas, and any other areas where heavy equipment and vehicle use are prevalent. Sediment and erosion controls include stabilization measures such as seeding, mulching, chemical stabilization, sodding, soil-retaining measures; and dust control and structural measures such as sediment traps, contouring, sediment basins, check dams, and silt fences. Chapters 12 and 13 provide further details on these measures.

Sector B: Paper and Allied Products Manufacturing

There are no requirements for sector B facilities beyond those described in Chap. 7.

Sector C: Chemical and Allied Products Manufacturing

All sector C facilities must prepare a drainage and site plan detailing the drainage patterns of the runoff and identifying the outfall and receiving water body.

The inventory of exposed materials as well as risk identification and summary of potential pollutant sources requirements were further defined to avoid confusion. In addition, the information submitted in the group application regarding pollutant sources and current management practices must be evaluated and considered when the plan is developed.

At a minimum, all facilities must consider establishing the following good housekeeping practices:

1. Schedule regular pickup and disposal of garbage and waste materials or other measures to dispose of waste. This schedule may be included in the plan. Individuals responsible for waste management and disposal should be informed of the procedures established under the plan.
2. Routinely inspect for leaks and conditions of drums, tanks, and containers. Ensure that spill cleanup procedures are understood by employees.
3. Keep an up-to-date inventory of all materials present at the facility. While the inventory is being taken, all containers should be clearly labeled. Hazardous containers that require special handling, storage, use, and disposal considerations should be clearly marked and readily recognizable.
4. Maintain clean ground surfaces by using brooms, shovels, vacuum cleaners, or cleaning machines.

All facilities must implement employee training to address procedures for equipment and container cleaning and washing. The training should emphasize the human hazards and the potential environmental impacts from the discharges of wash waters. The SWPPP for sector C facilities must identify periodic dates for such training of at least once per year. However, more frequent training may be necessary at facilities with high turnover of employees or where employee participation is essential to the SWPPP.

Qualified personnel must conduct quarterly inspections. A wet-weather inspection (during a rainfall event) must be conducted in the second (April to June) and third (July to September) quarters of each year. A dry-weather inspection (no precipitation) must be conducted in the first (January to April) and fourth (October to December) quarters. However, where a seasonal arid period is sustained for more than 3 months, a dry-weather inspection will satisfy the wet-weather inspection requirement. At least one inspection must be conducted every quarter.

Facilities should consider evaluating existing security systems such as fencing, lighting, vehicular traffic control, and securing of equipment and buildings, and should include existing and new systems in the plan to prevent accidental or intentional entry, which could cause a discharge of pollutants to waters of the United States.

For areas where liquid or powdered materials are stored, facilities must consider providing diking, curbing, or berms. For all other outside storage areas including storage of used containers, machinery, scrap and construction materials, and pallets, facilities must consider preventing or minimizing storm water run-on to the storage area by using curbing, culverting, gutters, sewers, or other forms of drainage control. For all storage areas, roofs, covers, or other forms of appropriate protection must be considered to prevent exposure to weather. In areas where liquid or powdered materials are transferred in bulk from truck or railcars, permittees must consider appropriate measures to minimize contact of material with precipitation. Permittees must consider providing for hose connection points at storage containers to be inside containment areas and drip pans to be used in areas which are not in a containment area, where spillage may occur (e.g., hose reels, connection points with railcars or trucks), or equivalent measures. In areas of transfer of contained or packaged materials and loading and in unloading areas, permittees must consider providing appropriate protection such as overhangs or door skirts to enclose trailer ends at truck loading and unloading docks, or an equivalent.

Contained areas should be restrained by valves or other positive means to prevent the discharge of a spill or leak. Containment units may be emptied by pumps or ejectors; however, these should be manually activated. Flapper-type drain valves should not be used to drain containment areas. Valves used for the drainage of containment areas should, as far as is practical, be of manual, open-or-closed design. If facility drainage is not engineered as above, the final discharge point of all in-facility sewers should be equipped to prevent the discharge in the event of an uncontrolled spill of materials.

The SWPPP must contain a description of storm water management practices used and/or to be used to divert, infiltrate, reuse, or otherwise manage storm water runoff in a manner that reduces pollutants in storm water discharges from the site.

For areas with a potential for significant soil erosion, the SWPPP should describe permanent stabilization practices to be used in order to stabilize disturbed areas.

A comprehensive site compliance evaluation must be conducted at least once every year. Members of the pollution prevention team or a qualified professional designated by the team must conduct the evaluation.

Sector D: Asphalt Paving and Roofing Materials Manufacturers and Lubricant Manufacturers

Comprehensive site compliance evaluations must be conducted at least once a year for asphalt facilities and lubricant manufacturers. The individual or individuals who will conduct the evaluations must be identified in the plan and should be members of the pollution prevention team. Inspection reports must be retained for at least 3 years after the date of the evaluation.

Comprehensive site compliance evaluations shall be conducted at least once a year at portable plant locations. Such evaluations shall be conducted at least once at portable plant locations that are not in operation a full year.

Based on the results of each evaluation, the description of potential pollution sources, and measures and controls, the plan must be revised as appropriate within 2 weeks after each evaluation. Changes in the measures and controls must be implemented on the site in a timely manner, but no later than 12 weeks after completion of the evaluation.

For portable plants, the plan must be revised as appropriate as soon as possible, but no later than 2 weeks after each evaluation. Two weeks is adequate time for portable plants to modify their plans due to the simpler and smaller nature of these operations in comparison to permanent facilities.

Sector E: Glass, Clay, Cement, Concrete, and Gypsum Product Manufacturing

Sector E facilities must identify on the site map the location of any baghouse or other air pollution control device, and any sedimentation or process wastewater recycling pond and the areas which drain to the pond. The site map for the facility must clearly indicate the portion of the facility which drains to sedimentation or recycle ponds that receive process wastewater. The site map must also indicate the outfall locations and the types of discharges contained in the drainage areas of the outfalls (e.g., storm water and air conditioner condensate). To increase the readability of the map, the inventory of the types of discharges contained in each outfall may be kept as an attachment to the site map. The site map for these facilities must also indicate the portion of the site where regular sweeping or other equivalent good housekeeping measures will be implemented to prevent the accumulation of spilled materials or settled dust.

The SWPPP for sector E facilities must specifically address measures to minimize the discharge of spilled cement, sand, kiln dust, fly ash, settled dust, or other significant materials in storm water from paved portions of the site that are exposed to storm water. Measures used to minimize the presence of these materials may include regular sweeping, or other equivalent measures. The plan shall indicate the frequency of sweeping or other measures. The frequency shall be determined based upon consideration of the amount of industrial activity occurring in the area and frequency of precipitation. The pollution prevention plan shall consider storing cement, fly ash, baghouse dust, and kiln dust in enclosed silos, hoppers, or other containers; in buildings; or in covered areas of the facility.

Facilities in the glass, clay, cement, concrete, and gypsum products industries are required to conduct self-inspections at a frequency which they determine to be adequate to ensure proper implementation of their pollution prevention plan, but not less frequently than once per month. The inspections must take place while the facility is in operation because this is the only time when potential pollutant sources (such as malfunctioning dust control equipment or non–storm water discharges from equipment-washing operations) may be evident. The inspectors must observe material handling areas, aboveground storage tanks, hoppers or silos, dust collection and containment systems, vehicle washing, and equipment-cleaning areas.

In addition to the requirements described in Chap. 7, the SWPPP for sector E facilities must require that the employee training program address procedures for equipment and vehicle washing. Training programs should focus on where and how equipment should be cleaned at the facility so that there will be no unpermitted discharge of wash water to the storm water conveyance system. Training should be conducted at least annually. However, more frequent training may be necessary at facilities with high turnover of employees or where employee participation is essential to the storm water pollution prevention plan.

Facilities engaged in the production of concrete products must include a description of measures which ensure that process wastewater which results from washing of trucks, mixers, transport buckets, forms, or other equipment is discharged in accordance with NPDES requirements or is recycled. These nonprocess wastewater discharges are common to this industry. However, these discharges are not eligible for coverage under this section, and it is necessary to assess the facility for the presence of these discharges so that steps may be taken to eliminate the discharges or to cover the process discharges with a separate permit.

A number of facilities in the concrete products industry maintain wash water recycle and retention ponds which receive the process wastewater from equipment-cleaning and other operations. These ponds may also receive a portion of or all the runoff from the industrial site. These facilities are required to provide an estimate of the depth of the 24-h-duration storm event that would be required to cause the recycle and retention pond to overflow and discharge to the waters of the United States. Methods to make this estimate can include, but are not limited to, the original design calculations for the recycle and retention pond or historical observation.

Pavement wash waters from sector E facilities may be discharged only after the accumulated fly ash, cement, aggregate, kiln dust, clay, concrete, or other dry significant materials handled at the facility have been removed from the pavement by sweeping, vacuuming, a combination thereof, or other equivalent measures; or the wash waters are conveyed into a BMP designed to remove solids prior to discharge, such as sediment basins, retention basins, and other equivalent measures. Where practicable, pavement wash water shall be directed to process wastewater treatment or recycling systems.

For facilities in the concrete product manufacturing industries, the annual site compliance evaluation must specifically address the following portions of the site: aboveground storage tanks, hoppers, or silos; dust collection and containment systems; truck washdown; and equipment-cleaning areas.

Sector F: Primary Metals

Sector F facilities must identify on the site map the location of any and all pollution control equipment such as baghouses, wet scrubbers, and electrostatic precipitators as well as any uncontrolled stack emissions which may be located on-site. The site map must also indicate the outfall locations and the types of discharges contained in the drainage areas of the outfalls (e.g., storm water and air conditioner condensate). To increase the readability of the map, the inventory of the types of discharges contained in each outfall may be kept as an attachment to the site map.

There are typically five types of activity and materials present at facilities in the primary metals industry that have potential impacts on storm water discharges. These include raw materials storage and handling; process activities related to furnace operations, casting, rolling, and extruding; waste material storage, handling, and disposal; erosion from unstabilized plant areas; and illicit discharges, spills, and leaks. Each of these areas that is applicable to a facility must be identified in the pollution prevention plan and evaluated with regard to the BMPs discussed.

Facilities must implement measures to limit the amount of spilled, settled, and leaked materials which are washed away by storm water. These materials include coal dust or coke breeze, metal fines from finishing operations, particulate emissions from furnaces and ovens, as well as dust and dirt from plant yards. In paved or other impervious areas, sweeping is an easy and effective way to reduce these pollutants. Sweeping frequency should be determined based on the rates of accumulation of a particular material and its potential impact on storm water discharges. Where significant particulates are generated in unstabilized areas of the plant, other measures may be necessary. Permittees must consider the storage of all such products under roof, in silos or covered hoppers, or under tarps, to minimize exposure of particulates to precipitation and wind-blown losses.

Unstabilized areas at a site which may be related to material handling and storage or vehicle and equipment traffic should be considered for paving. These areas can build up significant levels of particulates from materials and material handling as well as soil and dust particles. Paving these areas allows good housekeeping measures to be practiced and makes spills easier to clean up.

Permittees must consider preventive measures to minimize the exposure of significant materials to storm water. Measures include moving materials inside, under roof or cover, removing waste materials from the premises, and establishing scheduled removal of wastes to minimize storage on-site. Other measures to prevent runoff from contacting materials include swales, berms, dikes, or curbs to divert runoff away from significant materials or processes.

Facilities must incorporate into their plan the inspection and maintenance of all equipment which could lead to releases of pollutants. This includes all particulate emissions control equipment, storage tanks and piping systems, and any other material handling equipment which could fail and then release pollutants. All particulate pollution control equipment must be maintained to operate properly and effectively to control the settling of particulate matter.

Primary metals facilities are required to conduct self-inspections of all storage, process, and plant yard areas at least quarterly. These inspections will allow the effectiveness of the pollution prevention plan to be monitored. The potential for problems which could affect storm water is extremely varied and can have significant impacts over a short time period. These inspections are necessary to ensure that problems are identified and remedied as quickly as possible. Points of particular importance include pollution control equipment, material handling areas, and waste collection and disposal areas. Tanks, drums, silos, bins, and hoppers are other areas of potential concern.

Facilities must consider implementation of a range of management practices to control or treat storm water runoff. These include vegetative buffer strips or swales, filter fences and other types of filters, oil-water separators, and all types of settling basins and ponds. These practices allow the capture of pollutants from storm water before it leaves the site.

Due to the large size of many primary metals facilities, source controls may not be practical. In some cases, it may not be feasible to cover or otherwise protect large areas of material storage or exposed plant yards. Deposition of particulates from furnace or other process emissions may be relatively diffuse over a large area of the facility, and very difficult to control. In these cases management practices such as settling basins, retention or detention ponds, and recycle ponds can provide effective treatment of runoff. For smaller areas, filter fabric, booms, or other types of filters may be appropriate. In areas where oil and grease are a concern, oil-water separators may be appropriate and should be considered.

Primary metals facilities must certify that certain non–storm water discharges are not occurring at their facilities. A list of common non–storm water discharges that are not authorized by this section has been identified: waste discharges to floor drains or sinks connected to the facility's storm sewer or storm drainage system; water originating from vehicle and equipment washing; steam-cleaning wastewater; process wastewater; wash water originating from cleaning plant floor areas or material receiving areas; wastewater from wet scrubbers; boiler blowdown; contact or noncontact cooling water; discharges originating from dust control spray water; discharges originating from the cleaning out of oil-water separators or sumps; discharges from bermed areas with a visible oily sheen or other visible signs of contamination; discharges resulting from casting cleaning or casting quench operations; discharges from slag quench or slag rinsing operations; and discharges from wet sand reclamation operations.

Sector G: Metal Mining (Ore Mining and Dressing)

The SWPPP for sector G facilities must provide a narrative description of the mining and associated activities at the site which affect or may affect storm water runoff. The narrative description must report the total acreage within the mine site and an estimate of the acreage of land currently disturbed. A general description of the mining site relative to major transportation routes and communities must also be provided. Each SWPPP must describe activities, materials, and physical features of the facility that may contribute to storm water runoff or, during periods of dry weather, result in dry weather flows and mine pumpout.

Active facilities must also provide an estimate of the total acreage that will be disturbed throughout the life of the mine. Inactive facilities must report the approximate dates of operation and the activities (reclamation, etc.) that are currently taking place at the facility.

The SWPPP for all facilities must contain a map of the site that shows the pattern of storm water drainage, structural features that control pollutants in storm water runoff and process wastewater discharges (including mine drainage). Nonstructural features such as grass swales and vegetative buffer strips also should be shown. The plan must also show surface water bodies (including wetlands), places where significant materials[1] are exposed to rainfall and runoff. The site map must also indicate the outfall locations and the types of discharges contained in the drainage areas of the outfalls (e.g., storm water and air conditioner condensate). To increase the readability of the map, the inventory of the types of discharges contained in each outfall may be kept as an attachment to the site map.

The site maps for active sites must show the locations of major spills and leaks that occurred in the 3 years prior to the date of the submission of a Notice of Intent (NOI) to be covered under this permit, as well as areas where the following activities take place: fueling, vehicle and equipment maintenance and/or cleaning, loading and unloading, material storage (including tanks or other vessels used for liquid or waste storage), material processing, waste disposal, haul roads, access roads, and rail spurs.

The site maps for inactive sites must also show any remaining equipment storage, fueling, and maintenance areas; areas used for outdoor manufacturing, storage, or disposal of materials; the boundaries of former mining and milling sites; an outline of the portions of each outfall drainage area that are within the facility boundaries; tailings piles and ponds; mine drainage or any other process water discharge point; and an estimate of the direction of flow.

Operators of active facilities are required to carefully conduct an inspection of the site and related records to identify significant materials that are or may be exposed to storm water. The inventory must address materials that within 3 years prior to the date of the submission of a Notice of Intent to be covered under this permit have been handled, stored, processed, treated, or disposed of in a manner to allow exposure to storm water. Findings of the inventory must be documented in detail in the pollution prevention plan. At a minimum, the plan must describe the method and location of on-site storage or disposal; practices used to minimize contact of materials with rainfall and runoff; existing structural and nonstructural controls that reduce pollutants in storm water runoff; existing structural controls that limit process wastewater discharges; and any treatment that the runoff receives before it is discharged to surface waters or a separate storm sewer system. The description must be updated whenever there is a significant change in the types or amounts of materials, or material management practices, that may affect the exposure of materials to

---

[1]Significant materials commonly found at mining facilities include overburden; raw materials; waste rock piles; tailings; petroleum-based products; solvents and detergents; heap leach pads; tailings piles and ponds, both proposed and existing; and manufactured products, waste materials, or by-products used or created by the facility.

storm water. In addition, any existing ore or waste rock/overburden characterization data, including results of testing for acid rock generation potential, must be included in the pollution prevention plan.

Operators of active facilities are required to conduct monthly visual inspections of BMPs and designated equipment and mine areas. The inspections must include

1. An assessment of the integrity of storm water discharge diversions, conveyance systems, sediment control and collection systems, and containment structures
2. Visual inspections of vegetative BMPs, serrated slopes, and benched slopes to determine if soil erosion has occurred
3. Visual inspections of material handling and storage areas and other potential sources of pollution for evidence of actual or potential pollutant discharges of contaminated storm water

Owner/operators of temporarily inactive mining sites are required to conduct quarterly inspections, unless weather conditions make the mine site inaccessible. For remote and inaccessible sites, the annual site compliance evaluations may be conducted less frequently. However, a site compliance evaluation must in all cases be performed at least once every 3 years.

Facility operators are required to conduct employee training programs at least annually.

Each pollution prevention plan must include a certification, signed by an authorized individual, that discharges from the site have been tested or evaluated for the presence of non–storm water discharges, including discharges that are subject to 40 CFR Part 440. The certification must describe possible significant sources of nonstorm water, the results of any test and/or evaluation conducted to detect such discharges, the test method or evaluation criteria used, the dates on which tests or evaluations were performed, and the on-site drainage points directly observed during the test or evaluation. Pollution prevention plans must identify and ensure the implementation of appropriate pollution prevention measures for the non–storm water discharge.

EPA will allow non–storm water discharges that mix with storm water under this section, provided that the plan includes a certification that any non–storm water discharge which mixes with storm water is subject to a separate NPDES permit that applies applicable effluent limitations prior to the mixing of nonstorm water and storm water. In such cases, the certification must identify the non–storm water discharge(s), the applicable NPDES permit(s), the effluent limitations placed on the non–storm water discharge by the NPDES permit(s), and the point(s) at which the limitations are applied.

The storm water pollution prevention plan for inactive facilities must include, for each outfall, an inventory and narrative description of any significant materials that may still be at the site. The description and locations of the significant materials should be consistent with those shown on the site map. Findings of the inventory must be documented in detail in the pollution prevention plan. At a minimum, the plan must describe the method and location of on-site storage or disposal; practices used to minimize contact of materials with rainfall and runoff; existing structural and nonstructural controls that reduce pollutants in storm water runoff; existing structural controls that limit process wastewater discharges; and any treatment that the runoff receives before it is discharged to surface waters or a separate storm sewer system.

Sector H: Coal Mines and Coal Mining–Related Facilities

A site map, such as a drainage map required for SMCRA permits, must indicate drainage areas and storm water outfalls from potential pollutant sources. The map should provide, but not be limited to, the following information:

1. Drainage direction and discharge points from all applicable mining-related areas, including culvert and sump discharges from roads and rail beds and from equipment and vehicle maintenance areas, lubricants, and other potentially harmful liquids
2. Location of each existing erosion and sedimentation control structure and other control measures for reducing pollutants in storm water runoff
3. Receiving streams or other surface water bodies
4. Locations exposed to precipitation which contain acidic or metal ladened spoil, refuse, or unreclaimed disturbed areas
5. Locations where major spills or leaks of toxic or hazardous pollutants have occurred
6. Locations where liquid storage tanks containing potential pollutants, such as caustics, hydraulic fluids, and lubricants, are exposed to precipitation
7. Locations where fueling stations and vehicle and equipment maintenance areas are exposed to precipitation

The site map must indicate the outfall locations and the types of discharges contained in the drainage areas of the outfalls (e.g., storm water and air conditioner condensate). To increase the readability of the map, the inventory of the types of discharges contained in each outfall may be kept as an attachment to the site map.

SMCRA regulations require sediment and erosion control measures and practices for haul roads and most of the other active mining-related areas covered by this section. All such SMCRA requirements are also requirements of the pollution prevention plan and other applicable conditions of this section.

The purpose of good housekeeping practices is to remove or lessen the potential pollution sources before they come into contact with storm water. This includes collection and removal of waste oils collected in traps; cleaning up exposed maintenance areas of spilled lubricants and fuels, and similar measures; and preventing the off-site movement of dust by sweeping or by road watering.

A timely maintenance program should include inspections for preventing breakdowns, corrosion of tanks, and deterioration of pressure fuel or slurry pressure lines; periodic removal and disposal of accumulated solids in sediment traps; and replacement of straw bales and other control measures subject to weathering and deterioration.

For all SMCRA-regulated active mining-related sites, SMCRA authorities are required to conduct regular quarterly inspections. Coordinated inspections by the facility representative would be expected to take place before, during, or after the complete SMCRA inspections. In addition, sediment and erosion control measures should be evaluated at least once yearly during a storm period of at least 0.1-in rainfall, where effectiveness can be evaluated firsthand. Observations should also be made at this time of the resulting impact of any settled solids in the receiving stream.

Inactive coal mines should be inspected at least once yearly, except where very remote, to maintain an appraisal of sediment and erosion control measures, determine outstanding problem areas, and plan for improved measures.

Many inactive mines and portions of inactive mines are abandoned underground mines which have seeps or other discharges that are not due to storm events. These types of discharges from inactive mines are not covered by this section. In addition, the discharges from floor drains from maintenance buildings and other similar drains in mining and preparation plant areas may contain contaminants, and are not allowable discharges under the EPA multisector general permit.

The SWPPP must describe all sediment, erosion, and flow management controls used to control storm water discharges. The plan should also address the reasonableness and appropriateness of each sediment, erosion, and flow management control, and identify when they are required by state or federal SMCRA regulations. For the most part, these measures are best management practices expected of construction and other activities which are subject to storm runoff. However, construction activities are usually much more short-term than mining activities, so greater emphasis must be placed on implementing long-term measures for haul roads and other mining-related facilities.

The plan must be implemented, and where erosion control and pollution prevention measures in the plan are found to be deficient, the plan must be revised to include reasonable and appropriate control measures. Reports, including observations and incidences of noncompliance, must be prepared and kept on file for possible review.

Sector I: Oil and Gas Extraction

In its SWPPP description of potential pollutant sources, a sector I facility must include information about the RQ release which triggered the permit application requirements. Such information must include the nature of the release (e.g., spill of oil from a drum storage area); amount of oil or hazardous substance released; amount of substance recovered; date of the release; cause of the release (e.g., poor handling techniques, as well as lack of containment in area); area affected by release, including land and waters; procedure to clean up release; and remaining potential contamination of storm water from release.

The permittee must describe the measures taken to clean up RQ releases or related spills of materials, as well as measures proposed to avoid future releases of RQs. Such measures may include, among others, improved handling or storage techniques, containment around handling areas of liquid materials, and use of improved spill cleanup materials and techniques.

Vehicles and equipment associated with oil field activity are often coated with oil, oil field drilling muds, and the chemicals associated with drilling. These vehicles and equipment are a significant source of pollutants. The permittee must address these areas and institute practices to minimize pollutant runoff from this area.

The plan must describe measures that prevent or minimize contamination of the storm water runoff from all areas used for vehicle and equipment cleaning. The facility may consider performing all cleaning operations indoors, covering the cleaning operation, and/or collecting the storm water runoff from the cleaning area and providing treatment or recycling. These cleaning and maintenance activities can result in the exposure of cleaning solvents, detergents, oil and grease, and other chemicals to storm water runoff. The use of drip pans, maintaining an organized inventory of materials used in the shop, draining all parts of fluids prior to disposal, prohibiting the practice of hosing down the shop floor where the practice would result in the exposure of pollutants to storm water, using dry cleanup methods, and/or collecting the storm water runoff from the maintenance area and providing treatment or recycling may reduce the pollutants discharged in storm water runoff.

Labeling of all chemicals and materials storage containers helps facility personnel to respond effectively to spills or leaks. Additionally, covered storage of the materials and/or installation of berms and diking at the area can be effective BMPs.

The facility should consider covering the chemical mixing area, using spill and overflow protection, minimizing run-on of storm water to the mixing area, using dry cleanup methods, and/or collecting the storm water runoff and providing treatment or recycling. The facility should consider installation of berms and diking of the area. The wastewater pollutants associated with produced waters, drilling muds, drill cuttings, and produced sand from any source associated with onshore oil and gas production, field exploration, drilling, well completion, or well treatment are prohibited from being discharged [40 CFR 435.32].

All equipment and areas addressed in the pollution prevention plan must be inspected semiannually. Equipment and vehicles which store, mix, or transport hazardous materials must be inspected quarterly. Inspections must include the inspection of all on-site mixing tanks and equipment, and inspection of all vehicles which carry supplies and chemicals to oil field activities.

Sector J: Mineral Mining and Dressing

The permittee must evaluate, select, and describe the pollution prevention measures, best management practices, and other controls that will be implemented at the facility. The permittee must assess the applicability of the following BMPs for the site: discharge diversions, drainage and storm water conveyance systems, runoff dispersions, sediment control and collection mechanisms, vegetation and soil stabilization, and capping of contaminated sources. In addition, BMPs include processes, procedures, schedules of activities, prohibitions on practices, and other management practices that prevent or reduce the discharge of pollutants in storm water runoff.

Permittees are required to develop a preventive maintenance program that includes regular inspections and maintenance of storm water BMPs. The maintenance program requires periodic removal of debris from discharge diversions and conveyance systems. These activities should be conducted in the spring, after snow melt, and during the fall season. Permittees already controlling their storm water runoff frequently use impoundments or sedimentation ponds. Maintenance schedules for these ponds must be provided in the pollution prevention plan.

Operators of active facilities are required to conduct quarterly visual inspections of BMPs. Temporary and permanently inactive operations are required to perform annual inspections. The inspections shall include (1) an assessment of the integrity of storm water discharge diversions, conveyance systems, sediment control and collection systems, and containment structures; (2) visual inspections of vegetative BMPs, serrated slopes, and benched slopes to determine if soil erosion has occurred; and (3) visual inspections of material handling and storage areas and other potential sources of pollution for evidence of actual or potential pollutant discharges of contaminated storm water.

The inspection must be made at least once in each designated period during daylight hours. Inspections for active facilities must be conducted in each of the following periods: January through March, April through June, July through September, and October through December.

If BMPs are performing ineffectively, corrective action must be implemented. A set of tracking or follow-up procedures must be used to ensure that appropriate actions are taken in response to the inspections.

The permittee must describe procedures for developing and retaining records on the status and effectiveness of plan implementation. The plan must address spills, monitoring, and BMP inspection and maintenance activities. Ineffective BMPs must be reported and the date of their corrective action noted.

Permittees must indicate the location and design for proposed BMPs to be implemented prior to land disturbance activities. For sites already disturbed but without BMPs, the permittee must indicate the location and design of BMPs that will be implemented. The permittee is required to indicate plans for grading, contouring, stabilization, and establishment of vegetative cover for all disturbed areas, including road banks. Reclamation activities must continue until final closure notice has been issued.

The permittee must evaluate the appropriateness of each storm water BMP that diverts, infiltrates, reuses, or otherwise reduces the discharge of contaminated storm water. In addition, the permittee must describe the storm water pollutant source area or activity (i.e., loading and unloading operations, raw material storage piles, etc.) that is to be controlled by each storm water management practice.

Comprehensive site compliance evaluations should be conducted once per year. When annual comprehensive site compliance evaluations are shown in the plan to be impractical for inactive mining sites, due to remote location and inaccessibility, site evaluations must be conducted at least once every 3 years.

Sector K: Hazardous Waste Treatment Storage or Disposal Facilities

Most SWPPP requirements are already addressed by the RCRA program, and employed at hazardous waste treatment, storage, or disposal facilities. If RCRA does not address a particular condition which is stipulated in the storm water pollution prevention plan, the facility still must comply with that requirement of the plan.

Sector L: Landfills and Land Application Sites

Landfill permittees must identify on a site map the locations of active and closed cells or trenches, any known leachate springs or other areas where leachate may commingle with runoff, the locations of any leachate collection and handling systems, and the locations of stockpiles of landfill cover material. Land application site permittees must identify on their site maps the locations of active and inactive land application areas and the types of wastes applied in those areas, any known leachate springs or other areas where leachate may commingle with runoff, the locations of any leachate collection and handling systems, and the locations of temporary waste storage areas.

Owners or operators must summarize all available sampling data for storm water and leachate generated at the site and must identify any current NPDES-permitted discharges at their sites.

Good housekeeping practices are required for materials storage areas exposed to precipitation and for vehicle tracking of sediment and waste.

Frequent and thorough inspections are necessary to ensure adequate functioning of sediment and erosion controls, leachate collection systems, intermediate and final covers, and significant materials storage containers. Failure of any of the aforementioned items could cause contamination of storm water with sediment, leachate, or significant materials stored on-site. Inspections should be conducted both during storm events and during dry weather. Pollution prevention plans must address the specific inspection requirements for active and inactive landfills and land application sites.

Owners or operators must ensure the maintenance of material storage areas, to prevent leaking or rupture, and all elements of leachate collection and treatment systems, to prevent commingling of leachate with storm water. Pollution prevention plans must also describe measures to be taken to protect the integrity and effectiveness of any intermediate and final covers.

The SWPPP must address both stabilization and structural controls to reduce potential sediment loadings to surface waters.

Sector M: Automobile Salvage Yards

Specific requirements for a pollution prevention plan for automobile salvage yards are described below. These requirements must be implemented in addition to the baseline pollution prevention plan provisions discussed in Chap. 7.

Following completion of the source identification and assessment phase, the permittee must evaluate, select, and describe the pollution prevention measures, best management practices, and other controls that will be implemented at the facility. For the following areas at the site, the permittee must assess the applicability of the corresponding BMPs.

The plan must describe measures that prevent or minimize contamination of the storm water runoff from all areas used for vehicle dismantling and maintenance. The facility must consider draining and segregating all fluids from vehicles upon arrival at the site, or as soon as feasible thereafter. The facility must consider performing all maintenance activities indoors, maintaining an organized inventory of materials used in the shop, draining all parts' fluids prior to disposal, prohibiting the practice of hosing down the shop floor, using dry cleanup methods, and/or collecting the storm water runoff from the maintenance area and providing treatment.

Where dismantling and maintenance activities cannot take place indoors, facilities may consider methods for containing oil or other fluid spillage during parts removal. Drip pans, large plastic sheets, or canvas may be considered for placement under vehicles or equipment during maintenance and dismantling activities. Where drip pans are used, they should not be left unattended, to prevent accidental spills.

The storage of vehicles, parts, and equipment must be confined to designated areas (delineated on the site map). The plan must describe measures that prevent or minimize contamination of the storm water runoff from these areas. The facility must consider the use of drip pans, large sheets of plastic, canvas (or equivalent measures) under vehicles, parts, and equipment. Canvas or sheets of plastic may be used as temporary coverage of storage areas. Indoor storage of vehicles, parts, and equipment, as well as the installation of roofs, curbing, berms, and diking of these areas must be considered. Large plastic or metal bins with secure lids should be used to store oily parts (e.g., small engine parts). Used batteries should be stored within nonleaking secondary containment or by other equivalent means to prevent leaks of acid into storm water discharges.

As part of a good housekeeping program, consider labeling storage units of all materials (e.g., used oil, used oil filters, spent solvents, paint wastes, radiator fluids, transmission fluids, hydraulic fluids). Maintain such containers and units in good condition, so as to prevent contamination of storm water. The plan must describe measures that prevent or minimize contamination of the storm water runoff from such storage areas. The facility may consider indoor storage of the materials and/or installation of berms and diking of the area.

The plan must describe measures that prevent or minimize contamination of storm water from all areas used for vehicle, equipment, and parts cleaning. The facility must consider performing all cleaning operations indoors. In addition, the facility must consider covering or berming the cleaning operation area. Wash waters from vehicle, equipment, and parts cleaning areas are process wastewaters that are not authorized discharges under this section.

This section requires that, in addition to the comprehensive site evaluation of the permit, qualified facility personnel shall be identified to inspect upon arrival, or as soon as feasible thereafter, all vehicles for leaks; any equipment containing oily parts, hydraulic fluids, or any other fluids, at least quarterly for leaks; and any outdoor storage containers for liquids, including, but not limited to, brake fluid, transmission fluid, radiator water, and antifreeze, at least quarterly for leaks.

In addition, qualified facility personnel are required to conduct, at a minimum, quarterly visual inspections of BMPs. The inspections shall include (1) an assessment of the integrity of any flow diversion or source minimization systems and (2) visual inspections of dismantling areas; outdoor vehicle, equipment, and parts storage area; vehicle and equipment maintenance areas; vehicle, equipment, and parts washing areas; and liquid storage in aboveground containers. A set of tracking or follow-up procedures shall be used to ensure that appropriate actions are taken in response to the inspections.

The quarterly inspections must be made at least once in each of the following designated periods during daylight hours: January through March (storm water runoff or snow melt), April through June (storm water runoff), July through September (storm water runoff), October through December (storm water runoff).

Permittees are required to include a schedule for conducting training in the plan. EPA recommends that facilities conduct training annually at a minimum. However, more frequent training may be necessary at facilities with high turnover of employees or where employee participation is essential to the storm water pollution prevention plan. Employee training must, at a minimum, address the following areas when applicable to a facility: used-oil management; spill prevention and response; good housekeeping practices; used-battery management; and proper handling (i.e., collection, storage, and disposal) of all fluids.

Permittees must describe procedures for developing and retaining records on the status and effectiveness of plan implementation. The plan must address spills, monitoring, and BMP inspection and maintenance activities. Ineffective BMPs must be reported and the date of their corrective action noted.

The permittee must evaluate the appropriateness of each storm water BMP that diverts, infiltrates, reuses, or otherwise reduces the discharge of contaminated storm water. In addition, the permittee must describe the storm water pollutant source area or activity (i.e., loading and unloading operations, raw material storage piles, etc.) to be controlled by each storm water management practice.

Sector N: Scrap Recycling and Waste Recycling Facilities

The storm water pollution prevention plan is broken out into three subcategories: scrap recycling and waste recycling facilities (nonliquid materials), waste recycling facilities (liquid materials), and recycling facilities.

Scrap recycling and waste recycling facilities (nonliquid materials). The SWPPP for scrap recycling and waste recycling facilities (nonliquid recyclable wastes) shall include a recyclable and waste material inspection program to minimize the likelihood of receiving nonrecyclable materials (e.g., hazardous materials) that may be significant pollutant sources to storm water discharges. At a minimum, the plan shall address the following:

1. Information and education measures to encourage major suppliers of scrap and recyclable waste materials to drain residual fluids, whenever applicable, prior to their arrival at the facility. This includes vehicles and equipment engines, radiators, and transmissions; oil-filled transformers; white goods (appliances); and individual containers or drums.
2. Activities which accept scrap and materials that may contain residual fluids, e.g., automotive engines containing used oil and transmission fluids. Procedures shall be described to minimize the potential for these fluids to come into contact with either precipitation or runoff. The description shall also identify measures or procedures to properly store, handle, dispose of, and/or recycle these residual fluids.
3. Procedures pertaining to the acceptance of scrap lead-acid batteries. Additional requirements for the handling, storage, and disposal or recycling of batteries shall be in conformance with conditions for a scrap lead-acid battery program (see below).
4. A description of training requirements for those personnel engaged in the inspection and acceptance of inbound recyclable materials.
5. Liquid wastes, including used oil. These shall be stored in materially compatible and nonleaking containers, and disposed of or recycled in accordance with all requirements under the Resource Conservation and Recovery Act (RCRA), and other state or local requirements.

The plan shall address areas where significant materials are exposed to either storm water runoff or precipitation. The plan must describe those measures and controls used to minimize contact of storm water runoff with stockpiled materials. The plan should include measures to minimize the extent of storm water contamination from these areas. The operator shall consider (within the plan) the use of the following BMPs (either individually or in combination) or their equivalent to minimize contact with storm water runoff:

- Diversion devices or structures such as dikes, berms, containment trenches, culverts, and/or surface grading
- Media filtration such as catch basin filters and sand filters
- Silt fencing
- Oil-water separators, sumps, and dry adsorbents in stockpile areas that are potential sources of residual fluids, e.g., automotive engine storage areas

The operator may consider the use of permanent or semipermanent covers or other similar forms of protection over stockpiled materials, where the operator determines that such measures are reasonable and appropriate.

The operator may consider the use of sediment traps, vegetated swales, and/or vegetated strips to facilitate settling or filtering out of pollutants and sediment.

The plan shall address all areas where stockpiling of industrial turnings (previously exposed to cutting fluids) occurs. The plan shall implement those measures necessary to minimize contact of surface runoff with residual cutting fluids. The operator shall consider implementation of either of the following two alternatives or a combination of both or equivalent measures:

Alternative 1: Storage of all turnings previously exposed to cutting fluids under some form of permanent or semipermanent cover. Discharges of residual fluids from these areas to the storm sewer system in the absence of a storm event are prohibited. Discharges to the storm sewer system as a consequence of a storm event are permitted provided the discharge is first directed through an oil-water separator or its equivalent. Procedures to collect, handle, and dispose of or recycle residual fluids that may be present shall be identified in the plan.

Alternative 2: Establish dedicated containment areas for all turnings that have been exposed to cutting fluids where runoff from these areas is directed to a storm sewer system, to provide the following:

- Containment areas constructed of concrete, asphalt, or other equivalent type of impermeable material.
- A perimeter around containment areas to prevent runoff from moving across these areas. This would include the use of shallow berms or curbing, or constructing an elevated pad or other equivalent measure.
- A suitable drainage collection system to collect all runoff generated from within containment areas. At a minimum, the drainage system shall include a plate-type oil-water separator or its equivalent. The oil-water separator or its equivalent shall be installed according to the manufacturer's recommended specifications; whenever available, specifications will be kept with the plan.
- A schedule to maintain the oil-water separator (or its equivalent) to prevent the accumulation of appreciable amounts of fluids. In the absence of a storm event, no discharge from containment areas to the storm sewer system is permitted unless the discharge is covered by a separate NPDES permit.
- Identification of procedures for the proper disposal or recycling of collected residual fluids.

The plan shall address, at a minimum, measures and controls to minimize and, whenever feasible, eliminate residual liquids and particulate matter from materials stored indoors from coming in contact with surface runoff. The operator shall consider for inclusion in the plan good housekeeping measures to collect residual liquids from aluminum, glass, and plastic containers; prohibiting the practice of allowing wash water from tipping floors or other indoor processing areas from discharging to a storm sewer system; inspections to ensure that material stockpile areas with existing floor drains are not connected to the storm sewer system or any portion of the storm sewer system; and the disconnection of any floor drains to the storm drainage system.

The plan shall address areas where scrap and recyclable waste processing equipment are sited. This includes measures and controls to minimize surface runoff from coming in contact with scrap processing equipment. In the case of processing equipment that generates visible amounts of particulate residue, e.g., shredding facilities, the plan shall describe good housekeeping and preventive maintenance measures to minimize contact of runoff with residual fluids and accumulated particulate matter. At a minimum, the operator shall consider including the following: a schedule of periodic inspections of equipment for leaks, spills, and malfunctioning, worn, or corroded parts or equipment; a preventive maintenance program to repair and/or maintain processing equipment; measures to minimize shredder fluff from coming in contact with surface runoff; use of dry absorbents or other cleanup practices to collect and to dispose or recycle spilled or leaking fluids; and installation of low-level alarms or other equivalent protection devices on unattended hydraulic reservoirs with greater than 150-gal capacity. Alternatively, provide secondary containment with sufficient volume to contain the entire volume of the reservoir.

The operator shall consider using the following types of BMPs:

- Diversion structures such as dikes, berms, culverts, containment trenches, elevated concrete pads, and grading to minimize contact of storm water runoff with outdoor processing equipment
- Oil-water separators or sumps in processing areas that are potential sources of residual fluids and grease
- Permanent or semipermanent covers, or other similar measures
- Retention and detention basins or ponds, sediment traps, or vegetated swales and strips, to facilitate settling or filtering out of pollutants in runoff from processing areas
- Media filtration such as catch basin filters and sand filters

The plan shall address measures and controls for the proper receipt, handling, storage, and disposition of scrap lead-acid batteries (battery reclaiming is not eligible for coverage under this permit). The operator shall consider including procedures for accepting scrap batteries and describing how they will be segregated from other scrap materials; procedures for managing battery casings that may be cracked or leaking, including the proper handling and disposal of residual fluids; measures to minimize and, whenever possible, eliminate exposure of scrap batteries to either runoff or precipitation; the schedule for conducting periodic inspections of scrap battery storage areas and applicable source control measures; and measures to provide employee training on the management of scrap batteries.

The plan shall identify all areas associated with industrial activity that have a high potential for soil erosion and suspended solids loadings, i.e., areas that tend to accumulate significant particulate matter. Appropriate source control, stabilization measures, nonstructural and structural controls, or an equivalent shall be provided in these areas. The plan shall also contain a narrative discussion of the reason(s) for selected erosion and sediment controls. At a minimum, the operator shall consider in the plan, either individually or in combination, the following erosion and sediment control measures:

- Filtering or diversion practices, such as filter fabric, sediment filter boom, earthen or gravel berms, curbing, or other equivalent measure
- Catch basin filters, filter fabric, or equivalent measure, placed in or around inlets or catch basins that receive runoff from scrap, and waste storage areas and processing equipment

- Sediment traps, vegetative buffer strips, or equivalent, which effectively trap or remove sediment prior to discharge through an inlet or catch basin

In instances where significant erosion and suspended solids loadings continue after implementation of source control measures and nonstructural controls, the operator shall consider providing in the plan for a detention or retention basin or other equivalent structural control. All structural controls shall be designed using good engineering practice. All structural controls and outlets that are likely to receive discharges containing oil and grease must include appropriate measures to minimize the discharge of oil and grease through the outlet. This may include the use of an absorbent boom or other equivalent measure.

Where space limitations (e.g., obstructions caused by permanent structures such as buildings and permanently sited processing equipment, and limitations caused by a restrictive property boundary) prevent the siting of a structural control, i.e., retention basin, such a determination will be noted in the plan. The operator will identify in the plan what existing practices shall be modified or additional measures shall be undertaken to minimize erosion and suspended sediment loadings in lieu of a structural BMP.

To prevent or minimize storm water contamination at loading and unloading areas, and from equipment or container failures, the operator shall consider including in the plan the following practices:

- Description is given of spill prevention and response measures to address areas that are potential sources of leaks or spills of fluids.
- All significant leaks and spills should be contained and cleaned up as soon as possible. If malfunctioning equipment is responsible for the spill or leak, repairs should also be conducted as soon as possible.
- Cleanup procedures should be identified in the plan, including the use of dry absorbent materials or other cleanup methods. Where dry absorbent cleanup methods are used, an adequate supply of dry absorbent material should be maintained on-site. Used absorbent material should be disposed of properly.
- Drums containing liquids, including oil and lubricants, should be stored indoors; or in a bermed area; or in overpack containers or spill pallets; or in similar containment devices.
- Overfill prevention devices should be installed on all fuel pumps or tanks.
- Drip pans or equivalent measures should be placed under any leaking piece of stationary equipment until the leak is repaired. The drip

pans should be inspected for leaks and checked for potential overflow, and emptied regularly to prevent overflow; and all liquids will be disposed of in accordance with all requirements under RCRA.

- An alarm and/or pump shutoff system should be installed and maintained on all outside equipment with hydraulic reservoirs exceeding 150 gal (only those reservoirs not directly visible by the operator of the equipment), in order to prevent draining of the tank contents in the event of a line break. Alternatively, the equipment may have a secondary containment system capable of containing the contents of the hydraulic reservoir plus adequate freeboard for precipitation. Leaking hydraulic fluids should be disposed of in accordance with all requirements under RCRA.

A quarterly inspection shall include all designated areas of the facility and equipment identified in the plan. The inspection shall include a means of tracking and conducting follow-up actions based on the results of the inspection. The inspections shall be conducted by members of the storm water pollution prevention team. At a minimum, quarterly inspections shall include the following areas:

- All outdoor scrap processing areas
- All material unloading and loading areas (including rail sidings) that are exposed to either precipitation or storm water runoff
- Areas where structural BMPs have been installed
- All erosion and sediment BMPs
- Outdoor vehicle and equipment maintenance areas
- Vehicle and equipment fueling areas
- All areas where waste is generated, received, stored, treated, or disposed of and which are exposed to either precipitation or storm water runoff

If there is exposure to precipitation or storm water runoff, the inspection shall attempt to identify any corroded or leaking containers, corroded or leaking pipes, leaking or improperly closed valves and valve fittings, leaking pumps and/or hose connections, and deterioration in diversionary or containment structures. Spills or leaks shall be immediately addressed according to the facilities. A record of inspections shall be maintained with the plan.

The BMPs identified above have been employed by scrap recycling and waste recycling facilities and are believed to be appropriate, given the types of pollutants found in storm water discharges from these facilities. In addition, the diversity of options allows permittees to select those BMPs that are most applicable to the extent of the risk that exists at a particular facility. In instances where nonstructural measures are not sufficient, the conditions direct the permittee to more stringent requirements such as structural controls.

Waste recycling facilities (recyclable liquid wastes). SWPPPs for waste recycling facilities (recyclable liquid wastes) shall provide the following information: The operator shall consider including in the plan measures and controls to minimize residual liquids from waste materials stored indoors from coming into contact with surface runoff and provisions to maintain a sufficient supply of dry absorbent materials or a wet vacuum system or other equivalent measure to promptly respond to minor leaks or spills. Measures for secondary containment or its equivalent and procedures for proper material handling (including labeling and marking) and storage of containerized materials should be considered. Drainage from bermed areas should be discharged to an appropriate treatment facility or sanitary sewer system. Discharges from bermed areas should be covered by a separate NPDES permit or industrial user permit under the pretreatment program. The drainage system, where applicable, should include appropriate appurtenances such as pumps or ejectors and manually operated valves of the open-and-close design.

The plan will address areas where waste materials are exposed to either storm water runoff or precipitation. The plan must include measures to provide appropriate containment, drainage control, and/or other appropriate diversionary structures. The plan must describe those measures and controls used to minimize contact of storm water runoff with stored materials. The operator shall consider including in the plan the following preventive measures or an equivalent:

- An appropriate containment structure such as dikes, berms, curbing or pits, or other equivalent measure. The containment should be sufficient to store the volume of the largest single tank and should include sufficient freeboard for precipitation.
- A sufficient supply of dry absorbent materials or a wet vacuum system to collect liquids from minor spills and leaks in contained areas.
- Discharges of precipitation from containment areas containing used oil in accordance with applicable sections of 40 CFR Part 112.

The plan will describe measures and controls for truck and railcar loading and unloading areas. This includes appropriate containment and diversionary structures to minimize contact with precipitation and/or storm water runoff. The plan will also address measures to clean up minor spills and/or leaks originating from the transfer of liquid wastes. This may include dry cleanup methods, roof coverings, and other runoff controls.

The plan shall identify all areas associated with industrial activity that have a high potential for soil erosion. Appropriate stabilization measures, nonstructural and structural controls shall be provided in these areas. The plan shall contain a narrative consideration of the appropriateness for selected erosion and sediment controls. Where applicable, the facility shall consider the use of the following types of preventive measures: sediment traps; vegetative buffer strips; filter-fabric fence; sediment filtering boom; gravel outlet protection; or other equivalent measures that effectively trap or remove sediment prior to discharge through an inlet or catch basin.

The plan will address measures and procedures to address potential spill scenarios that could occur at the facility. This includes all applicable handling and storage procedures, containment, diversion controls, and cleanup procedures. The plan will specifically address all outdoor and indoor storage areas, waste transfer areas, material receiving areas (loading and unloading), and waste disposal areas.

Quarterly visual inspections shall be conducted by a member, or members, of the storm water pollution prevention team. If there is exposure to precipitation or storm water runoff, the inspection shall identify the presence of any corroded or leaking containers, corroded or leaking pipes, leaking or improperly closed valves and valve fittings, leaking pumps and/or hose connections, and deterioration in diversionary or containment structures. Spills or leaks shall be immediately addressed according to the facility's spill prevention and response procedures.

Recycling facilities. SWPPPs for recycling facilities [including Material Recovery Facilities (MRFs)] that receive only source-separated recyclable materials primarily from nonindustrial and residential sources shall include a recyclable material inspection program to minimize the likelihood of receiving nonrecyclable materials (e.g., hazardous materials) that may be a significant source of pollutants in surface runoff. At a minimum, the operator shall consider addressing in the plan the following:

- A description of information and education measures to educate the appropriate suppliers of recyclable materials on the types of recyclable materials that are acceptable and those that are not acceptable, e.g., household hazardous wastes
- A description of training requirements for drivers responsible for pickup of recyclable materials
- Clearly marked public dropoff containers showing what materials can be accepted

- Rejection of nonrecyclable wastes or household hazardous wastes at the source
- A description of procedures for the handling and disposal of nonrecyclable materials

The plan shall include BMPs to minimize or reduce the exposure of recyclable materials to surface runoff and precipitation. The plan, at a minimum, shall include good housekeeping measures to prevent the accumulation of visible quantities of residual particulate matter and fluids, particularly in high-traffic areas. The plan shall consider tarpaulins or their equivalent to be used to cover exposed bales of recyclable waste paper. The operator shall consider within the plan the use of the following types of BMPs (individually or in combination) or their equivalent:

- Provide totally enclosed dropoff containers for the public.
- Provide a sump and sump pump with each containment pit. Prevent the discharge of residual fluids to storm sewer system. Prevent discharging to the storm sewer system.
- Provide dikes and curbs around bales of recyclable waste paper.
- Divert surface runoff away from outside material storage areas.
- Provide covers over containment bins, dumpsters, and roll-off boxes.
- Store the equivalent of one day's volume of recyclable materials indoors.

The plan shall address BMPs to minimize the release of pollutants from indoor storage and processing areas to the storm sewer system. The plan shall establish specific measures to ensure that no floor drains discharge to the storm sewer system. The following BMPs shall be considered for inclusion in the plan:

- Schedule routine good housekeeping measures for all storage and processing areas.
- Prohibit the practice of allowing tipping-floor wash waters from draining to any portion of a storm sewer system.
- Provide employee training on pollution prevention practices.

The plan shall also provide for BMPs in those areas where vehicle and equipment maintenance is being done outdoors. At a minimum, the following BMPs shall be considered for inclusion in the plan:

- Prohibit vehicle and equipment wash water from discharging to the storm sewer system.
- Minimize or eliminate outdoor maintenance areas, wherever possible.

- Establish spill prevention and cleanup procedures in fueling areas.
- Provide employee training on not topping off fuel tanks.
- Divert runoff from fueling areas.
- Store lubricants and hydraulic fluids indoors.
- Provide employee training on proper handling and storage of hydraulic fluids and lubricants.

Scrap and waste recycling facilities that are typically classified in SIC 5093 (wholesale automotive wrecking for scrap) must certify that certain non–storm water discharges are not occurring at their facilities: waste discharges to floor drains or sinks connected to the facility storm sewer or storm drainage system; water originating from vehicle and equipment washing; steam-cleaning wastewater; process wastewaters; wash water originating from cleaning tipping-floor areas or material receiving areas that discharge to any portion of a storm sewer system; wastewater from wet scrubbers; boiler blowdown; noncontact and contact cooling water; discharges originating from dust control spray water; discharges from oil-water separators and sumps in the absence of a storm event; discharges originating from the cleaning out of oil-water separators or sumps; and non–storm water discharges from turnings containment areas.

Sector O: Steam Electric Generating Facilities

The areas required to be identified on the site map include the following: landfills, treatment ponds, scrap yards, general refuse areas, locations of short- and long-term storage of general materials, and the location of stockpile areas.

The plan must describe measures that prevent or minimize fugitive dust emissions from coal handling areas. The permittee shall consider establishing procedures to minimize off-site tracking of coal dust. To prevent off-site tracking, the facility may consider specially designed tires, or washing vehicles in a designated area before they leave the site and controlling the wash water.

Develop procedures for the inspection of delivery vehicles arriving on the plant site, and ensure overall integrity of the body or container.

Develop procedures to control leakage or spillage from vehicles or containers, and ensure that proper protective measures are available for personnel and the environment.

The plan must describe measures that prevent or minimize contamination of storm water runoff from fuel oil unloading areas. At a minimum the facility operator must consider using the following measures or an equivalent:

- Use containment curbs in unloading areas.
- During deliveries, station personnel familiar with spill prevention and response procedures must be present to ensure that any leaks or spills are immediately contained and cleaned up.
- Use spill and overflow protection. Drip pans, drip diapers, and/or other containment devices shall be placed beneath fuel oil connectors to contain any spillage that may occur during deliveries or due to leaks at such connectors.

The plan must describe measures that prevent or minimize the contamination of storm water runoff from chemical loading and unloading areas. At a minimum the permittee must consider using the following measures or an equivalent:

- Use containment curbs at chemical loading and unloading areas to contain spills.
- During deliveries, station personnel familiar with spill prevention and response procedures must be present to ensure that any leaks or spills are immediately contained and cleaned up.

Where practicable, chemical loading and unloading areas should be covered, and chemicals should be stored indoors.

The plan must describe measures that prevent or minimize the contamination of storm water runoff from loading and unloading areas. The facility may consider covering the loading area; minimizing storm water run-on to the loading area by grading, berms, or curbing the area around the loading area to direct storm water away from the area; or locating the loading and unloading equipment and vehicles so that leaks can be controlled in existing containment and flow diversion systems.

The plan must describe measures that prevent or minimize contamination of storm water runoff from aboveground liquid storage tanks. At a minimum the facility operator must consider employing the following measures or an equivalent:

- Use protective guards around tanks.
- Use containment curbs.
- Use spill and overflow protection (drip pans, drip diapers, and/or other containment devices shall be placed beneath chemical connectors to contain any spillage that may occur during deliveries or due to leaks at such connectors).
- Use dry cleanup methods.

The plan must describe measures that prevent or minimize contamination of storm water runoff from large bulk fuel storage tanks. The plan must describe measures to reduce the potential for an oil or chemical spill, or reference the appropriate section of the SPCC plan. At a minimum the structural integrity of all aboveground tanks, pipelines, pumps, and other related equipment shall be visually inspected on a weekly basis. All repairs deemed necessary based on the findings of the inspections shall be completed immediately, to reduce the incidence of spills and leaks occurring from such faulty equipment.

The plan must describe measures to reduce the potential for storm water contamination from oil-bearing equipment in switchyard areas. The facility may consider level grades and gravel surfaces to retard flows and limit the spread of spills; and collection of storm water runoff in perimeter ditches.

All residue-hauling vehicles shall be inspected for proper covering over the load, adequate gate sealing, and overall integrity of the body or container. Vehicles without load covers or adequate gate sealing, or with poor body or container conditions, must be repaired as soon as practicable.

Plant procedures shall be established to reduce and/or control the tracking of ash or residue from ash loading areas including, where practicable, requirements to clear the ash building floor and immediately adjacent roadways of spillage, debris, and excess water before each loaded vehicle departs. Reduce ash residue which may be tracked onto access roads traveled by residue trucks or residue handling vehicles. Reduce ash residue on exit roads leading into and out of residue handling areas.

The plan must address landfills, scrap yards, and general refuse sites. The BMPs for these activities were described previously in this chapter in the discussion of Sectors C and N.

For vehicle maintenance activities performed on the plant site, the permittee shall consider the applicable best management practices outlined in connection with sector P in the next section of this chapter.

The plan must describe measures that prevent or minimize contamination of storm water from material storage areas (including areas used for temporary storage of miscellaneous products and construction materials stored in lay-down areas). The facility operator may consider flat yard grades; runoff collection in graded swales or ditches; erosion protection measures at steep outfall sites (e.g., concrete chutes, riprap, stilling basins); covering lay-down areas; storing the materials indoors; and covering the material with a temporary covering made of polyethylene, polyurethane, polypropylene, or hypalon. Storm water run-on may be minimized by constructing an enclosure or building a berm around the area.

Qualified facility personnel must be identified to inspect designated equipment and areas of the facility on a monthly basis. The following areas must be included in the inspection: coal handling areas, fueling areas, loading and unloading areas, switchyards, bulk storage areas, ash handling areas, areas adjacent to disposal ponds and landfills, maintenance areas, liquid storage tanks, and long-term and short-term material storage areas. A set of tracking or follow-up procedures shall be used to ensure that appropriate actions are taken in response to the inspections. Records of inspections shall be maintained on-site.

Steam electric power generating facilities are required to identify periodic training dates in the pollution prevention plan, but in all cases training must be held at least annually.

Sector P: Land Transportation Facilities

The SWPPP must include storage areas for vehicles and equipment awaiting maintenance on the facility site map. The storage of vehicles and equipment with actual or potential fluid leaks must be confined to designated areas (delineated on the site map). The plan must describe measures that prevent or minimize contamination of the storm water runoff from these areas. The facility shall consider the use of drip pans under vehicles and equipment, indoor storage of the vehicles and equipment, installation of berms and diking of this area, use of absorbents, roofing or covering storage areas, cleaning of pavement surface to remove oil and grease, or other equivalent methods.

The plan must describe measures that prevent or minimize contamination of the storm water runoff from fueling areas. The facility shall consider covering the fueling area, using spill and overflow protection and cleanup equipment, minimizing run-on of storm water to the fueling area, using dry cleanup methods, collecting the storm water runoff and providing treatment or recycling, or other equivalent measures.

Storage units of all materials (e.g., used oil, used oil filters, spent solvents, paint wastes, radiator fluids, transmission fluids, hydraulic fluids) must be maintained in good condition, so as to prevent contamination of storm water, and plainly labeled (e.g., "used oil," "spent solvents"). The plan must describe measures that prevent or minimize contamination of the storm water runoff from such storage areas. The facility shall consider indoor storage of the materials, installation of berms and diking of the area, or other equivalent methods.

The plan must describe measures that prevent or minimize contamination of the storm water runoff from all areas used for vehicle and equipment cleaning. The facility shall consider performing all cleaning operations indoors, covering the cleaning operation, ensuring that all wash waters drain to the intended collection system (i.e., not the storm water drainage system unless NPDES-permitted), collecting the storm water runoff from the cleaning area and providing treatment or recycling, or other equivalent measures. The discharge of vehicle and equipment wash waters, including tank-cleaning operations, is not authorized by this section and must be covered under a separate NPDES permit, or discharge must be made to a sanitary sewer in accordance with applicable industrial pretreatment requirements.

The plan must describe measures that prevent or minimize contamination of the storm water runoff from all areas used for vehicle and equipment maintenance. The facility shall consider performing all maintenance activities indoors, using drip pans, maintaining an organized inventory of materials used in the shop, draining all parts of fluids prior to disposal, prohibiting the practice of hosing down the shop floor where the practice would result in the exposure of pollutants to storm water, using dry cleanup methods, collecting the storm water runoff from the maintenance area and providing treatment or recycling, or other equivalent measures.

The plan must describe measures that prevent or minimize contamination of the storm water runoff from areas used for locomotive sanding (including locomotive sanding). The facility shall consider covering sanding areas, minimizing storm water run-on and runoff, appropriate sediment removal practices to minimize the off-site transport of sanding material by storm water, or other equivalent measures.

A copy of any NPDES permit issued for vehicle and equipment wash waters or, if an NPDES permit has not yet been issued, a copy of the pending application must be attached to or referenced in the SWPPP. For facilities that discharge vehicle and equipment wash waters to the sanitary sewer system, the operator of the sanitary system and associated treatment plant must be notified. In such cases, a copy of the notification letter must be attached to the SWPPP. If an industrial user permit is issued under a pretreatment program, a copy of that permit must be attached to the SWPPP. In all cases, any permit conditions or pretreatment requirements must be considered in the SWPPP. If the wash waters are handled in another manner (e.g., hauled off-site), the disposal method must be described, and all pertinent documentation (e.g., frequency, volume, destination) must be attached to the SWPPP.

Qualified facility personnel must be identified to inspect designated equipment and areas of the facility quarterly, at a minimum. The following areas shall be included in all inspections: storage areas for vehicles and equipment awaiting maintenance, fueling areas, vehicle and equipment maintenance areas (both indoors and outdoors), material storage areas, vehicle and equipment cleaning areas, and loading and unloading areas.

Employees must be trained at least annually in at least the following topics:

- A summary of the facility's SWPPP requirements
- Used-oil management
- Spent-solvent management
- Spill prevention, response and control
- Fueling procedures
- General good housekeeping practices
- Proper painting procedures
- Used-battery management

Sector Q: Water Transportation Facilities

Permittees are required to include the location(s) on the facility site map where engine maintenance and repair work, vessel maintenance and repair work, and pressure-washing are performed. Include painting, blasting, welding, and metal fabrication areas. The location of liquid storage areas (i.e., paint, solvents, resins) and material storage areas (i.e., blasting media, aluminum, steel) must be shown.

When pressure-washing is used to remove marine growth from vessels, the discharge water must be permitted by an NPDES permit. It is not an allowable non–storm water discharge under the EPA multisector general permit. The plan must describe the measures to collect or contain the discharge from the pressure-washing area, detail the method for the removal of the visible solids, describe the method of disposal of the collected solids, and identify where the discharge will be released (i.e., the receiving water body, storm sewer system, sanitary sewer system).

The following non–storm water discharges are specifically prohibited: bilge and ballast water, pressure wash water, sanitary wastes, and cooling water originating from vessels.

The facility must consider containing all blasting and painting activities to prevent abrasives, paint chips, and overspray from reaching the receiving water or the storm sewer system. The plan must describe measures taken at the facility to prevent or minimize the discharge of spent abrasive, paint chips, and paint into the receiving water body and storm sewer system. The facility may consider hanging plastic barriers or tarpaulins during blasting or painting operations to contain debris. Where required, a schedule for cleaning storm systems to remove deposits of abrasive blasting debris and paint chips should be addressed within the plan. The plan should include any standard operating practices with regard to blasting and painting activities. Such included items may be the prohibition of performing uncontained blasting and painting over open water, or blasting and painting during windy conditions which can render containment ineffective.

All stored and containerized materials (fuels, paints, solvents, waste oil, antifreeze, batteries) must be stored in a protected, secure location away from drains and must be plainly labeled. The plan must describe measures that prevent or minimize contamination of the storm water runoff from such storage areas. The facility must specify which materials are stored indoors and must consider containment or enclosure for materials that are stored outdoors. Aboveground storage tanks, drums, and barrels permanently stored outside must be delineated on the site map with a description of the containment measures in place to prevent leaks and spills. The facility must consider implementing an inventory control plan to prevent excessive purchasing, storage, and handling of potentially hazardous materials. Those facilities where abrasive blasting is performed must specifically include a discussion of the storage and disposal of spent abrasive materials generated at the facility.

The plan must describe measures that prevent or minimize contamination of the storm water runoff from all areas used for engine maintenance and repair. The facility may consider performing all maintenance activities indoors, maintaining an organized inventory of materials used in the shop, draining all parts of fluids prior to disposal, prohibiting the practice of hosing down the shop floor, using dry cleanup methods, and/or collecting the storm water runoff from the maintenance area and providing treatment or recycling.

The plan must describe measures that prevent or minimize contamination of the storm water runoff from material handling operations and areas (i.e., fueling, paint and solvent mixing, disposal of process wastewater streams from vessels). The facility may consider covering fueling areas; using spill and overflow protection; mixing paints and solvents in a designated area, preferably indoors or under a shed; and minimizing run-on of storm water to material handling areas. Where applicable, the plan must address the replacement or repair of leaking connections, valves, pipes, hoses, and soil chutes carrying wastewater from vessels.

The plan must address the routine maintenance and cleaning of the drydock to minimize the potential for pollutants in the storm water runoff. The plan must describe the procedures for cleaning the accessible areas of the drydock prior to flooding and final cleanup after the vessel is removed and the dock is raised. Cleanup procedures for oil, grease, or fuel spills occurring on the drydock must also be included within the plan. The facility should consider items such as sweeping rather than hosing off debris and spent blasting material from the accessible areas of the drydock prior to flooding, and having absorbent materials and oil containment booms readily available to contain and clean up any spills.

The plan must include a schedule for routine yard maintenance and cleanup. Scrap metal, wood, plastic, miscellaneous trash, paper, glass, industrial scrap, insulation, welding rods, packaging, etc., must be routinely removed from the general yard area. The facility may consider such measures as providing covered trash receptacles in each yard, on each pier, and onboard each vessel being repaired.

Under the preventive maintenance requirements of the storm water pollution prevention plan elements, the plan specifically includes the routine inspection of sediment traps to ensure that spent abrasives, paint chips, and solids will be intercepted and retained prior to entering the storm drainage system. Because of the nature of operations such as abrasive blasting which occur at water transportation facilities, specific routine attention needs to be given to the collection and proper disposal of spent abrasive materials, paint chips, and other solids.

Under the inspection requirements of the storm water pollution prevention plan elements, qualified facility personnel shall be identified to inspect designated equipment and areas of the facility, at a minimum, on a monthly basis. The following areas shall be included in all inspections: pressure-washing area, blasting and painting areas, material storage areas, engine maintenance and repair areas, material handling areas, drydock area, and general yard area. A set of tracking or follow-up procedures shall be used to ensure that appropriate actions are taken in response to the inspections. Records shall be maintained.

The permittee is required to identify at least annual (once per year) dates for such training. Employee training must, at a minimum, address the following areas when applicable to a facility: used-oil management; spent-solvent management; proper disposal of spent abrasives; proper disposal of vessel wastewaters; spill prevention and control; fueling procedures; general good housekeeping practices; proper painting and blasting procedures; and used-battery management. Employees, independent contractors, and customers must be informed about BMPs and be required to perform in accordance with these practices. The facility must consider posting easy-to-read descriptions or graphic depictions of BMPs and emergency phone numbers in the work areas. Unlike some industrial operations, the industrial activities associated with water transportation facilities that may affect storm water quality require the cooperation of all employees. EPA, therefore, is requiring that employee training take place at least once per year to serve as (1) training for new employees, (2) a refresher course for existing employees, (3) training for all employees on any storm water pollution prevention techniques recently incorporated into the plan, and (4) a forum for the facility to invite independent contractors and customers to inform them on pollution prevention procedures and requirements.

Sector R: Ship and Boat Building or Repair Yards

When pressure-washing is used to remove marine growth from vessels, the discharge water must be collected or contained and disposed of as required by the NPDES permit for this process water, if the discharge is to waters of the United States or through a municipal separate storm sewer. The plan must describe the measures to collect or contain the discharge from the pressure-washing area, detail the method for the removal of the visible solids, describe the method of disposal of the collected solids, and identify where the discharge will be released (i.e., the receiving water body, storm sewer system, sanitary sewer system).

The facility must consider containing all blasting and painting activities to prevent abrasives, paint chips, and overspray from reaching a receiving water body or storm sewer system. The plan must describe measures taken at the facility to prevent or minimize the discharge of spent abrasive, paint chips, and paint into the receiving water body and storm sewer system. The facility may consider hanging plastic barriers or tarpaulins during blasting or painting operations to contain debris. Where appropriate, a schedule for cleaning storm water conveyances to remove deposits of abrasive blasting debris and paint chips should be addressed within the plan. The plan should include any standard operating practices with regard to blasting and painting activities. Such items may include the prohibition of performing uncontained blasting and painting over open water, or blasting and painting during windy conditions which can render containment ineffective.

All stored and containerized materials (fuels, paints, solvents, waste oil, antifreeze, batteries) must be stored in a protected, secure location away from drains and must be plainly labeled. The plan must describe measures that prevent or minimize contamination of the storm water runoff from such storage areas. The facility must specify which materials are stored indoors and consider containment or cover for materials that are stored outdoors. Aboveground storage tanks, drums, and barrels permanently stored outside must be delineated on the site map with a description of the containment measures in place to prevent leaks and spills. The facility must consider implementing an inventory control plan to prevent excessive purchasing, storage, and handling of potentially hazardous materials. Those facilities where abrasive blasting is performed must specifically include within the plan discussion of the storage and proper disposal of spent abrasive generated at the facility.

The plan must describe measures that prevent or minimize contamination of the storm water runoff from all areas used for engine maintenance and repair. The facility must consider performing all maintenance activities indoors, maintaining an organized inventory of materials used in the shop, draining all parts of fluids prior to disposal, prohibiting the practice of hosing down the shop floor where the practice would result in the exposure of pollutants to storm water, using dry cleanup methods, and/or collecting the storm water runoff from the maintenance area and providing treatment or recycling.

The plan must describe measures that prevent or minimize contamination of the storm water runoff from material handling operations and areas (i.e., fueling, paint and solvent mixing, disposal of process wastewater streams from vessels). The facility must consider covering fueling areas; using spill and overflow protection; mixing paints and solvents in a designated area, preferably indoors or under a shed; and minimizing runon of storm water to material handling areas. Where applicable, the plan must address the replacement or repair of leaking connections, valves, pipes, hoses, and soil chutes carrying wastewater from vessels.

The plan must address the routine maintenance and cleaning of the drydock to minimize the potential for pollutants in storm water runoff. The facility must describe the procedures for cleaning the accessible areas of the drydock prior to flooding and the final cleanup after the vessel is removed and the dock is raised. Cleanup procedures for oil, grease, or fuel spills occurring on the drydock must be included within the plan. The facility must consider items such as sweeping rather than hosing off debris and spent blasting material from the accessible areas of the drydock prior to flooding, and having absorbent materials and oil containment booms readily available to contain and clean up any spills.

The plan must include a schedule for routine yard maintenance and cleanup. Scrap metal, wood, plastic, miscellaneous trash, paper, glass, industrial scrap, insulation, welding rods, packaging, etc., must be routinely removed from the general yard area. The facility must consider such measures as providing covered trash receptacles in each yard, on each pier, and onboard each vessel being repaired.

The preventive maintenance requirements specifically include the routine inspection of sediment traps to ensure that spent abrasives, paint chips, and solids will be intercepted and retained prior to entering the storm drainage system. Because of the nature of operations occurring at ship and boat facilities, routine attention needs to be given to the collection and proper disposal of spent abrasives, paint chips, and other solids.

In addition to the comprehensive site evaluation, qualified facility personnel must be identified to inspect designated equipment and areas of the facility, at a minimum, on a monthly basis. The following areas shall be included in all inspections: pressure-washing areas, blasting and painting areas, material storage areas, engine maintenance and repair areas, material handling areas, drydock areas, and general yard areas. A set of tracking or follow-up procedures shall be used to ensure that appropriate actions are taken in response to the inspections. Records shall be maintained.

The permittee is required to identify annual (once-per-year) dates for employee training. Employee training must, at a minimum, address the following areas when applicable to a facility: used-oil management; spent-solvent management; proper disposal of spent abrasives; proper disposal of vessel wastewaters; spill prevention and control; fueling procedures; general good housekeeping practices; proper painting and blasting procedures; and used-battery management. Employees, independent contractors, and customers must be informed about BMPs and be required to perform in accordance with these practices. The permittee is required to consider posting easy-to-read or graphic depictions of BMPs that are included in the plan as well as emergency phone numbers in the work areas. This practice will enhance employees' understanding of the pollutant control measures. Unlike some industrial operations, the industrial activities associated with ship and boat building and repair facilities that may affect storm water quality require the cooperation of all employees. EPA, therefore, is requiring that employee training take place at least once per year to serve as (1) training for new employees, (2) a refresher course for existing employees, (3) training for all employees on any storm water pollution prevention techniques recently incorporated into the plan, and (4) a forum for the facility to invite independent contractors and customers to inform them of pollution prevention procedures and requirements.

Sector S: Air Transportation Facilities

In addition to the common pollution prevention plan requirements discussed in Chap. 7, the site map developed for an entire airport must identify the location of each tenant of the facility and describe his or her activities.

All aspects of aircraft deicing and anti-icing operations, including quantities used and stored, as well as application, handling, and storage procedures, are required to be addressed.

The plan must describe measures that prevent or minimize the contamination of storm water runoff from all areas used for aircraft, ground vehicle and equipment maintenance, and servicing. Management practices such as performing all maintenance activities indoors, maintaining an organized inventory of materials used, draining all parts of fluids prior to disposal, prohibiting the practice of hosing down the apron or hangar floor, using dry cleanup methods in the event of spills, and/or collecting the storm water runoff from maintenance and/or service areas and providing treatment or recycling should be considered.

The plan must describe measures that prevent or minimize the contamination of the storm water runoff from all areas used for aircraft, ground vehicle, and equipment maintenance. Management practices such as performing all cleaning operations indoors and/or collecting the storm water runoff from the area and providing treatment or recycling should be considered.

The storage of aircraft, ground vehicles, and equipment awaiting maintenance must be confined to designated areas (delineated on the site map). The plan must describe measures that prevent or minimize the contamination of storm water runoff from these areas. Management practices such as indoor storage of aircraft and ground vehicles, the use of drip pans for the collection of fluid leaks, and perimeter drains, dikes, or berms surrounding storage areas should be considered.

Storage units of all materials (e.g., used oils, hydraulic fluids, spent solvents, and waste aircraft fuel) must be maintained in good condition, so as to prevent contamination of storm water, and plainly labeled (e.g., "used oil," "contaminated jet A"). The plan must describe measures that prevent or minimize contamination of the storm water runoff from storage areas. Management practices such as indoor storage of materials, centralized storage areas for waste materials, and/or installation of berms and dikes around storage areas should be considered for implementation.

The plan must describe measures that prevent or minimize the discharge of fuels to the storm sewer resulting from fuel servicing activities or other operations conducted in support of the airport fuel system. Where the discharge of fuels into the storm sewer cannot be prevented, the plan shall indicate measures that will be employed to prevent or minimize the discharge of the contaminated runoff into receiving surface waters.

Where aboveground storage timers are present, pollution prevention plan (PPP) requirements shall be consistent with requirements established in 40 CFR 112.7 guidelines for the preparation and implementation of a spill prevention control and countermeasures (SPCC) plan. Where an SPCC plan already exists, the storm water pollution prevention plan may incorporate requirements into the PPP by reference.

Facilities which conduct aircraft and/or runway (including taxiways and ramps) deicing and anti-icing operations shall evaluate present operating procedures to consider alternative practices which would reduce the overall amount of deicing and anti-icing chemical used and/or lessen the environmental impact of the pollutant source.

With regard to runway deicing operations, operators should begin by evaluating present chemical application rates to ensure against excessive overapplication. Devices which meter the amount of chemical being applied to runways help to prevent overapplication. Operators should also emphasize anti-icing operations that would preclude the need to deice; less chemical is required to prevent the formation of ice on a runway than is required to remove ice from a runway. To further assist in implementing anti-icing procedures, operators should consider installing runway ice detection (RID) systems, otherwise known as *pavement sensors,* which monitor runway temperatures. Pavement sensors provide an indication of when runway temperatures are approaching freezing conditions, thus alerting operators to the need to conduct anti-icing operations. Deicing and anti-icing chemicals applied during extremely cold, dry conditions are often ineffective since they do not adhere to the ice surface and may be scattered as a result of windy conditions or aircraft movement. In an effort to improve the efficiency of the application, operators should consider prewetting the deicing chemical to improve adhesion to the iced surface.

With regard to substitute deicing chemicals for runway use, operators should consider using chemicals that have less of an environmental impact on receiving waters. Potassium acetate has a lower oxygen demand than glycol, is nontoxic to aquatic habitat or humans, and was approved by the FAA for runway deicing operations in November 1991 (AC No. 150/5200-30A CHG 1).

In considering alternative management practices for aircraft deicing operations, operators should evaluate present application rates to ensure against excessive overapplication. In addition, operators may consider pretreating aircraft with hot water or forced air prior to the application of chemical deicer. The goal of this management practice is to reduce the amount of chemical deicer used during the operation. This management practice alone is not sufficient since discharges of small concentrations of glycol can have significant effects on receiving waters. It is, however, an effective measure to reduce the amount of glycol needed per operation.

A number of reports including the EPA's Guidance for Issuing NPDES Storm Water Permits for Airports, September 28, 1991, and Federal Aviation Administration (FAA) Advisory Circular (AC 150-5320-15) indicate that the most common location for deicing and anti-icing aircraft at U.S. airports is along the apron areas, where mobile deicing vehicles operate from gate to gate. In a recent FAA survey of deicing and anti-icing operations at U.S. airports (June 1992), the majority of respondents indicated that spent deicer chemicals from aircraft deicing and anti-icing operations drain to the storm sewer system or open areas or are left to evaporate on the ramp.

Operators must provide a narrative description of BMPs to control or manage storm water runoff from areas where deicing and anti-icing operations occur, in an effort to minimize or reduce the amount of pollutants being discharged from the site. For example, when deicing and anti-icing operations are conducted on aircraft during periods of dry weather, operators should ensure that storm water inlets are blocked to prevent the discharge of deicing and anti-icing chemicals to the storm sewer system. Mechanical vacuum systems or other similar devices can then be used to collect the spent deicing chemical from the apron surface for proper disposal, to prevent those materials from later becoming a source of storm water contamination. Establishing a centralized deicing station would also provide better control over aircraft deicing and anti-icing operations in that it enables operators to readily collect spent deicing and anti-icing chemicals.

Once spent deicer and anti-icer chemicals have been collected, operators can select from various methods of disposal:

1. *Glycol treatment.* Because glycols are readily biodegradable, runoff can be treated along with sanitary sewage. The receiving treatment plant would, however, have to have the capacity to handle the hydraulic load as well as the additional biochemical oxygen demand associated with the deicing and anti-icing chemical. To lessen both the increased hydraulic and pollutant loads due to runoff from airport deicing and anti-icing operations, retention basins may be located at the airport facility.
2. *Glycol storage.* Conversion of suitable unused airport land into retention or detention basins allows for collection of large volumes of glycol waste from pavement surface runoff. The design capacity for such basins should at least handle surface runoffs for winter months, noting the decreased microbial activity during the winter season which is needed for biodegradation, plus additional capacity for runoff during thawing periods. Continuous aeration would supply required oxygen and allow for faster biodegradation and release of glycol waste, which may reduce capacity requirements. Metering the discharge of flow from an on-site basin allows the operator to better control the rate of flow during peak flight hours and to avoid BOD shock loadings to a sanitary treatment facility or a surface water.
3. *Glycol recycling.* Recycling provides operators with a chemical cost savings since recaptured glycol can be sold or reused for other non-aircraft applications (FAA AC 150-5320-15, February 1991). Studies indicate that collected deicing chemicals which have glycol concentrations ranging from 15% to 25% can be cost-effectively recycled. The optimal condition for collecting the highest concentration of glycol in spent deicing fluid is directly from the apron or centralized deicing sta-

tion when deicing operations are conducted during dry weather or light precipitation events. Deicing and anti-icing chemicals discharged to retention basins which are then allowed to mix with additional surface runoff typically result in glycol concentrations well below the acceptable range for recycling. There are, however, methods of physical separation presently available which increase the concentration of glycol and allow operators to recover a relatively reusable product.

In addition to the common pollution prevention plan requirements discussed in Chap. 7, qualified personnel must inspect equipment and areas involved in deicing and anti-icing operations on a weekly basis during periods when deicing and anti-icing operations are being conducted.

Training should address topics such as spill response, good housekeeping, material management practices, and deicing and anti-icing procedures. The pollution prevention plan must identify periodic dates for such training.

Sector T: Treatment Works

There are no additional requirements under this section, other than those described in Chap. 7 of this book.

Sector U: Food and Kindred Products

The following areas are also potential sources of pollutants in storm water from food and kindred products processing facilities: vents and stacks from cooking and drying operations and dry product vacuum transfer lines; animal holding pens; spoiled product and broken product container storage areas; and significant dust- or particulate-generating areas. The site map must identify all monitoring locations that must be sampled as part of the monitoring requirements of the permit.

In conducting the assessment of storm water pollution risks at the site, the facility operator must consider the following activities: loading and unloading areas; vehicle fueling; vehicle and equipment maintenance and/or cleaning areas; waste treatment, storage, and disposal locations; liquid storage tanks; vents and stacks from cooking and drying operations and dry product vacuum transfer lines; animal holding pens; out-of-date/spoiled product storage areas; and significant dust- or particulate-generating areas. The assessment must list any significant pollution sources at the site and identify the pollutant parameter or parameters (e.g., biochemical oxygen demand, oil and grease) associated with each source.

In addition to food and kindred products processing-related industrial activities, the plan must describe application and storage of pest control chemicals (e.g., rodenticides, insecticides, fungicides) used at the facility, including a discussion of application and storage procedures.

The permittee must evaluate, select, and describe the pollution prevention measures, BMPs, and other controls that will be implemented at the facility. EPA emphasizes the implementation of pollution prevention measures and BMPs that reduce possible pollutant discharges at the source. Source reduction measures include, among others, preventive maintenance, chemical substitution, spill prevention, good housekeeping, training, and proper materials management. Where source reduction is not appropriate, EPA supports the use of source control measures and BMPs such as material segregation or covering, water diversion, and dust control. If source reduction or source control is not possible, recycling or treatment is the remaining alternative. Recycling allows the reuse of storm water, while treatment lowers pollutant concentrations prior to discharge. Since the majority of food and kindred products processing is conducted indoors, the activities identified above are geared toward only those activities that may contribute pollutants to storm water. Also because of the relatively few activities that are conducted outdoors within this sector, pollution prevention measures, BMPs, and other controls should be relatively few and easy for any given permittee. Also, these measures are the most appropriate means to reduce pollutant loadings to storm water (as opposed to pollutant limitations) because of the relative ease and the significant reductions in pollutant loads that can be realized.

Permittees must describe protocols established to reduce the possibility of mishandling chemicals or equipment and training employees in good housekeeping techniques. Specifics of this plan must be communicated to appropriate plant personnel.

Areas and activities that typically pose a high risk for spills at food and kindred products processing facilities include raw material unloading and product loading areas, material storage areas, and waste management areas (e.g., dumpsters, compactors). These activities and areas, and their accompanying drainage points, must be described in the plan.

Inspections must be carried out by qualified facility personnel at least once each year.

Sector V: Textile Mills, Apparel, and Other Fabric Product Manufacturing

Permittees are required to include processing areas; loading and unloading areas; treatment, storage, and waste disposal areas; liquid storage tanks; and fueling areas on a site facility map.

All stored and containerized materials (fuels, petroleum products, solvents, dyes, etc.) must be stored in a protected area, away from drains, and clearly labeled. The plan must describe measures that prevent or minimize contamination of storm water runoff from such storage areas. The facility should specify which materials are stored indoors, and must provide a description of the contaminant area or enclosure for those materials which are stored outdoors. Aboveground storage tanks, drums, and barrels permanently stored outside must be delineated on the site map with a description of the appropriate containment measures in place to prevent leaks and spills. The facility may consider an inventory control plan to prevent excessive purchasing, storage, and handling of potentially hazardous substances. In the case of storage of empty chemical drums and containers, facilities should employ such practices as the triple-rinsing of containers. The discharge waters from such washings must be collected, contained, or treated, and facilities should identify where the discharge will be released.

The plan must describe measures that prevent or minimize contamination of the storm water runoff from materials handling operations and areas. The facility may consider the use of spill and overflow protection; covering fuel areas; and covering and enclosing areas where the transfer of materials may occur. Where applicable, the plan must address the replacement or repair of leaking connections, valves, transfer lines, and pipes that may carry chemicals, dyes, or wastewater.

The plan must describe measures that prevent or minimize contamination of the storm water runoff from fueling areas. The facility may consider covering the fueling area, using spill and overflow protection, minimizing run-on of storm water to the fueling area, using dry cleanup methods, and/or collecting the storm water runoff and providing treatment or recycling.

The plan must describe measures that prevent or minimize contamination of the storm water runoff from aboveground storage tank areas. The facility must consider storage tanks and their associated piping and valves. The facility may consider regular cleanup of these areas, preparation of a spill prevention control and countermeasures program, providing spill and overflow protection, minimizing run-on of storm water from adjacent facilities and properties, restricting access to the area, inserting filters in adjacent catch basins, providing absorbent booms in unbermed fueling areas, using dry cleanup methods, and permanently sealing drains within critical areas that may discharge to a storm drain.

Many facilities will find that management measures that have already been incorporated into the facility's operation, such as the installation of overfill protection equipment and labeling and maintenance of used-oil storage units, are already required under existing EPA programs and will meet the requirements of this section.

Under the preventive maintenance requirements, the plan specifically includes the routine inspection of sediment traps to ensure that solids will be intercepted and retained prior to entering the storm drainage system. Because of the nature of operations which occur at textile facilities, specific routine attention needs to be paid to the collection of solids.

In addition to the comprehensive site evaluation, qualified facility personnel must be identified to inspect designated equipment and areas of the facility, at a minimum, on a monthly basis.

The permittee is required to identify at least annual dates for employee training. EPA requires that facilities conduct training annually at a minimum. However, more frequent training may be necessary at facilities with high turnover of employees or where employee participation is essential to the storm water pollution prevention plan. Employee training must, at a minimum, address the following areas when applicable to a facility: use of reused and recycled waters; solvents management; proper disposal of dyes; proper disposal of petroleum products and spent lubricants; spill prevention and control; fueling procedures; and general good housekeeping practices. Employees, independent contractors, and customers must be informed about BMPs, and be required to perform in accordance with these practices. Copies of BMPs and any specific management plans, including emergency phone numbers, shall be posted in the work areas.

Sector W: Furniture and Fixtures

The permittee must assess the applicability of the following categories of BMPs for their site: discharge diversions, drainage and storm water conveyance systems, runoff dispersions, and good housekeeping measures. In addition, BMPs include processes, procedures, schedules of activities, prohibitions on practices, and other management practices that prevent or reduce the discharge of pollutants in storm water runoff.

Operators of furniture and fixture manufacturing facilities are required to conduct quarterly inspections. The inspections shall include (1) an assessment of the integrity of storm water discharge diversions, conveyance systems, sediment control and collection systems, and containment structures; (2) visual inspections of vegetative BMPs to determine if soil erosion has occurred; and (3) visual inspections of material handling and storage areas and other potential sources of pollution for evidence of actual or potential pollutant discharges of contaminated storm water.

Sector X: Printing and Publishing

For printing and publishing facilities, the SWPPP must describe protocols established to reduce the possibility of mishandling chemicals or equipment and training employees in good housekeeping techniques. Specifics of this plan must be communicated to appropriate plant personnel.

Sector Y: Rubber, Miscellaneous Plastic Products, and Miscellaneous Manufacturing Industries

For rubber manufacturers, permittees must develop specific BMPs to control discharges of zinc in storm water runoff. Rubber products manufacturers must review the possible sources of zinc at their facilities and include as appropriate the accompanying BMPs in their SWPPP.

Permittees are required to review the handling and storage of zinc bags at their facilities and to consider employee training regarding the handling and emptying of zinc bags, indoor storage of zinc bags, and thorough cleanup of zinc spills without washing the zinc into a storm drain. Facilities must also consider the use of 2500-lb sacks (from which spills are less likely) rather than 50- to 100-lb sacks.

Permittees must also consider providing a cover for the dumpster, moving the dumpster inside, or providing a lining for the dumpster.

Permittees must review dust collectors and baghouses as possible sources of zinc. Improperly operating dust collectors or baghouses must be replaced or repaired as appropriate; the plan must also provide for regular maintenance of these facilities.

Permittees must review dust generation from rubber grinding operations at their facility and, as appropriate, install a dust collection system.

The plan must include measures to prevent and/or clean up drips or spills of zinc stearate slurry that may be released to a storm drain. Alternate compounds to zinc stearate must be considered.

Sector Z: Leather Tanning and Finishing

Under the description of measures and controls in the SWPPP requirements, all areas that may contribute pollutants to storm water discharges shall be maintained in a clean, orderly manner. The following areas must be specifically addressed:

1. Pallets and/or bales of raw, semiprocessed, or finished tannery byproducts (e.g., splits, trimmings, shavings) that are stored where there is potential storm water contact must be stored indoors or protected by polyethylene wrapping, tarpaulins, roofed storage area, or other suit-

able means. Materials should be placed on an impermeable surface, the area should be enclosed or bermed, or other equivalent measures should be employed to prevent run-on or runoff of storm water.

2. Label storage units of all materials (e.g., specific chemicals, hazardous materials, spent solvents, waste materials). Maintain such containers and units in good condition. Describe measures that prevent or minimize contact with storm water. The facility must consider indoor storage and/or installation of berms and diking around the area to prevent run-on or runoff of storm water.

3. The plan must describe measures that prevent or minimize contamination of the storm water runoff with leather dust from buffing and shaving areas. The facility may consider dust collection enclosures, preventive inspection and maintenance programs, or other appropriate preventive measures.

4. The plan must describe measures that prevent or minimize contamination of the storm water runoff from receiving, unloading, and storage areas. Exposed receiving, unloading, and storage areas for hides and chemical supplies should be protected by a suitable cover, diversion of drainage to the process sewer, directing rain gutters away from loading and receiving areas, grade berming or curbing area to prevent run-on of storm water, or other appropriate preventive measures.

5. The plan must describe measures that minimize contact of storm water with contaminated equipment. Equipment should be protected by suitable cover, diversion of drainage to the process sewer, thorough cleaning prior to storage, or other appropriate preventive measures.

6. The plan must describe measures that prevent or minimize contamination of the storm water runoff from waste storage areas. The facility may consider inspection and maintenance programs for leaking containers or spills, covering dumpsters, moving waste management activities indoors, covering waste piles with temporary covering material such as tarpaulin or polyethylene, and minimizing storm water run-on by enclosing the area or building berms around the area.

7. Permittees must follow all applicable requirements described under Sector P in this chapter for controlling storm water discharges from vehicle maintenance and refueling areas.

8. The plan must describe measures which prevent and prohibit wash waters from processing areas from entering storm sewers. The facility must install safeguards against wash waters entering storm sewers and must train employees on proper disposal practices for disposal of all process waste materials.

Permittees using ponds to control their effluent limitation frequently use impoundments or sedimentation ponds as their BAT/BCT.

Maintenance schedules and maintenance measures for these ponds must be provided in the pollution prevention plan.

Qualified facility personnel must be identified to inspect designated areas of the facility, at a minimum of every 3 months. The individual or individuals who will conduct the inspections must be identified in the plan and should be members of the pollution prevention team. The following areas must be included in all inspections: storage areas for equipment and vehicles awaiting maintenance, facility yard area where outdoor storage occurs, receiving and unloading areas, and waste management areas. A set of tracking or follow-up procedures must be used to ensure that appropriate actions are taken in response to the inspections. Records of inspections shall be maintained and the pollution prevention plan modified where necessary.

In addition, qualified personnel must conduct quarterly visual inspections of all BMPs. The inspections must include an assessment of the effectiveness and need for maintenance of storm water roofing and covers, dikes and curbs, discharge diversions, sediment control and collection systems, and all other BMPs.

Quarterly visual inspections must be made at least once in each of the following designated periods during daylight hours: January to March (storm water runoff or snow melt), April to June (storm water runoff), July to September (storm water runoff), and October to December (snow melt runoff). Records must be maintained as part of the pollution prevention plan.

The permittee is required to identify annual (once-per-year) dates for training. Employee training must, at a minimum, address the following areas when applicable to a facility: general good housekeeping practices, spill prevention and control, waste management, inspections, preventive maintenance, detection of non–storm water discharges, and other areas. More frequent training may be necessary at facilities with high turnover of employees or where employee participation is essential to the storm water pollution prevention plan.

Permittees must describe procedures for developing and retaining records on the status and effectiveness of plan implementation. The plan must address spills, monitoring, and BMP inspection and maintenance activities. Ineffective BMPs must be reported and the date of their corrective action recorded. Employees must report incidents of leaking fluids to facility management, and these reports must be incorporated into the plan.

The permittee must evaluate the appropriateness of each storm water BMP that diverts, infiltrates, reuses, or otherwise reduces the discharge of contaminated storm water. In addition, the permittee must describe the storm water pollutant source or activity (i.e., loading and unloading operations, raw material storage piles, waste piles) to be controlled by each storm water management practice.

Sector AA: Fabricated Metal Products

The SWPPP for sector AA facilities should focus primarily on storage areas; unloading and loading areas; and any other area where outside operations occur. In particular; the SWPPP should address storage areas for raw metal; receiving, unloading, and loading areas; storage of heavy equipment; metal working fluid areas; unprotected liquid storage tanks; chemical cleaners and wastewaters; raw steel collection; paints and painting equipment; hazardous waste storage; chemical transportation; galvanized products; vehicle and equipment maintenance; wooden pallets and empty drums; and retention ponds.

Metal fabricating areas should be kept clean by frequent sweeping to avoid heavy accumulation of steel ingots, fines, and scrap. Dust is a byproduct of many processes in the fabricating areas and therefore should be absorbed through a vacuum system to avoid accumulation on rooftops and onto the ground. Tracking of metal dusts and metal fines outdoors may be minimized by employing these management practices: on a regular basis sweep all accessible paved areas; maintain floors in a clean and dry condition; remove and dispose of waste regularly; remove obsolete equipment expeditiously; sweep fabrication areas; and train employees in good housekeeping measures.

The storage of raw materials should be under a covered area whenever possible and protected from contact with the ground. The amount of material stored should be minimized to avoid corrosive activity from long-term exposed materials. Diking or berming the area to prevent or minimize run-on may be considered. Long-term exposure to weather conditions results in oxidation of the metals. Also, dirt, oil, and grease buildup on the metal are potential sources of pollutants. The following measures should be considered: Check raw metals for corrosion; keep area neat and orderly; stack neatly on pallets or off the ground; and cover exposed materials.

Receiving, unloading, and loading areas should be enclosed where feasible, using curbing, berming, diking, or other accepted containment systems in case of spills during delivery of chemicals such as lubricants, coolants, rust preventives, solvents, oil, sodium hydroxide, hydrochloric acid, calcium chloride, polymers, sulfuric acid, and other chemicals used in the metal fabricating processes. Directing roof downspouts away from loading sites and equipment and onto grassy or vegetated areas should help prevent storm water contamination by pollutants that have accumulated in these areas. The following measures should be considered: Clean up spills immediately; check for leaks and remedy problems regularly; and unload under covered areas when possible.

Vehicles should be stored indoors when possible. If they are stored outdoors, the use of gravel, concrete, or other porous surfaces should be considered to minimize or prevent heavy equipment from creating ditches or other conveyances that would cause sedimentation runoff and increase TSS loadings. Also, directing the flow toward the area by the use of grass swales or filter strips will reduce the runoff of materials. Directing drainage systems away from high-traffic areas into collection systems will help to reduce the TSS loadings caused by exposed and eroding open areas. The following measures should be considered: Clean prior to storage or store under cover; store indoors; and divert drainage to the grass swales, filter strips, retention ponds, or holding tanks.

Due to the toxicity of metal working fluids as well as the contamination of fluids by metal fines and dusts, spillage and loss of metal working fluids used to cleanse or prepare the steel components should be controlled throughout the process. Collection systems and storage areas need special consideration. The following measures should be considered: Store used metal working fluid with fine metal dust indoors; use tight-sealing lids on all fluid containers; use straw, clay absorbents, sawdust, or synthetic absorbents to confine or contain any spills, or other absorbent material; and establish recycling programs for used fluids when possible.

Storing these tanks indoors[2] will reduce potential waste or spills from contaminating storm water. Berming outdoor areas when unable to store inside will contain potential pollutants. Cleaning up spills is essential to minimizing buildup in these areas. The following measures should be considered: Cover all tanks whenever possible; berm tanks whenever possible; dike area or install grass filters to contain spills; keep area clean; and check piping, valves, and other related equipment on a regular basis.

Proper disposal and use of cleaners in various activities will minimize the amount of liquid exposed to storm water by reducing the need to store contaminated liquids for an extended time. Controlling potential contamination of pollutants by employing simple control devices during the activity will prevent potential contamination in storm water runoff. Recycling or reuse of these materials whenever possible serves as a source reduction by reducing the necessary amount of new materials. The following measures should be considered: Use drip pans and other spill devices to collect spills or solvents and other liquid cleaners; recycle waste water; store recyclable waste indoors or in covered containers; and substitute nontoxic cleaning agents when possible.

The collection areas must be kept clean. Materials should be kept in a covered storage bin or kept inside until pickup. The use of pitched

[2]Products that are gaseous at atmospheric pressure are not required to be stored indoors.

structures should be considered. The following measures should be considered: Collect scrap metals, fines, and iron dust; store under cover; and recycle.

Facilities use tarps, drip pans, or other spill collection devices to contain and collect spills of paints, solvents, or other liquid material. Blasting in windy weather increases the potential for runoff. Enclosing outdoor sanding areas with tarps or plastic sheeting contains the metal fines. Immediate collection of any waste and proper disposal may significantly contribute to the reduction of storm water runoff. Training employees to use the spray equipment properly may reduce waste and decrease the likelihood of accidents as well as reduce the amount of solvents needed to complete the job. The following measures should be considered: Paint and sand indoors when possible; avoid painting and sandblasting operations outdoors in windy weather conditions; if done outside, enclose sanding and painting areas with tarps or plastic sheeting; and use water-based paints when possible.

Changing of fluids or parts should be done indoors when possible. If maintenance is performed outdoors, fluids used in maintaining these vehicles should be contained in the area by using drip pans, large plastic sheets, canvas, or other similar controls under the vehicles, or by berming the area. Hydraulic fluids should be properly stored to prevent leakage and storm water contamination. The following measures should be considered: Berm area or use other containment device to control spills; use drip pans, plastic sheeting, and other similar controls; and discard fluids properly or recycle if possible.

All hazardous waste must be stored in sealed drums. Establishing centralized drum storage satellite areas throughout the complex to store these materials will decrease the potential for mishandling drums. Berming the enclosed structures is added protection in case of spills. Spills or leaks that are contained within an area are easier to contain and prevent storm water contamination or runoff. Checks for corrosion and leakage of storage containers are important. Proper labeling for proper handling should be considered. All other applicable federal, state, and local regulations must be followed. The following measures should be considered: Store indoors; label materials clearly; check for corrosion and leaking; properly dispose of outdated materials; dike or use grass swales, ditches, or other containment to prevent run-on or runoff in case of spills; post notices prohibiting dumping of materials into storm drains; store containers, drums, and bags away from direct traffic routes; do not stack containers in such a way as to cause leaks or damage to the containers; use pallets to store containers when possible; store materials with adequate space for traffic without disturbing drums; maintain low inventory level of chemicals based on need.

Proper handling of drums is needed to avoid damaging drums, causing leaks. Storage areas should be as close as possible to operational buildings. The following measures should be considered: Forklift operators should be trained to avoid puncturing drums; store drums as close to the operational building as possible; and label all drums with proper warning and handling instructions.

Improper storage of finished products can contribute pollutants to storm water discharges. Materials should be stored in such a way as to minimize contact with precipitation and runoff. The following measures should be considered: Store finished products indoors, on a wooden pallet, concrete pad, gravel surface, or other impervious surface; clean contaminated wooden pallets; cover empty drums; cover contaminated wooden pallets; store drums and pallets indoors; clean empty drums; and store pallets and drums on concrete pads.

Creating and maintaining retention ponds as a treatment system for settling out TSS would help to reduce the concentrations of these pollutants in storm water runoff. The following measures should be considered: Provide routine maintenance; remove excess sludge periodically; and aerate periodically to maintain pond's aerobic character and ecological balance.

In addition to the baseline inspection requirements, the required inspections for sector AA facilities must include periodic inspections of raw metal storage areas, finished product storage areas, material and chemical storage areas, recycling areas, loading and unloading areas, equipment storage areas, paint areas, fueling and maintenance areas, and waste management areas, as well as an annual comprehensive site compliance evaluation.

Sector AB: Transportation Equipment, Industrial or Commercial Machinery

The SWPPP for sector AB facilities should meet all the minimum SWPPP requirements discussed in Chap. 7. In addition, annual employee training must be required on specified topics. Good housekeeping practices and spill prevention and response procedures must be described and required for exposed areas. Annual inspections must be required for loading and unloading areas, storage areas, waste management units, and vents and stacks. An annual comprehensive site compliance evaluation is required.

The plan must contain a map of the site that shows the pattern of storm water drainage, structural and nonstructural features that control pollutants in storm water runoff and process wastewater discharges, surface water bodies (including wetlands), places where significant materials[3] are exposed to rainfall and runoff, and locations of major spills and leaks that occurred in the 3 years prior to the date of the submission of a Notice of Intent to be covered under this permit. The map must also indicate the direction of storm water flow. An outline of the drainage area for each outfall must be provided; and the location of each outfall and monitoring points must be indicated. An estimate of the total site acreage utilized for each industrial activity (e.g., storage of raw materials, waste materials, and used equipment) must be provided. These areas include liquid storage tanks, stockpiles, holding bins, used equipment, and empty drum storage. These areas are considered to be significant potential sources of pollutants at facilities which manufacture transportation equipment and industrial or commercial machinery. The site map must also indicate the outfall locations and the types of discharges contained in the drainage areas of the outfalls (e.g., storm water and air conditioner condensate). To increase the readability of the map, the inventory of the types of discharges contained in each outfall may be kept as an attachment to the site map.

Facility operators are required to carefully conduct an inspection of the site to identify significant materials that are or may be exposed to storm water discharges. The inventory must address materials that within 3 years prior to the date of the submission of a Notice of Intent to be covered under the permit have been handled, stored, processed, treated, or disposed of in a manner to allow exposure to storm water. Findings of the inventory must be documented in detail in the pollution prevention plan. At a minimum, the plan must describe the method and location of on-site storage or disposal; practices used to minimize contact of materials with precipitation and runoff; existing structural and nonstructural controls that reduce pollutants in storm water; existing structural controls that limit process wastewater discharges; and any treatment the runoff receives before it is discharged to surface waters or through a separate storm sewer system. The description must be updated whenever there is a significant change in the type or amounts of materials, or material management practices, that may affect the exposure of materials to storm water.

The plan must include a list of any significant spills and leaks of toxic or hazardous pollutants that occurred in the 3 years prior to the date of the submission of a Notice of Intent to be covered under this permit.

Each pollution prevention plan must include a certification, signed

[3]Significant materials commonly found at transportation equipment, industrial, or commercial machinery manufacturing facilities include raw and scrap metals; solvents; used equipment; petroleum-based products; and waste materials or by-products used or created by the facility.

by an authorized individual, that discharges from the site have been tested or evaluated for the presence of nonstorm water, the results of any test and/or evaluation conducted to detect such discharges, the test method or evaluation criteria used, the dates on which tests or evaluations were performed, and the on-site drainage points directly observed during the test or evaluation. Pollution prevention plans must identify and ensure the implementation of appropriate pollution prevention measures for any non–storm water discharges.

Any existing data describing the quality or quantity of storm water discharges from the facility must be summarized in the plan. The description should include a discussion of the methods used to collect and analyze the data. Sample collection points should be identified in the plan and shown on the site map.

The description of potential pollutant sources should clearly point to activities, materials, and physical features of the facility that have a reasonable potential to contribute significant amounts of pollutants to storm water. Any such activities, materials, or features must be addressed by the measures and controls subsequently described in the plan. In conducting the assessment, the facility operator must consider the following activities: raw materials (liquid storage tanks, stockpiles, holding bins), waste materials (empty drum storage), and used-equipment storage areas. The assessment must list any significant pollutant parameter(s) (i.e., total suspended solids, oil and grease, etc.) associated with each source.

Permittees must select, describe, and evaluate the pollution prevention measures, BMPs, and other controls that will be implemented at the facility. Source reduction measures include preventive maintenance, spill prevention, good housekeeping, training, and proper materials management. If source reduction is not an option, EPA supports the use of source control measures. These include BMPs such as material covering, water diversion, and dust control. If source reduction or source control is not available, then recycling and waste treatment are other alternatives. Recycling allows the reuse of storm water, while treatment lowers pollutant concentrations prior to discharge. Since the majority of transportation equipment, industrial or commercial machinery manufacturing occurs indoors, the BMPs identified above are geared toward only those activities occurring outdoors or that otherwise have a potential to contribute pollutants to storm water discharges.

Pollution prevention plans must discuss the reasons why each selected control or practice is appropriate for the facility and how each of the potential pollutant sources will be addressed. Plans must identify the time during which controls or practices will be implemented, as well the effect the controls or practices will have on storm water discharges from the site. At a minimum, the measures and controls must address the following components:

Permittees must describe protocols established to reduce the possibility of mishandling chemicals or equipment and to train employees in good housekeeping techniques. Specifics of this plan must be communicated to appropriate plant personnel.

Permittees are required to develop a preventive maintenance program that includes regular inspections and maintenance of storm water BMPs. Inspections should assess the effectiveness of the storm water pollution prevention plan. They allow facility personnel to monitor the components of the plan on a regular basis. The use of a checklist is encouraged, as it will ensure that all the appropriate areas are inspected and will provide documentation for record-keeping purposes.

Permittees are required to identify proper material handling procedures, storage requirements, containment or diversion equipment, and spill removal procedures to reduce exposure of spills to storm water discharges. Areas and activities which are high risks for spills at transportation equipment, industrial or commercial machinery manufacturing facilities include raw material unloading and product loading areas, material storage areas, and waste management areas. These activities and areas and their drainage points must be described in the plan.

Qualified personnel must inspect designated equipment and areas of the facility at the proper intervals specified in the plan. The plan should identify for periodic inspections areas which have the potential to pollute storm water. Records of inspections must be maintained on-site.

Permittees must describe a program for informing and educating personnel at all levels of responsibility in the components and goals of the storm water pollution prevention plan. A schedule for conducting this training should be provided in the plan. Where appropriate, contractor personnel must also be trained in relevant aspects of storm water pollution prevention. Topics for employee training should include good housekeeping, materials management, and spill response procedures. EPA recommends that facilities conduct training annually at a minimum. However, more frequent training may be necessary at facilities with high turnover of employees or where employee participation is essential to the storm water pollution prevention plan.

Permittees must describe procedures for developing and retaining records on the status and effectiveness of plan implementation. This includes the success and failure of BMPs implemented at the facility.

Permittees must identify areas due to topography, activities, soils, cover materials, or other factors that have a high potential for soil erosion. Measures to eliminate erosion must be identified in the plan.

Permittees must provide an assessment of traditional storm water management practices that divert, infiltrate, reuse, or otherwise manage storm water so as to reduce the discharge of pollutants. Based on this assessment, practices to control runoff from these areas must be identified and implemented as required by the plan.

The storm water pollution prevention plan must describe the scope and content of comprehensive site inspections that qualified personnel will conduct to (1) confirm the accuracy of the description of potential sources contained in the plan, (2) determine the effectiveness of the plan, and (3) assess compliance with the terms and conditions of this section. Comprehensive site compliance evaluations must be conducted once per year for transportation equipment, industrial or commercial machinery manufacturing facilities. The individual(s) who will conduct the evaluations must be identified in the plan and should be members of the pollution prevention team. Evaluation reports must be retained for at least 3 years after the date of the evaluation.

Based on the results of each evaluation, the description of potential pollution sources, and measures and controls, the plan must be revised as appropriate within 2 weeks after each evaluation. Changes in the measures and controls must be implemented on the site in a timely manner, never more than 12 weeks after completion of the evaluation.

Sector AC: Electronic, Electrical, Photographic, and Optical Goods

There are no additional requirements beyond those described in Chap. 7.

References

Environmental Protection Agency, 1982. *Final Development Document for Effluent Limitations Guidelines and Standards and Pretreatment Standards for the Steam Electric Point Source Category,* EPA-440/1-82/-29, November.

Government Printing Office: *Federal Register,* Office of the Federal Register, National Archives and Records Administration, Washington.

———, 1992. *Code of Federal Regulations,* Office of the Federal Register, National Archives and Records Administration, Washington, July 1.

Chapter

9

NPDES Industrial Storm Water Sampling and Monitoring

This is the fifth of five chapters which deal with the specific requirements for storm water discharges from industrial (nonconstruction) facilities. This chapter describes the requirements and procedures for storm water sampling and monitoring. Two types of requirements are described: the requirements for sampling data to be submitted as part of an application for an individual storm water discharge permit and the requirements for sampling data under the EPA multisector general permit.

Introduction to Storm Water Sampling

Sampling is the process of obtaining small quantities of discharge water and subjecting these to various analytical tests in order to characterize the pollutants within the discharge. National Pollutant Discharge Elimination System (NPDES) permits have traditionally included sampling requirements, so that the effectiveness of pollution control measures can be assessed. However, the EPA does not require sampling for all types of storm water discharges, as described later in this chapter.

Under the Clean Water Act (CWA), the EPA is authorized to require dischargers to provide information, monitoring, and record keeping to characterize discharges and to assess permit compliance. Discharge monitoring data can be used to assist in the evaluation of the risk of the discharge by indicating the types and the concentrations of pollutant parameters in the discharge. Monitoring of storm water from an industrial site can assist in evaluating sources of pollutants. Discharge monitoring data can be used in evaluating the potential of the discharge to cause or contribute to water quality impacts and water quality standards violations.

Discharge monitoring data can also be used to evaluate the effectiveness of controls on reducing pollutants in discharges. This monitoring function can be important in evaluating both the effectiveness of source control or pollution prevention measures and the operation of end-of-pipe treatment units.

The EPA has typically required that NPDES permittees [such as publicly owned treatment works (POTWs) and industrial dischargers] submit a *discharge monitoring report* (*DMR*) at least once per year. However, the amount of storm water sampling required for monitoring of multisector general permit compliance depends upon the type of facility. For all facilities, the EPA requires that dischargers conduct a comprehensive annual site compliance evaluation designed to evaluate the effectiveness of permit implementation and to identify pollutant sources. (See Chap. 7.)

The EPA is authorized under Section 308 of the CWA to require particular dischargers to submit sampling monitoring data necessary to carry out the objectives of the CWA. This authority can be used to obtain storm water monitoring data that are not otherwise required under the EPA multisector general permit.

Types of Storm Water Sampling

As described in Chap. 6, storm water sampling is required for individual discharge permit applications, but no sampling is required for initial coverage under the EPA multisector general permit for storm water discharges associated with industrial activity. Certain states or municipalities may have sampling and/or reporting requirements in addition to those described in this chapter. No sampling is required for discharges not subject to NPDES permit requirements, which are described in Chaps. 4 and 5.

As a part of the application for individual permit coverage, at least one *representative* storm event must be sampled. A representative storm event is defined as having the following characteristics:

- *More than 0.1 inch* (*in*) of *total rainfall.* This ensures that a significant volume of runoff will be generated by the storm event. Several EPA studies have established 0.1 in of rainfall as the minimum rainfall depth capable of producing the rainfall and runoff characteristics necessary to generate a sufficient volume of runoff for meaningful sample analysis.
- *More than 72 h since last event of more than 0.1 in of total rainfall.* This time interval allows pollutants to collect on the surface of the

drainage area and non–storm water discharges to build up within the drainage system. Therefore, the pollutants contributed by such sources will be measured.

- *Storm duration and total rainfall volume within 50% above or below the average rainfall event for the area of the facility.* This criterion is to be applied wherever feasible. The values in Table 9.1 may be used as estimates of the average rainfall depth and storm duration in each region of the continental United States. Figure 9.1 illustrates the boundaries of each region. Figures 9.2 and 9.3 compare the values for each region graphically.

In determining whether a particular storm meets these criteria, National Weather Service or other available meteorological data collected as close as possible to the facility may be used. However, it may be advantageous to install a nonrecording rainfall gauge at the site, if possible, to directly measure the depth and duration of rainfall and the interval between rainfall events. Later in this chapter, some guidelines are provided for situations in which it is difficult or impossible to find a storm event that meets the criteria listed above.

Snowmelt is acceptable as a source of runoff for sampling. However, snow melt runoff is typically more contaminated than rainfall runoff.

TABLE 9.1 Typical Rainfall Depths and Storm Durations

Region	Average duration, h	Acceptable range of durations, h	Average rainfall volume, in	Acceptable range of volumes, in
Northeast	11.2	5.6–16.8	0.50	0.25–0.75
Northeast, coastal	11.7	5.9–17.5	0.66	0.33–0.99
Mid-Atlantic	10.1	5.1–15.1	0.64	0.32–0.96
Central	9.2	4.6–13.8	0.62	0.31–0.93
North central	9.5	4.8–14.2	0.55	0.28–0.83
Southeast	8.7	4.4–13.0	0.75	0.38–1.13
East Gulf	6.4	3.2–9.6	0.80	0.40–1.20
East Texas	8.0	4.0–12.0	0.76	0.38–1.14
West Texas	7.4	3.7–11.1	0.57	0.29–0.85
Southwest	7.8	3.9–11.7	0.37	0.19–0.55
West inland	9.4	4.7–14.1	0.36	0.18–0.54
Pacific south	11.6	5.8–17.4	0.54	0.27–0.81
Northwest inland	10.4	5.2–15.6	0.37	0.19–0.55
Pacific central	13.7	6.9–20.5	0.58	0.29–0.87
Pacific northwest	15.9	8.0–23.8	0.50	0.25–0.75

SOURCE: Woodward-Clyde Consultants, 1990.

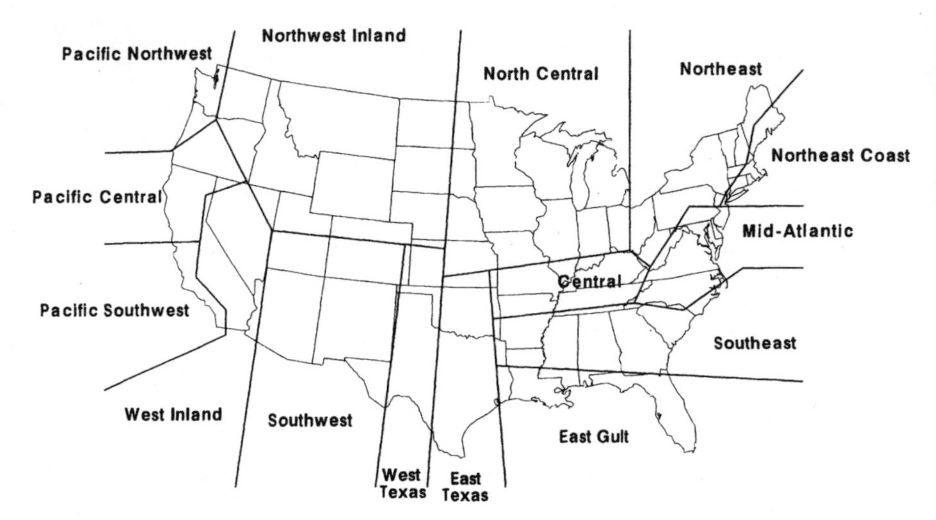

Figure 9.1 Rainfall zones of the contiguous United States. (*Source: Woodward-Clyde Consultants, 1990.*)

For individual permit applications, two types of samples must be obtained: a "first-flush" grab sample and a flow-weighted composite sample. Each of these different types of storm water discharge samples is described in the following sections.

First-flush grab samples

A *first-flush grab sample* is obtained within the first 30 minutes (min) of discharge. This is generally the most polluted portion of the dis-

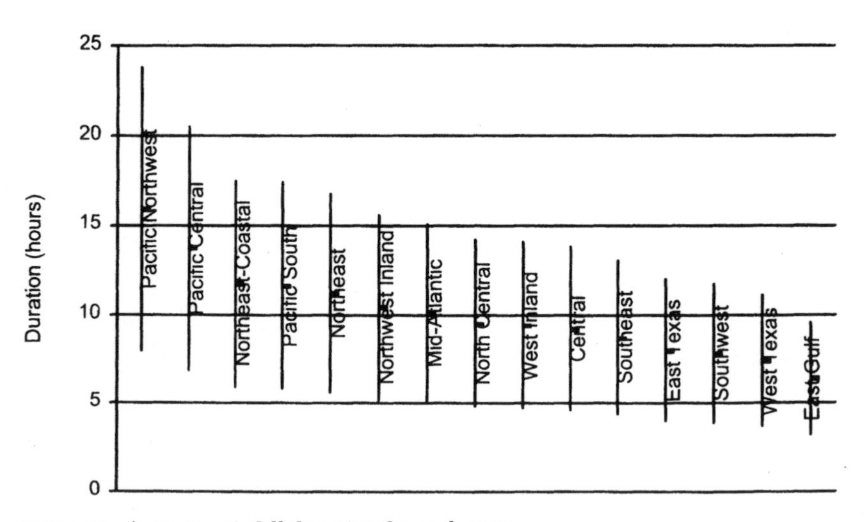

Figure 9.2 Average rainfall duration for each zone.

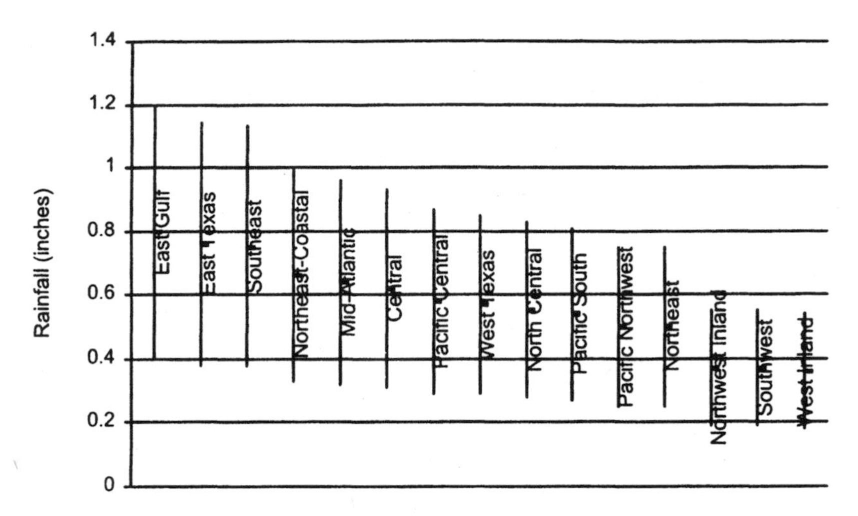

Figure 9.3 Average rainfall depth for each zone.

charge, because it may contain pollutants that lie on the surface of the drainage area. It may also contain a heavy concentration of non–storm water discharges, if these are present at the facility. The grab sample should be taken from a well-mixed portion of the flow stream, so that it is representative of the majority of discharge.

The first 30 min of discharge may not coincide with the first 30 min of rainfall. In fact, facilities equipped with detention or retention facilities may experience a considerable delay before the first discharge occurs as a result of a particular rainfall event.

If it is not possible to obtain a grab sample within the first 30 min of discharge, then the grab sample may be obtained within the first 1 h of discharge. However, the monitoring report should note this condition and explain why it was necessary.

Grab samples must be tested in the field for both pH and temperature. Testing for pH provides an indication of the potential availability of metals within the discharge. In some cases it will provide information regarding material management.

Grab samples should be tested in the laboratory for the following "conventional" pollutants:

- *Oil and grease* [in milligrams per liter (mg/L)]. This is a common component of storm water discharges that has serious effects on receiving waters. It can indicate materials management, housekeeping, and transportation activities. A value of 30 mg/L would be considered to be near the high end of the range of typical values for urban storm water runoff.

- *Total suspended solids* (*TSS*) (mg/L). The TSS are a common pollutant which can seriously affect receiving waters. TSS often indicate surface disturbances and material management practices. A value of 400 mg/L would be considered to be at the high end of typical values for urban storm water runoff.
- *The 5-day biochemical oxygen demand* (BOD_5) (mg/L). This is the most commonly used indicator of the oxygen depletion potential of the discharge. A value of 50 mg/L would be a relatively high value, when compared to typical urban storm water runoff.
- *Chemical oxygen demand* (*COD*) (mg/L). This is a more inclusive indicator of potential oxygen demand, especially where metals interfere with the BOD_5 test. COD is generally better suited for comparing the oxygen demand of a storm water discharge with that of other discharges.
- *Total Kjeldahl nitrogen* (*TKN*) (mg/L). This nutrient can degrade water quality.
- *Nitrate plus nitrite nitrogen* (mg/L). This is another form of nitrogen which can degrade water quality.
- *Total phosphorus* (mg/L). This is another type of nutrient which can degrade water quality.

In addition to these conventional pollutants, discharges should be analyzed for several other types of pollutants as applicable:

- Pollutants listed in a storm water effluent concentration guideline for the facility. Facilities with storm water effluent guidelines include cement manufacturing [40 CFR 411], feedlots [40 CFR 412], fertilizer manufacturing [40 CFR 418], petroleum refining [40 CFR 419], phosphate manufacturing [40 CFR 422], steam electric power generating [40 CFR 423], coal mining [40 CFR 434], ore mining and dressing [40 CFR 436], and asphalt emulsion [40 CFR 443].
- All components, chemicals, and pollutants limited by any existing NPDES permits for the facility. This may indicate the presence of any cross-contamination of storm water discharges by process water, or the presence of residual contamination which makes its way into the storm water discharge.
- Any other pollutant which may be present as a result of existing or previous operations at the facility. EPA Form 2F prescribes a list of toxic and hazardous substances which must be considered, but other pollutants must also be considered if applicable.

Laboratory testing may be done in-house or by a contract laboratory. Use of an EPA-certified laboratory is not required. However, all tests must be performed using the analytical protocols specified in 40 CFR

136. The analysis of one sample for conventional pollutants will cost about $200. The costs may total more than $1000 for a sample with several other pollutants, although volume discounts are often available.

Individual permit conditions and/or state and municipal requirements may involve additional laboratory tests. Also, some states require the use of state-certified laboratories.

Small businesses (with gross total annual sales of less than $100,000 per year for the past 3 years, in second-quarter 1980 dollars) and operators of small coal mines (with a probable annual production of less than 100,000 tons) are exempt from the reporting requirements for organic and toxic pollutants listed in table 2F-3. See the instructions for EPA Form 2F for further details.

Flow-weighted composite samples

In addition to a grab sample, the second type of sample required for each individual permit application is a *flow-weighted composite sample,* which is a combination of individual "aliquots" (samples) at least 15 min apart, or a minimum of 3 aliquots per hour for the entire discharge or for the first 3 h of discharge. Ideally, the aliquots are obtained at constant time intervals 20 min apart. The flow rate is measured at the time each aliquot is obtained.

The samples are combined in proportion to the flow rate at the time each sample was obtained. Along with the total volume of discharge, this provides a method of estimating the average pollutant concentration and total pollutant loading in the discharge. After the storm event ends, the maximum flow rate is identified, and the full volume [generally 1000 milliliters (mL)] of the corresponding aliquot is added to the composite sample container. For aliquots obtained at other flow rates, a lesser volume is added to the composite sample container. The volume in each case is determined by the following equation:

$$V = V_{\max}\left(\frac{Q}{Q_{\max}}\right)$$

where V = volume of individual aliquot added to composite sample container

Q = flow rate measured at time individual aliquot was obtained

$Q_{\max}$ = maximum flow rate observed during storm event

$V_{\max}$ = volume of aliquot obtained at time of maximum flow rate

Table 9.2 illustrates such a computation for a set of 9 aliquots. Note that the peak flow rate is 25 cubic feet per minute (ft^3/min), which occurred at the time the third aliquot was obtained. Therefore, the full volume of this aliquot (1000 mL) is added to the composite sample. The volume of each of the other aliquots added to the composite sample is

TABLE 9.2 Typical Example of Flow-Weighted Composite Sample

Sample	Flow rate, ft^3/min	Weight	Volume, mL	Adjusted volume, mL
1	10	10/25	400	80
2	20	20/25	800	160
3	25	25/25	1000	200
4	20	20/25	800	160
5	10	10/25	400	80
6	15	15/25	600	120
7	10	10/25	400	80
8	5	5/25	200	40
9	10	10/25	400	80
	Total volume		5000	1000

proportional to the flow rate measured at the time each aliquot was obtained. The Adjusted Volume column is the volume required to produce a total sample volume of 1000 mL. Figure 9.4 illustrates the relationship between the measured flow rate and computed sample volume for a flow-weighted composite sample.

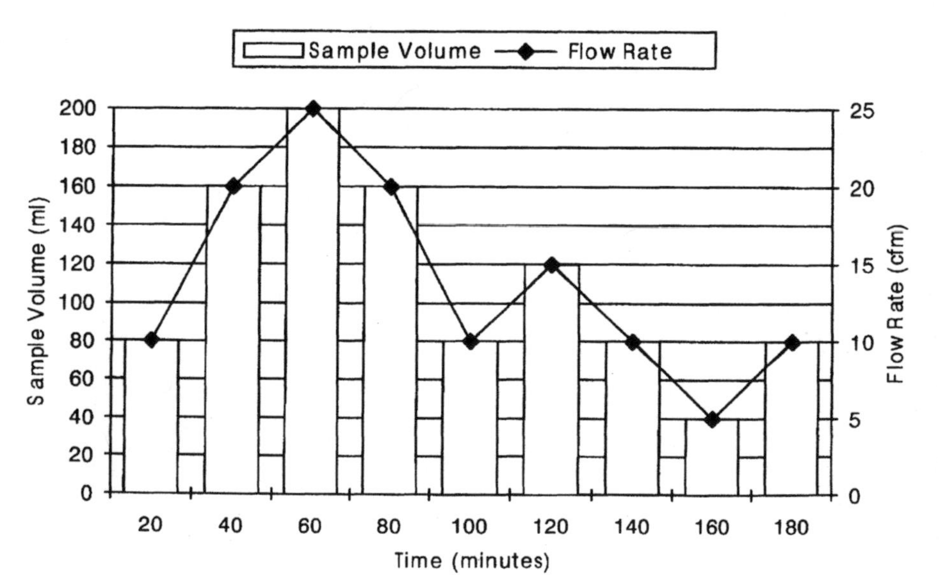

Figure 9.4 Flow-weighted composite sample.

Laboratory Analysis of Composite Samples

The analysis requirements for flow-weighted composite samples are almost identical to those for grab samples, except that there is no requirement for field measurement of temperature or pH and there is no requirement for laboratory analysis of oil and grease. Note that the EPA multisector general permit specifies minimum analysis requirements for certain types of facilities, as described later in this chapter. In addition, individual permits will specify the analysis requirements for the facility.

Reporting Requirements

Discharge monitoring reports and other information may be submitted to the appropriate EPA regional office or state agency, as specified by the permit for the discharge. In addition, copies must be provided to operators of municipal separate storm sewer systems (MS4s) if the facility discharges into a medium-size or large MS4, as defined in Chap. 1.

EPA Form 2F is used to report the results of storm water sampling in nonauthorized states. For all samples, the following items of information should be reported:

- Outfall locations
- Improvements
- Site drainage map
- Narrative description of pollutant sources
- Non–storm water discharge certification
- Significant leaks or spills
- Discharge information
- Biological toxicity testing data
- Contract analysis information
- Certification

At each outfall, the following data must be collected:

- Date and duration of storm event
- Time since last measurable event (hours)
- Flow measurements or estimates
- Total runoff volume for event
- Amount of precipitation (inches)
- Laboratory test results

Reducing Sampling Requirements

Because of the expense and time required to obtain and analyze samples and report the results, it is worthwhile to consider measures which may reduce the sampling requirements. Several such measures are available:

1. *Provide annual certification.* In lieu of sampling a particular outfall, the EPA multisector general permit specifies that the discharger may prepare an annual certification, under penalty of law, that material handling equipment or activities, raw materials, intermediate products, final products, waste materials, by-products, industrial machinery or operations, significant materials from past industrial activity, or airport deicing activities are not exposed to storm water within the drainage area of the outfall. This certification should be submitted to the EPA annually and should be included in the SWPPP. (However, the permittee must still comply with other applicable permit requirements which do not address sampling pollutants in discharges.) The EPA intends that this provision act as an incentive for permittees to eliminate exposure of potential pollutant sources to storm water, and the EPA will reward facilities that have already done so.
2. *Identify substantially identical outfalls.* The EPA multisector general permit allows sampling requirements to be waived where sampling results are available for another "substantially identical" discharge. In determining whether two or more discharges are substantially identical, the industrial activities, exposed materials, material management practices, and flow rates must all be substantially identical for each discharge. The EPA sampling guidance document (Environmental Protection Agency, 1992) describes an evaluation procedure which should be used to determine whether two or more outfalls are substantially identical. This procedure should be used, and this information should be submitted with any discharge monitoring report, and included in the SWPPP. (See Chap. 7.)
3. *Sample from a holding pond.* The EPA multisector general permit allows a single grab sample from the holding pond to be substituted for a flow-weighted or time-weighted composite sample, if the runoff from the site passes entirely through a holding pond with a retention time of greater than 24 h (i.e., if the pond volume is greater than the total storm water inflow for the first 24 h of the storm). The grab sample is obtained from the discharge of the holding pond at the end of the runoff event (after essentially all runoff has entered the holding pond). The pond volume averages the pollutant concentrations from all inflows, producing practically the same results as would be obtained by taking a flow-weighted composite sample of the discharge. Note that the holding time is computed for the actual storm event, and not for a particular "design" storm event such as the 2-year, 10-year, or 100-year storm.

4. *Combine outfalls.* It may be possible to construct a temporary or permanent diversion system which diverts flow from one outfall to combine with another. This eliminates an outfall which would otherwise require sampling.

Sampling Equipment and Laboratories

It is advisable to coordinate the sampling effort in advance with the laboratory which will perform the analyses. The laboratory will generally be able to provide decontaminated sample containers with a shipping container and other necessary supplies. The following equipment is recommended:

- *Weather radio.* The National Weather Service maintains a network of radio stations which provide broadcasts of weather forecasts and warnings. Special weather radio receivers are available at low cost at most consumer electronics stores. These receivers are tuned to the correct frequency to receive the National Weather Service broadcasts.
- *Sample containers.* Plastic sample containers are safer and more convenient, but glass containers may be required for grab samples, because they must be tested for oil and grease. Sample containers should be prelabeled and numbered, so as to avoid any confusion during or after the storm event.
- *Sample scoop.* A wide, flat plastic scoop is used to obtain storm water samples when the depth of flow is not sufficient to fill sample containers directly.
- *Sample logbook with waterproof markers.* This book is used to record sample times, container numbers, and flow rates occurring at each sample time. A survey field book (available at survey supply stores and most engineering reprographics suppliers) serves as a good sample logbook.
- *Ice chest with ice* (*or refrigerator*). This is used to lower the temperature of samples to about 4°C until they reach the laboratory. All storm water samples must be stabilized so that the chemical composition does not change before the laboratory tests can be performed. Generally, thermal stabilization (refrigeration) is best. However, in some circumstances, when the samples will not reach the testing laboratory within 24 h, chemical stabilization may have to be used. If this may be the case, check with the laboratory for instructions.
- *Thermometer.* It is used to perform field tests on grab samples. These items are readily available from any scientific supply store. A good-quality digital electronic thermometer is recommended.

- *A pH meter.* This is also used for field tests on grab samples and is available from a scientific supply store. A good-quality electronic meter with glass electrodes is required to meet EPA test requirements [40 CFR 136.3] of accuracy to within ± 0.1 pH unit. Pocket-size battery-operated digital pH meters are available for less than $100. For best results, however, a self-calibrating digital pH meter is recommended. These are available for about $500. A pH "strip kit" (pH paper) is not acceptable for pH measurements because of the lack of precision. Other potential problems with pH paper are that it is judged subjectively by comparison to a color chart; is affected by humidity, age, and method of storage; and may be affected by turbidity or suspended solids. In spite of these difficulties, however, the use of a strip kit is recommended as a field check of electronic meter readings. Suppliers of strip kits and pH meters can also supply buffered solutions that may be used to confirm the accuracy of pH meters and test procedures.
- *Container of distilled water.* This is used to clean the pH meter, temperature probe, and sample scoop after each use.
- *Watch or stopwatch.* This is needed to record sample times and/or sample intervals.
- *Rain gear.* This is used to protect employees during field work. Flashlights and other protective equipment may also be necessary.

The chain of custody from the person obtaining the samples to the laboratory should be clearly recorded in writing.

Manual and Automatic Sampling Methods

An automatic continuous sampler can be used as an alternative to a series of manually collected samples. Manual and automatic sampling methods each have advantages and disadvantages. Manual methods have the advantage of being appropriate for all types of pollutants and requiring minimal equipment. However, manual grab samples are labor-intensive, may involve exposing personnel to hazardous situations, and are subject to human error. Manual flow-weighted composites require that flow measurements be taken during sampling. In addition, it may not be possible to get the sampling personnel and equipment to the outfall location(s) within the required 30 min.

Automatic sampling methods minimize labor requirements, reduce the risk of human error, reduce personnel exposure to hazardous situations, and can be triggered remotely or initiated according to present conditions. The equipment may be battery-operated for remote operation and may include refrigeration systems to stabilize storm water samples.

Automatic sampling methods may eliminate the need for manual compositing of aliquots for flow-weighted composite samples. However, automatic sampling methods may not be representative for oil and grease parameters, and cannot be used for analysis of volatile organic carbons. The samples may not be appropriate for pH and temperature tests, or for other parameters with short holding times (such as fecal streptococcus, fecal coliform, and chlorine). There is also the potential for cross-contamination of the aliquot if the tubing and sample containers are not properly washed. This is especially true for certain pollutants such as pH, temperature, cyanide, total phenols, residual chlorine, oil and grease, fecal coliform, and fecal streptococcus.

Automatic samplers can also be costly ($7000 or more each for battery-powered units). The automated equipment requires installation, maintenance, and operator training, and an accurate flow measurement device must be linked to the automatic sampler.

Common Problems with Storm Water Sampling

Storm water sampling is considerably more difficult than the sampling of process water discharges, because of the lack of control over sampling times and conditions. Rainfall events do not follow a predetermined schedule. They often occur at night, on weekends and holidays, and with little advance notice.

The EPA multisector general permit has an "adverse climatic conditions" provision allowing a discharger to temporarily waive sampling and to submit a description of why samples could not be collected. *Adverse climatic conditions* include insurmountable weather conditions that create dangerous conditions for personnel (such as local flooding, flash floods, high winds, hurricane, tornadoes, and electrical storms) or otherwise make the collection of a sample impracticable (drought, extended frozen conditions, and so on). These events tend to be isolated incidents and should not be used as an excuse for not conducting sampling under more favorable conditions associated with other storm events. This waiver may be used only once during each 2-year period.

The sampling waiver is not intended to apply to difficult logistical conditions, such as remote facilities with few employees, or discharge locations which are difficult to access. "Ordinary" difficulties of the following sorts do not relieve the discharger from the responsibility to collect storm water samples:

- *Lack of rainfall.* In arid regions or regions experiencing a period of unusually low rainfall, it is advisable to submit whatever samples are available, even if fewer than 9 aliquots have been obtained.

Document the conditions under which the available samples have been obtained. In the case of difficult climatic conditions, the EPA will accept sampling data from any storm event of at least 0.1-in total rainfall.

- *Too much rainfall.* Some regions may experience such frequent rainfall events that there is never a period of 72 h or more between storms. Again, it is advisable to submit the best samples available, even if the 72-h criterion cannot be met. Document the conditions under which the samples were obtained.
- *Short rainfall events.* If rainfall begins and then ends before at least 3 h of rainfall has occurred, then it is advisable to go ahead and perform the laboratory tests on whatever samples were obtained, provided that at least 3 aliquots are available. Be sure to document the conditions under which the samples were obtained when you submit the test results.
- *Start-and-stop rainfall.* If rainfall ends before the 3-h duration and then begins again later, it is advisable to continue sampling, provided that no more than a "reasonable" period has passed before the rainfall resumes. What constitutes a reasonable period is left to the judgment of the person obtaining the samples. However, it should not be more than a few hours. As with the other conditions noted above, this condition should be documented when the test results are submitted.
- *Inaccessible outfalls.* Some outfalls are inaccessible, especially during storm events. For example, a storm sewer outfall may be submerged during storm conditions. In such cases, it may be necessary to obtain samples at an upstream location, such as the next manhole or inspection point upstream. (*Note:* Before you sample within manholes, read the safety warnings later in this chapter.) If another manhole or inspection point is not available, tap into the pipe or sample at several locations to best represent the total site runoff.
- *Multiple outfalls.* If one person is responsible for sampling at multiple outfall points or multiple sites, it may not be possible to obtain a first-flush grab sample for each outfall for a single storm event. In such cases, it may be necessary to obtain the first-flush sample for different storm events at each outfall or site. For most parameters, automatic samplers may be used to collect samples within the first 30 min, triggered by the amount of rainfall, depth of flow, flow volume, or time.
- *Numerous small point discharges.* If it is not feasible to sample all discharges, then impound the discharge channel or combine the discharges by building a weir or digging a ditch to collect discharges at a

low point for sampling. This artificial collection point should be lined with plastic to prevent infiltration and/or high levels of sediment.

- *Sampling in manholes.* Sample in manholes only when necessary. Any sampling performed in confined spaces should meet the "SET" test: *S*upported, *E*quipped, and *T*rained. Anyone sampling in a confined space should have a support person monitoring performance and safety. The person doing the sampling should be properly equipped for the task and should be trained in confined-space entry.
- *Run-on from other property.* If possible, estimate the volume of off-site run-on contributions and off-site run-on sources of pollutants to perform a mass-balance calculation. The mass-balance calculation simply provides an estimate of the total volume of run-on times the concentration of various pollutants in the run-on, compared with the total volume and pollutant concentrations of site discharges. Include this information in the permit application. If this estimation is not possible, provide a narrative discussion of the upstream site (e.g., the type of facilities located on the upstream site and the types of pollutants that may be present in discharges).
- *Commingled discharges.* If process water is known to be mixed with the storm water at a particular outfall, attempt to sample the storm water discharge before it mixes with the non–storm water discharge. If this is impossible, perform dry-weather sampling of the process water discharge and provide both sets of data to the permitting authority. This will give an indication of the contribution of pollutants from each source.

Safety Issues in Storm Water Sampling

By definition, storm water sampling is performed under adverse conditions. These conditions can be merely uncomfortable, or they may be hazardous. Sampling in confined spaces, such as in storm sewer manholes, can be hazardous because of vapors which may collect in such locations. Physical hazards such as lightning and slick surfaces can create significant danger. In addition, there are biological hazards (sometimes known as the "three R's"—roaches, rodents, and reptiles). Finally, sampling personnel may be subjected to crime while outside or in parked vehicles during the 3-h sampling period required for a composite sample. As noted above, the EPA multisector general permit has a temporary sampling waiver for unusually dangerous conditions.

Occupational Safety and Health Administration (OSHA) safety rules should be followed, along with any applicable state and local regulations.

In addition to personal safety and health issues, there are property hazards that must be considered when designing a storm water sampling program. Expensive equipment such as automatic samplers may be subject to theft and vandalism. Appropriate measures should be taken to safeguard such equipment.

Estimating Flow Rates

It is necessary to be able to estimate the flow rate in order to assemble a flow-weighted composite sample and to fulfill all reporting requirements. The most important thing to understand about flow measurements is that *highly precise primary flow measurement devices such as flumes and weirs are not required by the EPA* (*although they are acceptable if used*). The EPA has clearly and specifically stated that most dischargers should not have to install expensive flow measurement equipment.

Acceptable flow rate estimates can generally be prepared by secondary methods, using simple principles of hydrology and hydraulics. *Hydrology* is the study of runoff resulting from precipitation. *Hydraulics* is the study of fluid (mainly water) flow.

Estimating flow rates by hydrologic methods

The most common method for computing the peak rate of runoff from a small drainage area is the *rational method:*

$$Q = CiA$$

where Q = peak flow rate, ft^3/s, approximate
C = runoff coefficient of drainage area
= fraction of rainfall which appears as runoff at discharge point
i = average rainfall intensity during time of concentration, in/h
A = drainage area, acres

Chapter 10 provides further information on methods of estimating the drainage area and runoff coefficient. The rainfall intensity can be estimated during the actual storm event by reading a rainfall gauge periodically.

Estimating flow rates by hydraulic methods

For all types of conveyances, the flow rate equals the product of the average flow velocity and the cross-sectional area of flow. The cross-sectional area of flow in a channel or culvert may be measured. Then the flow velocity can be measured or computed in one of two ways:

1. *Direct measurement.* One way of measuring flow velocity in a channel or culvert is by the *float method,* in which an object such as an orange is dropped into flowing water. A stopwatch is used to measure the elapsed time required for the orange to move a known distance. Thus the average flow velocity can be easily computed. An orange makes a good "float" for this purpose because it floats very low in the water, so that its velocity represents the flow velocity some distance below the surface, where most of the flow is occurring.

2. *Hydraulic analysis.* The principles of conservation of energy or momentum may be used to relate flow velocity, flow depth, and the characteristics of the conveyance. For open channels, the flow capacity is generally computed by using the *Manning equation:*

$$Q = \frac{1.486}{n} A \left(\frac{A}{\text{WP}}\right)^{2/3} \sqrt{S}$$

where Q = flow rate, ft^3/s
n = Manning roughness coefficient (depends upon condition of channel)
A = cross-sectional area of flow, ft^2
WP = wetted perimeter of flow in channel, ft
S = slope of channel, ft/ft

Existing hydraulic structures such as culverts and inlets can also be used to compute flow rates. Charts or equations are generally available for different types of culverts and inlets, relating the depth of flow to the flow rate. Culvert flow is generally analyzed by using charts or equations available from the Federal Highway Administration or various state and local transportation agencies.

Special devices such as flumes and weirs are available to facilitate flow rate measurements. These devices are convenient because they have simple and unique relationships between flow rate and depth of flow. Therefore, it is only necessary to measure the depth of flow, and then the flow rate may be determined from a simple equation or chart.

Even more complex devices are available. Floats may be used to measure the depth of flow. Used in conjunction with a weir or flume, a float makes the use of such a device even more convenient. Ultrasonic transducers are also available to measure the flow depth and velocity.

Requirements for EPA Multisector General Permit Compliance

The multisector general permit contains three different types of monitoring provisions:

1. *Quarterly visual examinations.* These are the least burdensome type of monitoring requirement under the permit. Almost all industrial activities are required to perform visual examinations of their storm water discharges when they are occurring on a quarterly basis.
2. *Analytical monitoring.* This involves laboratory chemical analyses of samples collected by the permittee. The results of the analytical monitoring are quantitative concentration values for different pollutants, which can be easily compared to the results from other sampling events, other facilities, or national benchmarks.
3. *Annual compliance monitoring.* This ensures that discharges subject to numerical effluent limitations under the storm water effluent limitation guidelines are in compliance with those limitations.

The monitoring requirements vary among industry subsectors. Monitoring is required primarily to provide facilities with a means for assessing their storm water contamination and evaluating the performance of their storm water pollution prevention plans.

Representative storm event

Grab samples may be used for all monitoring, unless otherwise stated. All such samples shall be collected from the discharge resulting from a storm event that is greater than 0.1 in in magnitude and that occurs at least 72 h from the previously measurable (greater than 0.1-in rainfall) storm event. The required 72-h storm event interval may be waived by the permittee where the preceding measurable storm event did not result in a measurable discharge from the facility. The 72-h requirement may also be waived by the permittee where the permittee documents that less than a 72-h interval is representative for local storm events during the season when sampling is being conducted.

The grab sample must be taken during the first 30 min of the discharge. If the collection of a grab sample during the first 30 min is impracticable, a grab sample can be taken during the first hour of the discharge, and the discharger must submit with the monitoring report a description of why a grab sample during the first 30 min was impracticable. A minimum of one grab is required. Where the discharge to be sampled contains both storm water and nonstorm water, the facility shall sample the storm water component of the discharge at a point upstream of the location where the nonstorm water mixes with the storm water, if practicable.

Compliance monitoring

In addition to the analytical monitoring requirements for certain sectors, the EPA multisector general permit contains monitoring requirements for discharges which are subject to effluent limitations. These discharges must be sampled annually and tested for the parameters which are limited by the permit. Discharges subject to compliance monitoring include coal pile runoff, contaminated runoff from phosphate fertilizer manufacturing facilities, runoff from asphalt paving and roofing emulsion production areas, material storage pile runoff from cement manufacturing facilities, and mine dewatering discharges from crushed stone, construction sand and gravel, and industrial sand mines located in Texas, Louisiana, Oklahoma, New Mexico, and Arizona. All samples are to be grabs taken within the first 30 min of discharge where practicable, but in no case later than the first hour of discharge. Where practicable, the samples shall be taken from the discharges subject to the numeric effluent limitations prior to mixing with other discharges.

Permittees subject to compliance monitoring requirements are required to submit all compliance monitoring results annually on October 28th. Compliance monitoring results must be submitted on signed DMR forms. For each outfall, one discharge monitoring report form must be submitted for each storm event sampled. Permittees must retain all records for a minimum of 3 years from the date of the sampling, examination, or other activity that generated the data.

Quarterly visual examination of storm water

Almost all facilities covered by the multisector general permit are required to perform visual examination of their storm water discharges on a quarterly basis. The visual examination is to be performed by the operator, who must examine a sample of the discharge and note the color, odor, clarity, floating solids, settled solids, suspended solids, foam, oil sheen, or other obvious indicators of storm water pollution. No analytical tests are required to be performed on these samples.

EPA recommends that the quarterly visual examination be conducted at different times from the analytical or compliance monitoring, but this is not required. The quarterly visual examination is intended to be performed by members of the pollution prevention team. To be most effective, the personnel conducting the quarterly visual examination should be fully knowledgeable about the storm water pollution prevention plan, the sources of contaminants on the site, the industrial activities conducted exposed to storm water, and the day-to-day operations that may cause unexpected pollutant releases. Whenever practicable, the same individual should carry out the collection and examination of discharges throughout the life of the permit, to ensure the greatest degree of consistency possible in recording observations.

The examination shall be made during daylight hours unless there is insufficient rainfall or snowmelt to produce a runoff event. In any event, the examination must be conducted in a well-lighted area. Examinations shall be conducted in each of the following periods for the purposes of inspecting storm water quality associated with storm water runoff and snowmelt: January through March, April through June, July through September, October through December. Reports of the visual examination include the examination date and time, examination personnel, visual quality of the storm water discharge, and probable sources of any observed storm water contamination. The visual examination reports must be maintained on-site with the pollution prevention plan.

The quarterly visual examination is not required if there is insufficient rainfall or snowmelt to run off or if hazardous conditions prevent sampling. Inactive, unstaffed facilities can exercise a waiver of the requirement to conduct quarterly visual examination, as described later in this chapter. The representative discharge provision, also described later in this chapter, may be used to reduce the requirements for visual examinations.

A facility is not required to submit the visual examination results unless required to do so by the EPA. Visual examination reports must be maintained on-site with the pollution prevention plan. Records of visual examination of storm water discharge need not be lengthy. They can be simple typed or hand-written reports using forms or tables which may be created for the facility. The report should include the date and time of examination; the identity of the person(s) performing the examination; the nature of the discharge (i.e., runoff or snowmelt); the visual quality of the storm water discharge (including observations of color, odor, clarity, floating solids, settled solids, suspended solids, foam, oil sheen, and other obvious indicators of storm water pollution); and the probable sources of any observed storm water contamination.

When conducting a storm water visual examination, the pollution prevention team member(s) should attempt to relate the results of the examination to potential sources of storm water contamination on the site. For example, if the visual examination reveals an oil sheen, the facility personnel (preferably members of the pollution prevention team) should conduct an inspection of the area of the site draining to the examined discharge, to look for obvious sources of spilled oil, leaks, etc. If a source can be located, then this information allows the facility operator to immediately conduct a cleanup of the pollutant source and/or to design a change to the pollution prevention plan to eliminate or minimize the contaminant source in the future.

If the visual examination results in an observation of floating solids, personnel should carefully examine the solids to see if they are raw materials, waste materials, or other known products stored or used at the site. If an unusual color or odor is sensed, personnel should attempt to compare the color or odor to the colors or odors of known chemicals and of other materials used at the facility. If the examination reveals a large amount of settled solids, personnel may check for unpaved, unstabilized areas or areas of erosion. If the examination results in a cloudy sample that is very slow to settle out, personnel should evaluate the site draining to the discharge point for fine particulate material, such as dust, ash, or other pulverized, ground, or powdered chemicals. Other observations may prompt other investigations.

If best management practices (BMPs) are performing ineffectively, corrective action must be implemented. A set of tracking or follow-up procedures must be used to ensure that appropriate actions are taken in response to the examinations.

The multisector general permit allows the requirement for a quarterly visual examination to be waived for facilities that are both inactive and unstaffed. This waiver is intended to apply to these types of facilities only when the ability to conduct visual examinations would be severely hindered and would result in an inability to meet the time and representative rainfall sampling specifications. This sampling waiver is not intended to apply to remote facilities that are active and staffed, or to typical difficult logistical conditions. When a discharger is unable to perform quarterly visual examinations as specified in the permit, the discharger shall maintain on-site with the pollution prevention plan a certification stating that the facility is unstaffed and inactive and that the ability to perform visual examinations within the specifications is not possible. Permittees are not required to obtain advance approval for visual examination waivers.

Analytical monitoring requirements

A number of facilities are required to perform analytical monitoring of their discharges by the multisector general permit during the second year of permit coverage.

Analytical monitoring is required only for the industry sectors or subsectors which are determined to have a high potential to discharge a pollutant at concentrations of concern. To identify these industries, the EPA analyzed the sampling data submitted by the group application members. The data were divided by industry sector or subsector and statistically analyzed. The median concentration value and other statistics were calculated for each pollutant within each subsector. The median value was compared to a pollutant benchmark value (typically derived from the national water quality criteria).

A *benchmark* is the concentration at which a storm water discharge could potentially impair, or contribute to impairing, water quality or affect human health from ingestion of water or fish. If a facility is discharging pollutants at levels below the established benchmarks, then the facility represents little potential for water quality concern, and would not warrant further analytical monitoring. Therefore, the benchmarks help to determine whether a facility's storm water pollution prevention measures are being successfully implemented. The benchmark concentrations are not effluent limitations and should not be interpreted or adopted as such. These values are merely levels which the EPA has used to determine whether a storm water discharge from any given facility merits further monitoring to ensure that the facility has been successful in implementing a storm water pollution prevention plan. As such, these levels represent a target concentration for a facility to achieve through implementation of pollution prevention measures at the facility. Tables 9.3 through 9.6 list the parameter benchmark values for various types of pollutants.[1]

Benchmark concentrations were determined based upon a number of existing standards or other sources. The EPA has also sought to develop values which can realistically be measured and achieved by industrial facilities. The EPA selected the median concentration from the National Urban Runoff Program as the benchmark for total suspended solids and for nitrate plus nitrite as nitrogen. EPA selected the storm water effluent limitation guideline for petroleum refining facilities as the benchmark for oil and grease. The concentration value for phosphorus was designed to prevent eutrophication of freshwater bodies from storm water runoff.

Where the discharge to be sampled contains both storm water and nonstorm water, the facility shall sample the storm water component of the discharge at a point upstream of the location where the nonstorm water mixes with the storm water, if practicable.

Analytical monitoring is not required during the first, third, and fifth years of the permit. Analytical monitoring is required to be performed by the permittees on a quarterly basis during the second year of permit coverage (after the facility has implemented the storm water pollution prevention plan). Monitoring must be conducted for the same storm water discharge outfall in each sampling period. Where a given storm water discharge is addressed by more than one sector's or subsector's monitoring requirements, the monitoring requirements for the applicable sec-

---

[1] These assumptions are made for all values in all tables: Receiving water temperature of 20°C, receiving water pH of 7.8, receiving water hardness $CaCO_3$, of 100 mg/L, receiving water salinity of 20 grams per kilogram (g/kg), acute-to-chronic ratio (ACR) of 10.

TABLE 9.3 Parameter Benchmark Values for Conventional Pollutants

Parameter	Benchmark level	Source
5-Day biochemical oxygen demand	30 mg/L	Secondary treatment regulations [40 CFR 133]
Chemical oxygen demand	120 mg/L	Factor of 4 times BOD, concentration—North Carolina benchmark
Oil and grease	15 mg/L	Median concentration of Storm Water Effluent Limitation Guideline [40 CFR Part 419]
Total suspended solids	100 mg/L	National Urban Runoff Program (NURP) median concentration
Nitrate + nitrite nitrogen	0.68 mg/L	National Urban Runoff Program median concentration
Total phosphorus	2.0 mg/L	North Carolina storm water benchmark derived from North Carolina water quality standards
pH	6.0–9.0 standard units	Secondary treatment regulations [40 CFR 133]

tor's or subsector's activities are cumulative. Therefore, if a particular discharge fits under more than one set of monitoring requirements, the facility must comply with all sets of sampling requirements. Monitoring requirements must be evaluated on an outfall-by-outfall basis.

Analytical monitoring is designed to provide some feedback for a facility operator to assess the effectiveness of the pollution prevention plan. An incentive is also built in to stimulate the implementation of more effective pollution prevention measures. At the end of the second year of permit coverage, the average concentration for each required monitoring parameter must be computed, using all samples collected. If the average concentration for a pollutant parameter is less than or equal to the benchmark value, then analytical monitoring is not required for that pollutant during the fourth year of the permit. If, however, the average concentration for a pollutant is greater than the benchmark value, then quarterly monitoring is required for that pollutant during the fourth year of permit coverage. The exclusion from analytical monitoring in the fourth year of the permit is conditional on the facility's maintaining industrial operations and BMPs that will ensure a quality of storm water discharges consistent with the average concentrations recorded during the second year of the permit.

TABLE 9.4 Parameter Benchmark Values for Metals

Parameter	Benchmark level, mg/L	Source
Cadmium, total	0.0159	3.18 × highest method detection limit (MDL)
Copper, total	0.0636	3.18 × highest MDL
Aluminum, total (pH 6.5–9)	0.75	"EPA Recommended Ambient Water Quality Criteria," acute aquatic life freshwater
Lead, total	0.0816	"EPA Recommended Ambient Water Quality Criteria," acute aquatic life freshwater
Mercury, total	0.0024	"EPA Recommended Ambient Water Quality Criteria," acute aquatic life freshwater
Nickel, total	1.417	"EPA Recommended Ambient Water Quality Criteria," acute aquatic life freshwater
Zinc, total	0.117	"EPA Recommended Ambient Water Quality Criteria," acute aquatic life freshwater
Iron, total	1.0	"EPA Recommended Ambient Water Quality Criteria," chronic aquatic life freshwater
Antimony, total	0.636	3.18 × highest MDL
Selenium, total	0.2385	3.18 × highest MDL *Note:* Limit established for oil and gas exploration and production facilities only
Silver, total	0.0318	3.18 × highest MDL
Manganese	1.0	Colorado—chronic aquatic life freshwater, water quality criteria

All analytical monitoring results must be submitted to the appropriate EPA regional office within 3 months of the conclusion of the second and fourth years of coverage of the permit. For each outfall, one discharge monitoring report form must be submitted per storm event sampled. All records must be retained for a minimum of 3 years from the date of the sampling, examination, or other activity that generated the data.

Deferral for adverse weather conditions. The permit allows for temporary waivers from sampling based on adverse climatic conditions. This temporary sampling waiver is intended to apply only to insurmountable weather conditions such as drought or dangerous conditions such

TABLE 9.5 Parameter Benchmark Values for Carcinogens

Parameter	Benchmark level, mg/L	Source
Arsenic, total	0.16854	3.18 × highest method detection limit (MDL)
PCB-1016	0.000127	3.18 × Highest MDL
PCB-1232	0.000318	3.18 × Highest MDL
PCB-1248	0.002544	3.18 × Highest MDL
PCB-1260	0.000477	3.18 × Highest MDL
PCB-1221	0.10	Laboratory-derived minimum level (ML)
PCB-1242	0.00020	Laboratory-derived ML
PCB-1254	0.10	Laboratory-derived ML
Pyrene polynuclear aromatic hydrocarbon	0.01	Laboratory-derived ML
Trichloroethylene	0.0027	"EPA Recommended Ambient Water Quality Criteria," human health criteria for consumption of water and organisms
Acrylonitrile	7.55	"EPA Recommended Ambient Water Quality Criteria," lowest observed effect level (LOEL) acute freshwater
Beryllium, total	0.13	"EPA Recommended Ambient Water Quality Criteria," LOEL acute freshwater

TABLE 9.6 Parameter Benchmark Values for Other Pollutants

Parameter	Benchmark level, mg/L	Source
Ammonia	19	"EPA Recommended Ambient Water Quality Criteria," acute aquatic life freshwater
Chloride	860	"EPA Recommended Ambient Water Quality Criteria," acute aquatic life freshwater
Toluene	10.0	"EPA Recommended Ambient Water Quality Criteria," human health criteria for consumption of water and organisms
Butylbenzyl phthalate	3	"EPA Recommended Ambient Water Quality Criteria," human health criteria for consumption of water and organisms
Ethylbenzene	3.1	"EPA Recommended Ambient Water Quality Criteria," human health criteria for consumption of water and organisms
Fluoranthene	0.042	"EPA Recommended Ambient Water Quality Criteria," human health criteria for consumption of water and organisms
Dimethyl phthalate	1.0	Discharge limitations and compliance data
Phenols, total	1.0	Discharge limitations and compliance data
Benzene	0.01	Laboratory-derived minimum level
Fluoride	1.8	North Carolina storm water benchmark derived from North Carolina water quality standards

as lightning, flash flooding, or hurricanes. These events tend to be isolated incidents and should not be used as an excuse for not conducting sampling under more favorable conditions associated with other storm events.

The sampling waiver is not intended to apply to difficult logistical conditions, such as remote facilities with few employees or discharge locations that are difficult to access. When a discharger is unable to collect samples within a specified sampling period due to adverse climatic conditions, a substitute sample should be collected from a separate qualifying event in the next sampling period, as well as a sample for the routine monitoring required in that period. Both samples should be analyzed separately, and the results should be submitted to EPA. Permittees are not required to obtain advance approval for sampling waivers.

Waiver for inactive and unstaffed facilities. If a facility is inactive and unstaffed, it may be difficult to collect storm water discharge samples when a qualifying event occurs. Therefore, inactive, unstaffed facilities can exercise a waiver of the requirement to conduct analytical monitoring. This waiver is intended to apply to these types of facilities only when the ability to conduct sampling would be severely hindered and would result in an inability to meet the time and representative rainfall sampling specifications. This sampling waiver is not intended to apply to remote facilities that are active and staffed or to typical difficult logistical conditions. Permittees are not required to obtain advance approval for this waiver.

When a discharger is unable to collect samples for analytical or compliance monitoring as specified in the permit, the discharger shall certify in the DMR that the facility is unstaffed and inactive and that the ability to conduct samples within the specifications is not possible.

When a discharger is unable to perform analytical monitoring as specified in the permit, the discharger shall maintain on-site with the pollution prevention plan a certification stating that the facility is unstaffed and inactive and that the ability to perform analytical monitoring within the specifications is not possible.

Waiver for low concentration. The results of the quarterly grab samples collected in the second year of permit coverage must be averaged and compared to the benchmark values for the pollutants. Discharges with an average pollutant concentration of less than the benchmark concentration for that pollutant are not subject to monitoring requirements for the remainder of the permit coverage.

However, facilities with average pollutant concentrations that are higher than the benchmark must review and revise their pollution prevention plan and must monitor their discharges on a quarterly basis during the fourth year of permit coverage. This incentive-based monitoring is intended to provide feedback to the permittee on the effectiveness of the storm water pollution prevention plan.

Waiver for no exposure ("alternative certification"). A facility which is in an industry sector or subsector subject to analytical monitoring requirements can obtain an exemption from the monitoring for any particular pollutant if the facility operator can certify that there is no source of that pollutant which is exposed or is expected to be exposed to storm water.

A discharger is not subject to the monitoring requirements, provided the discharger makes a certification for a given outfall, or on a pollutant-by-pollutant basis in lieu of monitoring, under penalty of law, that material handling equipment or activities, raw materials, intermediate products, final products, waste materials, by-products, industrial machinery or operations, and significant materials from past industrial activity that are located in areas of the facility that are within the drainage area of the outfall are not presently exposed to storm water and will not be exposed to storm water for the certification period. Such certification must be retained in the storm water pollution prevention plan, and submitted to EPA instead of the monitoring reports that would otherwise be required. If the permittee cannot certify for an entire period, she or he must submit the date that exposure was eliminated and any monitoring required up until that date. EPA does not expect facilities to be able to exercise this certification for indicator parameters, such as TSS and BOD.

The permit does not allow facilities with discharges subject to numeric effluent limitations to submit alternative certification in lieu of the compliance monitoring requirements. The permit also does not allow air transportation facilities subject to the analytical monitoring requirements to exercise an alternative certification.

It is permissible to exercise the alternative certification in the fourth year of permit coverage, even if the facility performed sampling in the second year of permit coverage and had pollutants that exceeded the benchmark levels. In fact, EPA encourages dischargers to eliminate the exposure of industrial activities and significant materials to storm water wherever practicable.

Representative discharge. The permit allows permittees to use the substantially identical outfalls to reduce their monitoring burden. This representative discharge provision provides facilities with multiple storm water outfalls, a means for reducing the number of outfalls that must be sampled and analyzed. This may result in a substantial reduction of the resources required in order for a facility to comply with analytical monitoring requirements.

For a facility having two or more outfalls that—based on a consideration of industrial activity, significant materials, and management practices and activities within the area drained by the outfall—the permittee reasonably believes to discharge substantially identical effluents, the permittee may test the effluent of one of such outfalls and report that the quantitative data also apply to the substantially identical outfalls, provided that the permittee includes in the storm water pollution prevention plan a description of the location of the outfalls and explains in detail why the outfalls are expected to discharge substantially identical effluent. In addition, for each outfall that the permittee believes to be representative, an estimate of the size of the drainage area (in square feet) and an estimate of the runoff coefficient of the drainage area [e.g., low (under 40%), medium (40 to 65%) or high (above 65%)] shall be provided in the plan.

Facilities that select and sample a representative discharge are prohibited from changing the selected discharge in future monitoring periods unless the selected discharge ceases to be representative or is eliminated. Permittees do not need EPA approval to claim that discharges are representative, provided they have documented their rationale within the storm water pollution prevention plan. However, the director may determine that the discharges are not representative and require sampling of all nonidentical outfalls.

The representative discharge provision in the permit is available to almost all facilities subject to the analytical monitoring requirements, but is not applicable to compliance monitoring for effluent guideline limit compliance purposes, or to facilities subject to visual examination requirements.

Table 9.7 lists all of the sampling requirements for the EPA multisector general permit.

Reporting requirements

Permittees are required to submit all monitoring results obtained during the second and fourth year of permit coverage within 3 months of the conclusion of each year. For each outfall, one signed DMR form must be submitted to the director per storm event sampled. For facilities conducting monitoring beyond the minimum requirements, an additional signed DMR form must be filed for each analysis. The permittee must include a measurement or estimate of the total precipitation, volume of runoff, and peak flow rate of runoff for each storm event sampled.

TABLE 9.7 Summary of Sampling Requirements for EPA Multisector General Permit

	Visual examination	Analytical monitoring	Compliance monitoring
Frequency during first year of permit coverage	Quarterly	None	Annual
Frequency during second year of permit coverage	Quarterly	Quarterly	Annual
Frequency during third year of permit coverage	Quarterly	None	Annual
Frequency during fourth year of permit coverage	Quarterly	Quarterly (if necessary)	Annual
Frequency during fifth year of permit coverage	Quarterly	None	Annual
Type of sample	First-flush grab sample	First-flush grab sample	First-flush grab sample
Report form	Simple user-designed form	EPA Discharge Monitoring Report (DMR)	EPA Discharge Monitoring Report
Report submittal	Stored with SWPPP	Submitted to EPA within 3 months of end of sampling year	Submitted to EPA on October 28th
Deferral for adverse weather conditions	Yes	Yes	Yes
Waiver for inactive and unstaffed facilities	Yes	Yes	Yes
Waiver for no exposure	Yes	Yes (except airports)	No
Waiver for low concentration	No	Fourth year only	No

Numeric effluent limitations

The EPA multisector general permit contains numeric effluent limitations for phosphate fertilizer manufacturing facilities, asphalt emulsion manufacturers, cement manufacturers, coal pile runoff from steam electric power generating facilities, and sand, gravel, and crushed stone quarries. These limitations are required under EPA's storm water effluent limitation guidelines in the Code of Federal Regulations at 40 CFR Parts 418, 443, 411, 423, and 436.

Coal pile runoff

The EPA multisector general permit establishes effluent limitations of 50 mg/L total suspended solids and a pH range of 6.0 to 9.0 for coal pile runoff. Any untreated overflow from facilities designed, constructed, and operated to treat the volume of coal pile runoff associated with a 10-year, 24-h rainfall event is not subject to the 50 mg/L limitation for total suspended solids. Steam electric generating facilities must comply with these limitations upon submittal of the NOI.

Sector A: Timber products

All facilities, unless inactive and unstaffed, must conduct quarterly visual examinations of storm water discharges. In addition, analytical monitoring is required.

The modified MSGP authorizes nonstorm water discharges resulting from the spray down of lumber and wood products in storage yards (wet decking) provided that no chemicals are used in the spray and no chemicals are applied to the wood during storage. The numerical limits which apply to these nonstorm water discharges are: (1) no debris shall be discharged (debris refers to woody materials such as bark, twigs, branches) that does not pass through a 2.5-cm × 1.0-inch opening, and (2) pH shall range from 6.0 to 9.0.

Subsector 1, SIC Code 2421. *General sawmills and planing mills* must collect quarterly grab samples for the following parameters: chemical oxygen demand (COD), total suspended solids (TSS), and total recoverable zinc during the second and fourth years of permit coverage.

Subsector 2, SIC Code 2491. *Wood preserving facilities* must collect quarterly grab samples for total recoverable arsenic and total recoverable copper during the second and fourth years of permit coverage.

Subsector 3, SIC Code 2411. *Log storage and handling facilities* must collect quarterly grab samples for total suspended solids during the second and fourth years of permit coverage.

Subsector 4. *Mills, wood containers, and other wood products facilities* must collect quarterly grab samples for chemical oxygen demand and TSS during the second and fourth years of permit coverage.

All facilities may exercise the low-concentration waiver, inactive-and-unstaffed waiver, or alternative certification in lieu of analytical monitoring.

Sector B: Paper and allied products manufacturing

All facilities, unless inactive and unstaffed, must conduct quarterly visual examinations of storm water discharges. In addition, paperboard mills must collect quarterly grab samples for COD during the second and fourth years of permit coverage. If the average concentrations of these pollutants exceed the benchmark values discussed earlier in this chapter, then sampling must also be conducted during the fourth year of permit coverage.

All facilities may exercise the low-concentration waiver, inactive-and-unstaffed waiver, or alternative certification in lieu of analytical monitoring.

Sector C: Chemical and allied products manufacturing

All facilities must conduct quarterly visual examinations of storm water discharges unless inactive and unstaffed.

Subsector 1, SIC Code 281X. *Industrial inorganic chemicals facilities* must collect quarterly grab samples for total recoverable aluminum, total recoverable iron, and nitrate + nitrite nitrogen during the second and fourth years of permit coverage.

Subsector 2, SIC Code 282X. *Facilities of plastics materials and synthetic resins, synthetic rubber, cellulosic, and other manmade fibers except glass* must collect quarterly grab samples for total recoverable zinc during the second and fourth years of permit coverage.

Subsector 4, SIC Code 284X. *Facilities using soaps, detergents, and cleaning preparations as well as perfumes, cosmetics, and other toilet preparations* must collect quarterly grab samples for total recoverable zinc and nitrate + nitrite nitrogen during the second and fourth years of permit coverage.

Subsector 7, SIC Code 287X. *Facilities using nitrogenous and phosphatic basic fertilizers, mixed fertilizers, pesticides, and other agricultural chemicals* must collect quarterly grab samples for total recoverable lead, total recoverable iron, total recoverable zinc, phosphorus, and nitrate + nitrite nitrogen during the second and fourth years of permit coverage.

All facilities may exercise the low-concentration waiver, inactive-and-unstaffed waiver, or alternative certification in lieu of analytical monitoring.

The multisector general permit establishes limits on the contaminated storm water at phosphate fertilizer manufacturing facilities. *Contaminated storm water* is any precipitation runoff which during manufacturing or processing comes into incidental contact with any raw materials, intermediate product, finished product, by-products, or waste product. These limits are equivalent to 40 CFR 418. The concentration of total phosphorus in storm water discharges cannot exceed the following effluent limitations:

Daily maximum = 105.0 mg/L

Average of daily values for 30 consecutive days = 35.0 mg/L

The concentration of fluoride in storm water discharges cannot exceed the following effluent limitations:

Daily maximum = 75.0 mg/L

Average of daily values for 30 consecutive days = 25.0 mg/L

Phosphate fertilizer manufacturing facilities must be in compliance with these effluent limitations upon commencement of coverage and for the entire term of this permit. Compliance monitoring is required at least annually for discharges subject to effluent limitations, and no waivers are available. However, if the analytical monitoring methods are consistent with the requirement of the compliance monitoring, these results of the analytical monitoring may be used to satisfy the compliance monitoring requirements.

Sector D: Asphalt paving and roofing materials manufacturers and lubricant manufacturers

All facilities, unless inactive and unstaffed, must conduct quarterly visual examinations of storm water discharges.

Subsector 1, SIC Code 295X. *Asphalt paving and roofing materials manufacturing facilities* must collect quarterly grab samples for total suspended solids during the second and fourth years of permit coverage.

All facilities may exercise the low-concentration waiver, inactive-and-unstaffed waiver, or alternative certification in lieu of analytical monitoring.

The multisector general permit includes limits for storm water discharges from asphalt paving and roofing emulsion facilities. These effluent limitations are in accordance with 40 CFR 443.12 and 40 CFR 443.13, effluent guidelines and standards, paving and roofing materials point source category, asphalt emulsion subcategory. These limitations represent the degree of effluent reduction attainable by the application of best practicable control technology and best available technology.

The concentration of TSS in storm water discharges cannot exceed the following effluent limitations:

Daily maximum = 23 mg/L

Average of daily values for 30 consecutive days = 15 mg/L

The concentration of oil and grease in storm water discharges cannot exceed the following effluent limitations:

Daily maximum = 15 mg/L

Average of daily values for 30 consecutive days = 10 mg/L

The pH must be within the range of 6.0 to 9.0 (standard units).

Facilities must be in compliance with these effluent limitations upon commencement of coverage and for the entire term of this permit. Compliance monitoring is required at least annually for discharges subject to effluent limitations, and no waivers are available. However, if the analytical monitoring methods are consistent with the requirement of the compliance monitoring, these results of the analytical monitoring may be used to satisfy the compliance monitoring requirements.

Sector E: Glass, clay, cement, concrete, and gypsum product manufacturing

All facilities, unless inactive and unstaffed, must conduct quarterly visual examinations of storm water discharges.

Subsector 3, SIC Code 325X. *Structural clay products facilities* must collect quarterly grab samples for total recoverable aluminum during the second and fourth years of permit coverage.

Subsector 4, SIC Code 327X (except 3274). *Facilities of concrete, gypsum, and plaster products (except lime)* must collect quarterly grab samples for total suspended solids and total recoverable iron during the second and fourth years of permit coverage.

All facilities may exercise the low-concentration waiver, inactive-and-unstaffed waiver, or alternative certification in lieu of analytical monitoring.

Storm water discharges from storage areas for materials used or produced at cement manufacturing facilities may not exceed a maximum TSS concentration of 50 mg/L. The pH of the discharges from these areas must be within the range of 6.0 to 9.0. Untreated discharges from the facility which are a result of a storm with a rainfall depth greater than the 10-year, 24-h storm event are not subject to this limitation. These effluent limitations are in accordance with 40 CFR 411.32 and 40 CFR 411.37.

Facilities must be in compliance with these effluent limitations upon commencement of coverage and for the entire term of this permit.

Compliance monitoring is required at least annually for discharges subject to effluent limitations, and no waivers are available. However, if the analytical monitoring methods are consistent with the requirement of the compliance monitoring, these results of the analytical monitoring may be used to satisfy the compliance monitoring requirements.

Sector F: Primary metals

All facilities, unless inactive and unstaffed, must conduct quarterly visual examinations of storm water discharges.

Subsector 1, SIC Code 331X. *Steel works, blast furnaces, and rolling and finishing mills* must collect quarterly grab samples for total recoverable aluminum and total recoverable zinc during the second and fourth years of permit coverage.

Subsector 2, SIC Code 332X. *Iron and steel foundries* must collect quarterly grab samples for total recoverable copper, total recoverable zinc, total recoverable iron, total recoverable aluminum, and total suspended solids during the second and fourth years of permit coverage.

Subsector 5, SIC Code 335X. *Facilities for rolling, drawing, and extruding of nonferrous metals* must collect quarterly grab samples for total recoverable copper and total recoverable zinc during the second and fourth years of permit coverage.

Subsector 6, SIC Code 336X. *Nonferrous foundries (castings)* must collect quarterly grab samples for total recoverable copper and total recoverable zinc during the second and fourth years of permit coverage.

All facilities may exercise the low-concentration waiver, inactive-and-unstaffed waiver, or alternative certification in lieu of analytical monitoring.

Sector G: Metal mining (ore mining and dressing)

All facilities, unless inactive and unstaffed, must conduct quarterly visual examinations of storm water discharges.

Subsector 2, SIC Code 102X. *Active copper ore mining and dressing facilities* must collect quarterly grab samples for chemical oxygen demand, total suspended solids, and nitrate + nitrite nitrogen during the second and fourth years of permit coverage.

All facilities may exercise the low-concentration waiver, inactive-

and-unstaffed waiver, or alternative certification in lieu of analytical monitoring.

Sector H: Coal mines and coal mining–related facilities

All facilities, unless inactive and unstaffed, must conduct quarterly visual examinations of storm water discharges. Visual examinations are not required at inactive areas not under SMCRA bond. Active areas under SMCRA bond that are located in areas with an average annual precipitation greater than 20 in must perform the visual examinations quarterly. Active areas under SMCRA bond with an average annual precipitation less than or equal to 20 in are required to perform visual examinations on a semiannual basis.

Coal mines and coal mining–related facilities must collect quarterly grab samples for TSS, total recoverable aluminum, and total recoverable iron during the second and fourth years of permit coverage.

All facilities may exercise the low-concentration waiver, inactive-and-unstaffed waiver, or alternative certification in lieu of analytical monitoring.

Sector I: Oil and gas extraction

All facilities, unless inactive and unstaffed, must conduct quarterly visual examinations of storm water discharges. No analytical monitoring is required.

Sector J: Mineral mining and dressing

All facilities, unless inactive and unstaffed, must conduct quarterly visual examinations of storm water discharges.

Subsector 1. *Dimension stone, crushed stone, and nonmetallic minerals except fuels mining and processing facilities* must collect quarterly grab samples for TSS during the second and fourth years of permit coverage.

Subsector 2, SIC Code 144X. *Sand and gravel mining and processing facilities* must collect quarterly grab samples for total suspended solids and nitrate + nitrite nitrogen during the second and fourth years of permit coverage.

All facilities may exercise the low-concentration waiver, inactive-and-unstaffed waiver, or alternative certification in lieu of analytical monitoring.

Mine dewatering discharges from construction sand and gravel, and crushed stone mine facilities must be monitored for the presence of pH. Facilities must be in compliance with these effluent limitations upon commencement of coverage and for the entire term of this permit. Compliance monitoring is required at least annually for discharges subject to effluent limitations, and no waivers are available. However, if the analytical monitoring methods are consistent with the requirement of the compliance monitoring, these results of the analytical monitoring may be used to satisfy the compliance monitoring requirements.

Sector K: Hazardous waste treatment storage or disposal facilities

All facilities, unless inactive and unstaffed, must conduct quarterly visual examinations of storm water discharges.

Treatment, storage, and disposal facilities must collect quarterly grab samples for ammonia, total recoverable magnesium, chemical oxygen demand, total recoverable arsenic, total recoverable cadmium, total cyanide, total recoverable lead, total recoverable mercury, total recoverable selenium, and total recoverable silver during the second and fourth years of permit coverage.

All facilities may exercise the low-concentration waiver, inactive-and-unstaffed waiver, or alternative certification in lieu of analytical monitoring.

Sector L: Landfills and land application sites

All facilities, unless inactive and unstaffed, must conduct quarterly visual examinations of storm water discharges.

Landfills and land application sites must collect quarterly grab samples for total recoverable iron and total suspended solids during the second and fourth years of permit coverage. Municipal solid waste landfills closed in accordance with 40 CFR 258.60 are not required to monitor total recoverable iron.

All facilities may exercise the low-concentration waiver, inactive-and-unstaffed waiver, or alternative certification in lieu of analytical monitoring.

Sector M: Automobile salvage yards

All facilities, unless inactive and unstaffed, must conduct quarterly visual examinations of storm water discharges. Automobile salvage yards must collect quarterly grab samples for total recoverable iron, total recoverable aluminum, total recoverable lead, and total suspended solids during the second and fourth years of permit coverage.

All facilities may exercise the low-concentration waiver, inactive-and-unstaffed waiver, or alternative certification in lieu of analytical monitoring.

Sector N: Scrap recycling facilities

All facilities, unless inactive and unstaffed, must conduct quarterly visual examinations of storm water discharges. Scrap and waste material processing and recycling (nonliquid) facilities must collect quarterly grab samples for total recoverable copper, total recoverable aluminum, total recoverable iron, total recoverable lead, total recoverable zinc, chemical oxygen demand, and total suspended solids during the second and fourth years of permit coverage.

All facilities may exercise the low-concentration waiver, inactive-and-unstaffed waiver, or alternative certification in lieu of analytical monitoring.

Sector O: Steam electric generating facilities

All facilities, unless inactive and unstaffed, must conduct quarterly visual examinations of storm water discharges. Steam electric generating facilities must collect quarterly grab samples for total recoverable iron during the second and fourth years of permit coverage.

All facilities may exercise the low-concentration waiver, inactive-and-unstaffed waiver, or alternative certification in lieu of analytical monitoring.

Facilities with coal pile runoff associated with steam electric power generation are to monitor for the presence of total suspended solids and pH at least annually.

Sector P: Land transportation facilities

All facilities, unless inactive and unstaffed, must conduct quarterly visual examinations of storm water discharges. No analytical monitoring is required.

Sector Q: Water transportation facilities

All facilities, unless inactive and unstaffed, must conduct quarterly visual examinations of storm water discharges. Water transportation facilities must collect quarterly grab samples for total recoverable aluminum, total recoverable iron, total recoverable lead, and total recoverable zinc during the second and fourth years of permit coverage.

All facilities may exercise the low-concentration waiver, inactive-and-unstaffed waiver, or alternative certification in lieu of analytical monitoring.

Sector R: Ship and boat building or repairing yards

All facilities, unless inactive and unstaffed, must conduct quarterly visual examinations of storm water discharges. No analytical monitoring is required.

Sector S: Air transportation facilities

All facilities that use more than 100,000 gal of glycol-based deicing/anti-icing chemicals and/or 100 tons or more of urea on an average annual basis must prepare estimates of annual pollutant loadings resulting from discharges of spent deicing/anti-icing chemicals from the facility. The loading estimates must reflect the amounts of deicing/anti-icing chemicals discharged to separate storm sewer systems or surface waters, prior to and after implementation of the facility's storm water pollution prevention plan. These estimates must be reviewed and certified by an environmental professional (engineer, scientist, etc.) with experience in storm water pollution prevention. The environmental professional need not be certified or registered. However, experience with development of storm water pollution prevention plans and with airport operations is essential to the preparation of accurate estimates. The environmental professional, having examined the facility's deicing/anti-icing procedures and proposed control measures described in the storm water pollution prevention plan, must certify that the loading estimates have been accurately prepared.

At a minimum, storm water discharges from airport facilities that use 100,000 gal or more of glycol-based deicing/anti-icing chemicals and/or 100 tons or more of urea on an average basis must be monitored four times during the second year of permit coverage when deicing/anti-icing activities are occurring (from December through February) and from outfalls that receive storm water runoff from those areas. The required monitoring parameters include biochemical oxygen demand (BOD_5), chemical oxygen demand, ammonia, and pH.

The alternative certification provisions discussed in other industry sectors are not applicable to discharges resulting from deicing/anti-icing operations.

A minimum of one grab and one flow-weighted composite sample must be taken from each outfall that collects runoff from areas where deicing/anti-icing activities occur. The required 72-h storm event interval is waived where the preceding measurable storm event did not result in a measurable discharge from the facility. The required 72-h storm event interval may also be waived where the permittee documents that less than a 72-h interval is representative for local storm events during the season when sampling is being conducted. The grab sample is intended to provide information on the maximum expected concentrations of BOD_5, COD, and ammonia as a result of deicing/anti-icing chemicals discharged during the precipitation event. The composite sample is intended to provide a measure of the BOD_5, COD, and ammonia loadings for the entire precipitation event as a result of the discharge of deicing/anti-icing chemicals.

Sector T: Treatment works

All facilities, unless inactive and unstaffed, must conduct quarterly visual examinations of storm water discharges. No analytical or compliance monitoring is required.

Sector U: Food and kindred products

All facilities, unless inactive and unstaffed, must conduct quarterly visual examinations of storm water discharges.

Subsector 4, SIC Code 204X. *Grain mill products facilities* must collect quarterly grab samples for total suspended solids during the second and fourth years of permit coverage.

Subsector 7, SIC Code 207X. *Fats and oils facilities* must collect quarterly grab samples for BOD, chemical oxygen demand, total suspended solids, and nitrate + nitrite nitrogen during the second and fourth years of permit coverage.

All facilities may exercise the low-concentration waiver, inactive-and-unstaffed waiver, or alternative certification in lieu of analytical monitoring.

Sector V: Textile mills, apparel, and other fabric product manufacturing

All facilities, unless inactive and unstaffed, must conduct quarterly visual examinations of storm water discharges. No analytical or compliance monitoring is required.

Sector W: Furniture and fixtures

All facilities, unless inactive and unstaffed, must conduct quarterly visual examinations of storm water discharges. No analytical or compliance monitoring is required.

Sector X: Printing and publishing

All facilities, unless inactive and unstaffed, must conduct quarterly visual examinations of storm water discharges. No analytical or compliance monitoring is required.

Sector Y: Rubber, miscellaneous plastic products, and miscellaneous manufacturing industries

All facilities, unless inactive and unstaffed, must conduct quarterly visual examinations of storm water discharges.

Subsector 1, SIC Codes 301X, 302X, 305X, 306X. *Rubber products manufacturing facilities* must collect quarterly grab samples for total recoverable zinc during the second and fourth years of permit coverage.

All facilities may exercise the low-concentration waiver, inactive-and-unstaffed waiver, or alternative certification in lieu of analytical monitoring.

Sector Z: Leather tanning and finishing

All facilities, unless inactive and unstaffed, must conduct quarterly visual examinations of storm water discharges. No analytical or compliance monitoring is required.

Sector AA: Fabricated metal products

All facilities, unless inactive and unstaffed, must conduct quarterly visual examinations of storm water discharges.

Subsector 1, SIC Codes 3429, 3441, 3442, 3443, 3444, 3451, 3452, 3462, 3471, 3494, 3496, 3499, and 391X. *Fabricated metal products except coating manufacturing facilities* must collect quarterly grab samples for total recoverable iron, total recoverable aluminum, total recoverable zinc, and nitrate + nitrite nitrogen during the second and fourth years of permit coverage.

Subsector 2, SIC Code 3479. *Coating, engraving, and allied services facilities* must collect quarterly grab samples for total recoverable zinc and nitrate + nitrite nitrogen during the second and fourth years of permit coverage.

All facilities may exercise the low-concentration waiver, inactive-and-unstaffed waiver, or alternative certification in lieu of analytical monitoring.

Sector AB: Transportation equipment, industrial, or commercial machinery

All facilities, unless inactive and unstaffed, must conduct quarterly visual examinations of storm water discharges. No analytical or compliance monitoring is required.

Sector AC: Electronic, electrical, photographic, and optical goods

All facilities, unless inactive and unstaffed, must conduct quarterly visual examinations of storm water discharges. No analytical or compliance monitoring is required.

References

Environmental Protection Agency, 1974a. *Development Document for Effluent Limitation Guidelines and New Source Performance Standards for the Cement Manufacturing Point Source Category,* EPA/440/1-79-005-a.

———, 1974b. *Development Document for Effluent Limitations Guidelines and New Source Performance Standards—Feedlots Point Source Category,* EPA-440/1/74-004-a.

———, 1992. *NPDES Storm Water Sampling Guidance Document,* EPA 833-B-92-001, July.

———, 1993. *Guidance Specifying Management Measures for Sources of Non-point Pollution in Coastal Waters,* Office of Water, EPA 840-B-92-002, January.

Woodward-Clyde Consultants, Inc., 1990. *Urban Targeting and BMP Selection: An Information and Guidance Manual for State Nonpoint Source Program Staff Engineers and Managers,* prepared for Region V, Water Division, Watershed Management Unit, Environmental Protection Agency, Chicago, and Office of Water Regulations and Standards, Office of Water Enforcement and Permits, Environmental Protection Agency, Washington, distributed by The Terrene Institute, Washington, November.

Chapter

10

Construction Storm Water Permitting

This is the first of five chapters which deal with the specific requirements for storm water discharges from construction facilities. This chapter introduces the requirements for construction storm water permitting and describes the EPA general permit for storm water discharges from construction facilities.

Chapter 11 describes how to prepare a storm water pollution prevention plan (SWPPP) for a construction site. Chapter 12 provides details on stabilization measures (such as vegetation) for erosion and sediment control and Chap. 13 on structural measures (such as sediment basins) for erosion and sediment control. Chapter 14 describes other pollution prevention measures for construction sites.

There is additional information on construction permit requirements for specific states in App. A. Appendix B contains a glossary of terms.

Why EPA Regulates Construction Storm Water Discharges

Construction activities produce many different kinds of pollutants which may cause storm water contamination problems. Grading activities remove grass, rocks, pavement, and other protective ground covers, resulting in the exposure of underlying soil to the elements. Because the soil surface is unprotected, soil and sand particles are easily picked up by wind and/or washed away by rain or snowmelt. This process is called *erosion*. The water carrying these particles eventually reaches a stream, river, or a lake where the water slows down, allowing the particles to fall onto the bottom of the streambed or lake. This process is called *sedimentation*. Gradually, layers of these clays and silt build up in the stream beds, choking the river and stream channels and covering the areas where fish spawn and plants grow. These particles also cloud waters, causing aquatic respiration problems, and can kill fish and plants growing in the river stream.

Sediment runoff rates from construction sites are typically 10 to 20 times those of agricultural lands and 1000 to 2000 times those of forestlands. Even a small amount of construction may have a significant negative impact on water quality in localized areas. Over a short time, construction sites can contribute more sediment to streams than was deposited previously over several decades. Figure 10.1 illustrates the effect of construction activity on sediment concentrations.

In addition, the construction of buildings and roads may require the use of toxic or hazardous materials such as petroleum products, pesticides, fertilizers, and herbicides, and building materials such as asphalt, sealants, and concrete which may pollute storm water running off the construction site. These materials can be toxic to aquatic organisms and can degrade water for drinking and water-contact recreational purposes.

Regulated Discharges

The first step in complying with EPA or state regulations on construction storm water discharges is to determine if a particular discharge is required to obtain NPDES permit coverage. To make this determination, it is important to understand exactly how terms such as *storm water, discharge, point source,* and *waters of the United States* are defined. Chapter 3 provides detailed definitions and a discussion of these terms.

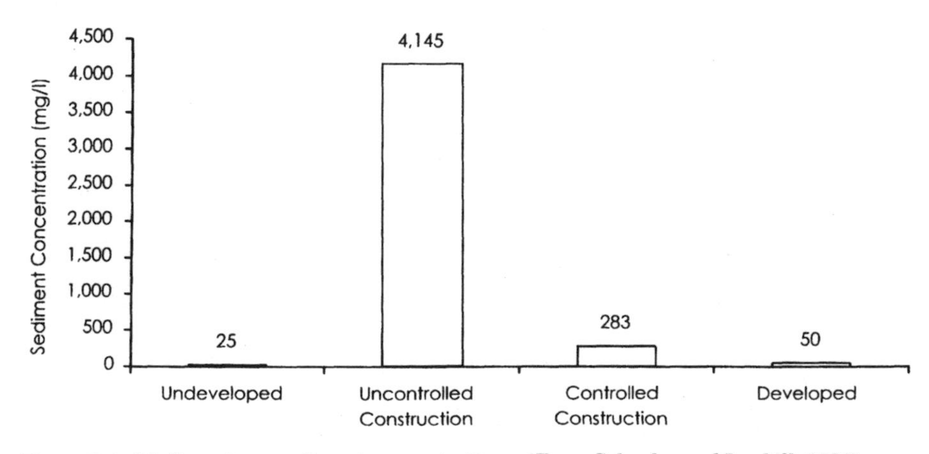

Figure 10.1 Median storm sediment concentrations. (*From Schueler and Lugbill, 1990*)

Under the National Storm Water Program, the EPA regulates storm water discharges from construction sites, including clearing, grading, and excavation activities, if the disturbed land area is 5 acres (2 ha) or more. Construction sites with disturbed areas of less than 5 acres (2 ha) may also be regulated, if they are part of a "larger common plan of development or sale." A detailed discussion of this requirement is presented later in this chapter.

The location of the construction activity and the ultimate land use at the site do not influence the decision of whether permit coverage is required, except that no permit is required for agricultural or silvicultural activities, including the clearing of land for agricultural purposes, because these are exempt from NPDES permit requirements under 40 CFR 122.4 and CWA Section 502(14). Regulated construction activities may include road building; construction of residential houses, office buildings, or industrial buildings; demolition; and other activities. At a demolition site, disturbed areas might include where building materials, demolition equipment, or disturbed soil is situated, which may alter the surface of the land (Environmental Protection Agency, 1993).

If a construction activity is undertaken at an industrial facility that already holds a permit for industrial storm water discharges, a separate permit must be obtained for the construction activity if it meets the 5-acre (2 ha) limitation. For construction activities on less than 5 acres (2 ha), no construction permit is required, but the industrial operator should modify the storm water pollution prevention plan to address all new sources of pollution and runoff from construction activities (Environmental Protection Agency, 1993).

Repaving of roads is not regulated under the storm water program unless 5 acres (2 ha) or more of underlying and/or surrounding soil is cleared, graded, or excavated as part of the repaving operation. Maintenance activities for flood control channels or roadside ditches (such as removal of vegetation) must obtain storm water discharge permit coverage if they involve grading, clearing, or excavation activities that disturb 5 acres (2 ha) or more, either individually or as part of a long-term maintenance plan (Environmental Protection Agency, 1993).

As discussed in Chap. 3, the Transportation Act of 1991 deferred NPDES storm water discharge permit requirements for construction projects owned by small municipalities (less than 100,000), in spite of whether the projects meet the other criteria for regulation described above. If the construction activity is owned by a small municipality, the activity is exempt from permit coverage, even though it may be operated by a private contractor. Permit coverage may still be required if the project is a significant contributor of pollutants to waters of the United States or if it contributes to a violation of a water quality standard.

This deferral of permit requirements is scheduled to end on August 7, 2001 [63 FR 01635 (Jan. 9, 1998)]. After that date, all storm water discharges from construction projects that meet the applicable minimum size requirements must be permitted, even if the project owner is a small municipality.

General Permit Coverage

There are two types of NPDES permits available for storm water discharges from construction sites: an individual permit and a general permit. Many states have developed their own general permits for construction storm water discharges; review App. A for further details on state requirements.

As noted in Chap. 1, a series of court rulings have established the authority of the EPA to utilize innovative concepts such as general permits to regulate discharges under the Clean Water Act. The EPA has used general permits as a tool to accommodate the large number of dischargers included in the National Storm Water Program. Most traditional NPDES permits have been individual permits, with permit requirements which were specific to the facility named in the permit. In addition, the traditional NPDES permit gives numeric effluent limitations on various pollutants and specifies a minimum discharge sampling interval.

The EPA issued its baseline general permit for storm water discharges associated with construction activity in 1992. This baseline general permit was reissued, with some revisions, in 1998. It applies directly to nonauthorized states and also serves as a model for other general permits issued by authorized states.

The EPA baseline general permit requires the submittal of a Notice of Intent, which states the permittee's desire to discharge according to the terms and provisions of the general permit. Compliance with the provisions of the general permit involves the preparation and maintenance of an SWPPP (as described in Chap. 11).

The EPA does not currently assess permit fees for general permit coverage, but several states do. The EPA has worked on a federal permit fee system which would impose fees on permit holders in states without NPDES permit authority. This fee system was developed under the 1990 Budget Reconciliation Act, which required the EPA to develop a system of user fees to pay for "services rendered." It is reasonable to expect the EPA to implement a fee system for storm water discharge permit applications.

Almost all construction activities which require NPDES permit coverage for storm water discharges can be covered under the general permit. Therefore, this book strongly emphasizes the general permit requirements. However, the EPA or state government administering the NPDES permit program has the right to require that a construction project submit an application for an individual permit under certain conditions. These conditions are most likely to be encountered in large, complex, and controversial construction projects, such as new airports, freeways, and other major facilities. These types of facilities often face some opposition from environmental groups, which already use existing permit requirements (such as the requirement for a Section 404 wetlands permit) as a means of delaying, altering, or killing the project. The requirement for a permit to discharge storm water provides a new legal means for such groups to oppose such major projects.

EPA definition of *operator*

The "operator" of a discharge of storm water associated with construction activity is required to submit the NOI and obtain coverage under an NPDES permit. Therefore, it is important to understand that EPA defines *operator* to mean any party associated with a construction project that meets either of the following two criteria:

1. Anyone who has operational control over construction plans and specifications, including the ability to make modifications to those plans and specifications, is an operator. Since the owner of the site generally has the ability to change the construction specifications, this definition tends to include the owner.
2. Anyone who has day-to-day operational control of those activities at a project which are necessary to ensure compliance with a storm water pollution prevention plan for the site or other permit conditions (e.g., he or she is authorized to direct workers at a site to carry out activities required by the storm water pollution prevention plan or comply with other permit conditions) is an operator. Since the general contractor usually maintains day-to-day control of the site activities, this definition tends to include the general contractor.

The following parties generally would not meet either of the two definitions of operator given above, and therefore would not be required to obtain permit coverage in most cases:

- Subcontractors hired by, and working under the supervision of, the owner or a general contractor.
- Utility service companies and their subcontractors, who may carry out activities that disturb the soil on the construction site. However, there may be circumstances in which utility service companies may be considered to be operators, as described later in this chapter.

- Individuals who hire a general contractor to construct a home for their personal use (e.g., not those homes to be sold for profit or used as rental property). The general contractor, being a professional in the building industry, is better equipped to meet the requirements of both applying for permit coverage and developing and properly implementing a SWPPP. However, individuals would meet the definition of operator in instances where they performed the general contracting duties for construction of their personal residences.

There may be more than one party at a site performing the tasks relating to "operational control" as defined above. Depending on the site and the relationship between the parties (e.g., owner, developer), either there can be a single party acting as site operator and consequently responsible for obtaining permit coverage, or there can be two or more operators with all needing permit coverage. There are many possible scenarios, including the following:

1. *Owner as sole permittee.* The property owner designs the structures for the site, develops and implements the SWPPP, and serves as general contractor (or has an on-site representative with full authority to direct day-to-day operations). In this case, the owner may be the only party that needs a permit. All others on the site may be considered subcontractors and may not need permit coverage. Only one NOI form may need to be submitted. Permittees who intend to act as the sole "overall" operator need to comply with both the "plans and specifications" and "implementation" requirements of the SWPPP.
2. *Contractor as sole permittee.* The property owner hires a construction company to design the project, prepare the SWPPP, and supervise "turnkey" (project). Here, the contractor would be the only party needing a permit. It is under this scenario that an individual having a personal residence built for her or his own use (e.g., not to be sold for profit or used as rental property) would not be considered an operator. Only one NOI form may need to be submitted.
3. *Owner and contractor as copermittees.* The owner retains control over any changes to site plans, SWPPPs, or storm water conveyance or control designs; but the contractor is responsible for overseeing actual earth-disturbing activities and daily implementation of SWPPP and other permit conditions. In this case, both parties may need coverage, and more than one NOI form may need to be submitted.

An operator with control over only a portion of a project is only responsible for a corresponding portion of the permit compliance activities. An operator must also take care not to damage or disrupt another operator's pollution controls. Operators must either implement their portion of a joint SWPPP or develop and implement their own individual SWPPP.

Several states have different requirements concerning the operator. Some states name the owner as the operator while others name the general contractor; some follow the EPA example and require them to become copermittees.

For purposes of this permit and determining who is an operator, *owner* refers to the party that owns the structure being built, but not necessarily the land on which the construction is occurring. For example, a landowner whose property is being disturbed by construction of a gas pipeline would not usually be considered an operator. Likewise, if the erection of a structure has been contracted for, but possession of the title or lease to the land or structure is not to occur until after construction, the would-be owner may not be considered an operator. An example of this situation would be an individual having a house built by a residential home builder.

The EPA definition of *operator* does not include government agencies or other organizations that develop or enforce building codes or other common design standards. These standards, and those that develop and enforce them, do not have "control" over a construction project's plans and specifications in the same sense that the owner has such control. Instead, these standards direct or limit a project operator's latitude when drafting or modifying a particular aspect of the project's plans and specifications.

Common plan of development or sale

As described earlier in this chapter, an NPDES permit to discharge storm water associated with construction activity is only needed when a "larger common plan of development or sale" will disturb 5 acres (2 ha) or more. The simple case is when the "common plan" involves the construction of a single building, etc., for a single owner. The more complicated case is when the common plan consists of several smaller construction projects that cumulatively will disturb 5 ac (2 ha) or more, but may or may not be under construction at the same time. Residential development, with houses being built by several home builders in a master-planned subdivision, is an excellent example of this second case. Another example would be a new airport runway, where the construction of the associated taxiways, additional hangers, terminals, parking lots, etc., at the airport would also be part of the same common plan of development.

For sites disturbing less than 5 acres (2 ha), the first two steps in deciding whether a permit is needed for storm water discharges associated with construction activity are to determine the following:

1. Is there a "common plan of development or sale" tying individual sites together? For example, if the lots are part of a subdivision plan

filed with the local land-use planning authority, then this would be a "common plan."

2. Will the total area disturbed by all the individual sites (including the cumulative total disturbance necessary to completely build out the subdivision) come to 5 acres (2 ha) or more?

If the answer to both questions is "no," a storm water discharge permit is not needed—unless the EPA determines that discharges contribute to a violation of water quality standards or are a significant contributor of pollutants to waters of the United States—and specifically requests a permit application.

The *plan* in a common plan of development or sale is broadly defined as any announcement or piece of documentation (including a sign, public notice or hearing, sales pitch, advertisement, drawing, permit application, zoning request, computer design, etc.) or physical demarcation (including boundary signs, lot stakes, surveyor markings, etc.) indicating that construction activities may occur on a specific plot. However, a city's land-use master plan or zoning map would not be considered a "plan" that would create a storm water discharge permit requirement for every construction project in the city.

The concept of the larger common plan has to be applied with some common sense, and should not be taken to extremes. One of the primary criteria that may be useful in determining whether there is a larger common plan of development or sale is whether the "plan" has been developed by or for one of the operators of the project (probably the owner). A plan developed by or for one of the operators is much more likely to be considered a larger common plan of development or sale than a plan developed by or for someone else, such as a government agency that may have the authority to regulate or influence the development, but does not propose to actually undertake the construction of the development.

Construction activities that could result in a regulated discharge of storm water include all earth-disturbing activities necessary to complete the planned project, such as grading lots, installing utilities, building roads, and preparing storm water control structures. Various support activities, such as exposed materials storage and equipment staging areas, would also generate regulated discharges.

Once the extent of the common plan has been determined, the total area to be disturbed must be calculated. A single ¼-acre (0.1-ha) lot is not large enough by itself to require a permit. However, if 100 such lots are present in a subdivision and each one will be fully disturbed during the process of home building, then the total disturbed area is 25 acres (10 ha), and permit coverage is needed.

The EPA general permit is designed to provide coverage for all the permittee's activities on the common plan. It is not necessary (and is, in fact, counterproductive) to submit separate NOI forms and develop a separate storm water pollution prevention plan for each individual lot.

Installation of utility service lines

In many areas utility service companies (companies that install distribution facilities for natural gas, electricity, telephone, or cable television) will not need to obtain separate permit coverage for installing utility service lines. Often, utility companies are not responsible for overall SWPPP compliance at a project. Typically, a project's general contractor has overall responsibility for SWPPP implementation and compliance, and the utility service companies tend to act more as subcontractors. As with any other party involved in a construction project, permit coverage will be required for utility service companies only when they meet the definition of *operator,* as discussed earlier in this chapter.

To the extent that a utility service company needs to develop its own site-specific plans and specifications for a service installation at a project requiring storm water permit coverage, the utility will be considered to meet the definition of *operator* and must allow for appropriate storm water control measures, either by designing and implementing controls themselves or by ensuring that another project operator has designed and will implement storm water controls for the area disturbed by the utility service installation.

Other examples in which a service line installation would require construction storm water permit coverage are that the activity disturbed 5 acres (2 ha) or more [40 CFR 122.26(b)(14)(x)], or was designated by the EPA or state to obtain coverage for another reason [40 CFR 122.26(a)(1)(v), 122.26(a)(9) or 122.26(g)(1)(i)]. Other utility company activities, such as the installation of main transmission lines or a large transformer station, should likewise be reviewed to see if permit coverage is required.

Future construction

Once a residence or commercial building has been completed and occupied by the owner (or tenant), future activities by the owner on the individual's property are not considered part of the original common plan of development. For example, after a home is occupied by the homeowner or a tenant, future construction activity on that particular lot is considered a new and distinct project and is compared to applicable disturbed-area limits for permit applicability. If a homeowner installs a swimming pool after occupying the house, only the disturbed area on the lot—not the total area of the subdevelopment—is considered for determining whether a permit is needed. Likewise, demolition and reconstruction of individual houses originally built as part of a common plan of development, including those destroyed or damaged by fire or natural disasters, are also considered to be "new" plans of development or redevelopment, and not part of any larger common plan.

In many cases, a common plan of development or sale consists of many small construction projects that collectively come to 5 acres (2 ha) of total disturbed land. For example, an original common plan of development for a residential subdivision might lay out the streets, house lots, and areas for parks, schools, and commercial development that the developer plans to build or sell to others for development. All these areas would remain part of the common plan of development or sale until the intended construction occurred. After this initial plan has been completed for a particular parcel, any subsequent development or redevelopment of that parcel will be regarded as a new plan of development, and will then be subject to the 5-acre (2-ha) cutoff for storm water permitting purposes.

Infill development

Infill development occurs after most of a subdivision or similar development has been completed, but when isolated residential or commercial lots remain vacant. If the total area of these remaining lots that will be disturbed is less than 5 acres (2 ha), then no permit coverage is required for storm water discharges occurring during construction on these lots. However, if the total area of expected disturbance on the remaining lots is greater than 5 acres (2 ha), then construction on any of the remaining lots must be covered by a storm water discharge permit.

A home builder can discontinue permit coverage once the total area of lots to be disturbed by future construction has dropped below 5 acres (2 ha). However, the project still has to meet the standard requirements for submittal of a Notice of Termination, as described later in this chapter. Basically, this means that if the home builder has completed construction on several lots in a subdivision, stabilized the lots, and delivered the finished houses to the owners, then the home builder can complete additional construction without permit coverage, provided that the total area remaining to be disturbed is now less than 5 acres (2 ha). This does not imply that the home builder can terminate permit coverage and discontinue all other permit compliance activities on lots that are still under construction. As long as there is still construction continually under way, the permit requirement remains in effect. The permit requirement is removed only if the construction is temporarily completed and the portion of the site remaining to be disturbed is less than 5 acres (2 ha).

A common plan of development must at least theoretically be capable of having 5 acres (2 ha) or more of land disturbed at one time in order to trigger the need for a permit. Requiring that all parts of the project, including unbuilt portions of the larger common plan of development, have achieved final stabilization before the total disturbed area can be "recalculated" ensures that there is a period of time during which all discharges of storm water associated with construction activity from the common plan of development or sale have ceased. The requirement to compare disturbed area to the total remaining unbuilt area of the larger common plan protects against attempts to artificially divide a project in such a way as to avoid having to provide environmental controls for construction activities.

Support activities at off-site areas

Off-site areas are commonly used for storage of fill material or soil excavated from the construction site, borrow areas to obtain fill material, storage of building materials, concrete or asphalt batch plants, or storage of construction equipment. Where activities at off-site locations would not exist without the construction project, discharges of pollutants in storm water from these areas must be controlled. Off-site storage areas, support bases, disposal areas, and borrow areas used for a construction project are considered to be part of the larger common plan of development or sale, as described earlier in this chapter. The pollution prevention plan for the construction project must include controls for all off-site areas directly supporting the construction project.

If the off-site location is a permanent facility used to support multiple construction projects or customers, then the off-site location does not have to be covered by the storm water pollution prevention plan for the construction project. For example, a fixed base of operations, such as a construction company's home office, warehouse, commercial warehouse, landfill, equipment yard, etc. used for all construction projects, would not require permit coverage under a construction permit (although it might meet the requirements for industrial permit coverage, as described in Chap. 5).

Allowing such off-site locations to be included under the construction permit for the construction site avoids the need for a separate permit for the remote location. Where the same operator uses a temporary off-site location to support construction activities at several nearby locations, permit coverage may be obtained by identifying the site and including controls for this common site in at least one of the pollution prevention plans for the individual construction projects. For example, a common support area for three highway projects could be permitted by identifying the site, including appropriate controls in at least one of the three pollution prevention plans for the separate projects, and ensuring that an NOI is not submitted until the support area is finally stabilized.

EPA General Permit NOI Requirements

The NOI must be postmarked at least 2 days before permit coverage is required. For owners (who have control over the plans and specifications for the project), the NOI should be filed at least 2 days before any work commences on the project. For contractors (who have day-to-day operational control), the NOI should be filed at least 2 days before work commences on that contractor's portion of the project. If a different operator assumes control of the site, the new operator must file a new NOI at least 2 days before beginning work on the site.

The EPA will accept a late NOI, but the authorization covers discharges only from 2 days after the postmarked date. The authorization does not retroactively apply to any prior, unpermitted discharges. The EPA may choose to take enforcement action for any unpermitted discharges of pollutants to waters to the United States.

The address for submitting the NOI is found in the instruction portion of the NOI form. Copies may have to be sent to a state or Indian tribe. Only NOI forms provided by the EPA (or photocopies thereof) are valid. It is acceptable to fill in information that will be the same for every project (e.g., a company's name, address) and make copies of the partially completed form for future use.

Only one NOI is required to cover all the activities of one operator on any one common plan of development or sale. For example, a home builder in a residential development should submit only one NOI to cover all lots, even if they are on opposite sides of the development.

Before the NOI form can be submitted, the SWPPP must be completed to ensure that appropriate controls to meet ESA and National Historic Preservation Act certification requirements, if needed, are included to avoid or mitigate adverse effects to listed endangered or threatened species, critical habitat, or historic properties. The ESA and NHPA requirements are described later in this chapter.

Storm water discharges are authorized 2 days after the date on which the NOI is postmarked, unless otherwise notified by the EPA. Permittees must implement their SWPPP, or their portion of the plan, as soon as they begin work on site.

The NOI form requires the following information:

- The operator's (applicant's) name, address, and telephone number and whether it is a federal, state, tribal, public, or private entity.
- The street address (description of location if street address is unavailable), county, and latitude and longitude of the approximate center of

the construction site. The use of latitude and longitude allows the EPA to use the NOI data effectively with state Geographic Information Systems (GIS) systems, thus enhancing the ability to deal with projects on a watershed basis. Help in finding the latitude and longitude is provided in the instructions to the NOI form. Portable global positioning system (GPS) units are available at a relatively low cost to provide readouts of the latitude and longitude.

- Whether the construction project is located on an Indian country land.
- The name of the receiving water(s); or, if the discharge is through a municipal separate storm sewer, the name of the municipal operator of the storm sewer and the receiving water(s).
- An estimate of the project start date and completion date, and an estimate of the size of the area on the site on which soil will be disturbed. Note that the project start and stop dates need not be exact. The EPA recognizes that many factors, often beyond the permittee's control, contribute to whether a project will actually start or end on the estimated dates.
- Whether the SWPPP has been prepared and (optional) the location of where the plan can be viewed, if different from the project address.
- Whether any endangered species are in proximity to the construction project and which of the listed options enables the operator to claim eligibility for permit coverage.

A signature block is provided following a certification statement that everything on the NOI form is correct. The person signing the NOI must have a sufficient level of authority and responsibility within the organization to help ensure compliance with the terms and conditions of the permit. For a sole proprietorship, the proprietor must sign the NOI. For a partnership, a general partner must sign.

For a corporation, the person signing the NOI must be a responsible corporate officer, including the president, secretary, treasurer, or vice president of the corporation in charge of a principal business function or any other person who performs similar policy- or decision-making functions for the corporation.

A corporate plant or facility manager may sign the NOI only under certain conditions. The plant or facility must employ more than 250 persons or have gross annual sales or expenditures exceeding $25 million (in second-quarter 1980 dollars). In addition, the authority to sign documents must have been assigned or delegated to the plant or facility manager in accordance with corporate procedures.

For a municipality, state, federal, or other public agency, the person signing the NOI must be a principal executive officer or ranking elected official. A principal executive officer of a federal agency includes the chief executive officer of the agency and a senior executive officer having responsibility for the overall operations of a principal geographic unit of the agency (e.g., regional administrators of the EPA). Within the military, the commanding officer of the facility represents the appropriate level of authority.

The person signing the NOI makes the following certification:

> I certify under penalty of law that this document and all attachments were prepared under my direction or supervision in accordance with a system designed to assure that qualified personnel properly gather and evaluate the information submitted. Based on my inquiry of the person or persons who manage the system, or those persons directly responsible for gathering the information, the information submitted is, to the best of my knowledge and belief, true, accurate, and complete. I am aware that there are significant penalties for submitting false information, including the possibility of fine and imprisonment for knowing violations.

Endangered Species Act (ESA) Certification

The EPA is required to comply with the Endangered Species Act (ESA) of 1966. Part of EPA's obligations under the ESA is to consider the effects of any permits that are issued by EPA on endangered or threatened species (known as *listed species*) and their critical habitat. This type of consideration must be done on a site-specific basis. Therefore, anyone who applies for coverage of a storm water discharge under the EPA construction general permit must supply the EPA with site-specific information that will provide some assurance that the possible effects of the discharge on listed species and their critical habitat have been adequately considered. Impacts to listed species and critical habitat can occur from development and construction even on fully developed sites. Often, the impacts occur at the point of discharge into surface waters. Therefore, even those projects that are located on fully developed sites such as residential lots must meet the endangered-species certification requirements.

Depending on the results of the assessment, the SWPPP may need to include specific measures to eliminate, reduce, or mitigate the adverse effects on listed species and their critical habitat. If the adverse effects cannot be adequately reduced or mitigated, general permit coverage may not be available. An individual storm water discharge permit, along with additional consideration of ESA requirements, may be required.

In cooperation with the U.S. Fish and Wildlife Service (FWS) and the National Marine Fisheries Service (NMFS), the EPA has developed a six-step procedure for addressing the possible effects of a construction project on listed species and their critical habitat. The following sections of this chapter describe each of these six steps and give appropriate guidance on carrying out each step.

Because compliance with ESA requirements can take a long time, it is very important to address this issue early in the project planning process. If the ESA-related issues are fully addressed and adequately documented early in the project planning stages, then the contractors, home builders, utility service companies, and others who will become involved in the project later can operate under the initial assessment, and the project should not experience any additional delays. However, the use of earlier assessments by others is subject to limitations and obligations as described under step 6 below.

It is important to note that the EPA/FWS/NMFS six-step procedure is designed primarily to identify and address the adverse effects of storm water discharges and associated activities. Therefore, this procedure may not adequately address other effects of the proposed project on listed species or their critical habitat. It may be necessary to obtain an ESA Section 10 permit or to request a formal consultation under ESA Section 7, to address these additional effects. These procedures are described under step 6 below.

Many of the six steps in the ESA assessment procedure are best performed with the assistance of a qualified biologist or other environmental professional, working as a consultant to the operator or designer of the construction project. However, the nearest FWS and/or NMFS office, or a state or tribal natural heritage center, can also provide very useful information and some assistance.

Step 1: Is the construction site within a critical habitat area?

Some, but not all, listed species have designated critical habitat. Exact locations of such habitat are provided in the service regulations at 50 CFR Parts 17 and 226. If the construction site is located in a designated critical habitat, then it is necessary to consider impacts to critical habitat when following steps 2 through 6 below. It is important to note that protection of listed species and protection of critical habitat are two distinct considerations, and different protection measures may be required for each.

Step 2: Are there listed species in the project county(ies)?

Lists of endangered species, arranged by state and county, are available from several sources:

- The EPA Office of Wastewater Management's Internet Web site http://www.epa.gov/owm.
- The Fish and Wildlife Service's Internet Web site http://www.fws.gov.
- The National Marine Fisheries Service's Internet Web site http://www.nmfs.gov.
- The Code of Federal Regulations [50 CFR 17 and 50 CFR 226], which is available in printed form at U.S. government bookstores and many libraries, or in electronic form from the Government Printing Office Internet Web site http://www.gpo.gov.
- The Biodiversity Heritage Centers of the Natural Heritage Network, which also maintains an Internet Web site at the address http://www.heritage.tnc.org.

Where a facility is located in more than one county, the lists for all counties should be reviewed. Where a facility discharges into a water body which serves as a border between counties or which crosses a county line that is in the immediate vicinity of the point of discharge, applicants should also review the species list for the county that lies immediately downstream or is across the water body from the point of discharge.

The EPA does not require permit applicants to consider impacts to state-protected species and designated critical habitat that are not protected under the federal ESA. However, states have the authority to impose their own requirements under state law to protect federally or state-protected species from construction activities.

After a review of the available information from the sources mentioned above, if no listed species are located in a facility's county or if a facility's county is not listed, and the construction site is not located in critical habitat as described under step 1, the proposed construction storm water discharge is eligible for EPA construction general permit coverage without further inquiry into the presence of, or effects on, listed species. Otherwise, the procedure must continue with step 3.

Step 3: Are listed species present in the project area?

The project area consists of

- The areas on the construction site where storm water discharges originate and flow toward the point of discharge into the receiving waters (including all areas where excavation, site development, or other ground disturbance activities occur) and the immediate vicinity. Examples include (1) where bald eagles nest in a tree that is on or bordering a construction site and could be disturbed by the construction activity and (2) where grading causes storm water to flow into a small wetland or other habitat that is on the site which contains listed species.

- The areas where storm water discharges flow from the construction site to the point of discharge into receiving waters, e.g., where storm water flows into a ditch, swale, or gully which leads to receiving waters and where listed species (such as amphibians) are found in the ditch, swale, or gully.
- The areas where storm water from construction activities discharges into receiving waters and the areas in the immediate vicinity of the point of discharge, e.g., where storm water from construction activities discharges into a stream segment that is known to harbor listed aquatic species.
- The areas where storm water BMPs will be constructed and operated, including any areas where storm water flows to and from BMPs, for example, where a storm water retention pond would be built.

The project area will vary with the size and structure of the construction activity, the nature and quantity of the storm water discharges, the storm water discharge–related activities, and the type of receiving water. In many cases, the project area will encompass an entire construction site. However, there could be situations where the project area may encompass a portion of the site (e.g., where the actual construction disturbs only a portion of a land development project).

Applicants should use an appropriate method that allows them to determine, to the best of their knowledge, whether listed species are located in their project area. These methods may include

- Conducting visual inspections. This method may be particularly suitable for construction sites that are smaller or located in nonnatural settings such as highly urbanized areas or industrial parks where there is little or no natural habitat, or for construction activities that discharge directly into municipal storm water collection systems.
- Contacting local and regional conservation groups or the state or tribal natural heritage centers.
- Submitting a data request to a natural heritage center. Many of these centers will provide site-specific information on the presence of listed species in a project area. Some of these centers will charge a fee for researching data requests.
- Conducting an environmental assessment under the National Environmental Policy Act (NEPA).

As noted previously, the help of a qualified environmental consultant may be extremely useful. If no species are found in the project area, the proposed construction project is eligible for EPA construction general permit coverage. Otherwise, the location and nature of this presence should be documented in the storm water pollution prevention plan, and then the assessment should continue with step 4.

Step 4: Are listed species or critical habitat likely to be adversely affected?

If listed species or critical habitat is present in the project area, it is necessary to assess whether either is likely to be adversely affected by storm water discharges or storm water discharge–related activities. *Storm water discharge–related activities* include

- Activities which cause, contribute to, or result in point source storm water pollutant discharges, including but not limited to excavation, site development, grading, and other surface disturbance activities
- Measures to control storm water discharges including the siting, construction, and operation of best management practices (BMPs) to control, reduce, or prevent storm water pollution

Potential adverse effects from storm water discharges and storm water discharge–related activities include

- *Hydrological effects.* Storm water discharges may cause siltation or sedimentation or may induce other changes in receiving waters such as changes in temperature, salinity, or pH. These effects will vary with the amount of storm water discharged and the volume and condition of the receiving water. Where a storm water discharge constitutes a minute portion of the total volume of the receiving water, adverse hydrological effects are less likely. Construction activity itself may also alter drainage patterns on a site where construction occurs, which can impact listed species or critical habitat.
- *Habitat effects.* Excavation, site development, grading, and other surface disturbance activities from construction activities, including the installation or placement of storm water BMPs, may adversely affect listed species or their habitat. Storm water may drain or inundate listed species habitat.
- *Toxicity effects.* In some cases, pollutants in storm water may have toxic effects on listed species.

The scope of effects to consider will vary with each site. If adverse effects are not likely, the proposed construction project is eligible for EPA construction general permit coverage. If adverse effects are likely, applicants must follow step 5.

Step 5: Can adverse effects be avoided?

If adverse effects are likely, the proposed construction project can still receive coverage under the EPA construction general permit if appropriate measures are undertaken to avoid or eliminate the likelihood of adverse effects prior to applying for permit coverage. These measures may involve relatively simple changes to construction activities, such as rerouting a storm water discharge to bypass an area where species are located, relocating BMPs, or changing the "footprint" of the construction activity.

It is important to note that any measures that are undertaken to avoid or eliminate adverse affects must continue to remain fully in effect throughout the course of permit coverage. These measures must be described in the storm water pollution prevention plan and may be enforceable as permit conditions. If adequate measures cannot be identified, the ESA assessment must continue with step 6.

Step 6: Can the proposed project meet minimum eligibility requirements?

Section 9 of the ESA prohibits any person from "taking" a listed species (e.g., harassing or harming it) unless (1) the taking is authorized through an "incidental take statement" as part of undergoing ESA Section 7 formal consultation; (2) an incidental take permit is obtained under ESA Section 10 (which requires the development of a habitat conservation plan); or (3) the person is otherwise authorized or exempted under the ESA. This prohibition applies to all entities including private individuals, businesses, and governments.

After completion of steps 1 to 5, it may be apparent that the proposed construction project is likely to result in adverse effects on a listed species or its critical habitat. This could trigger the penalty provisions in Section 9 of the ESA. Therefore, it is necessary to contact the EPA and FWS/NMFS and discuss the situation. The project may still be eligible for coverage under the EPA construction general permit if the adverse effects can be addressed through meeting one of three minimum criteria that are available:

1. An ESA Section 7 consultation
2. An ESA Section 10 "incidental take" permit
3. Reliance on a previous ESA certification

ESA Section 7 consultations. Section 7 of the ESA provides for a process of consultation with FWS and/or NMFS to address possible effects on listed species or their critical habitat. Section 7 consultations may be either formal or informal. The formal consultation must result in either a "no-jeopardy opinion" or a "jeopardy opinion" that identifies reasonable and prudent alternatives to avoid jeopardy which are to be implemented by the applicant. The informal consultation must result in a written concurrence by FWS and/or NMFS on a finding that the storm water discharge(s) and storm water discharge–related activities are not likely to adversely affect listed species or critical habitat [50 CFR 402]. Most consultations are informal. The EPA has automatically designated applicants for the construction general permit for storm water discharges as nonfederal representatives for the purpose of conducting informal consultations.

The Section 7 storm water consultation may be combined with consideration of other federal permits or actions on the same project. For example, an ESA Section 7 consultation may also be required for the issuance of a wetlands dredge and fill permit for the project, or where an NEPA review is performed for the project which incorporates a Section 7 consultation. Any terms and conditions developed through consultations to protect listed species and critical habitat must be incorporated into the SWPPP.

Whether ESA Section 7 consultation must be performed with the FWS, NMFS, or both services depends on the listed species that may be affected by the applicant's activity. In general, NMFS has jurisdiction over marine, estuarine, and anadromous species. Applicants should also be aware that while formal Section 7 consultation provides protection from incidental-takings liability, informal consultation does not.

ESA Section 10 permits. Section 10 of the ESA provides for a permit process that involves the development of a habitat conservation plan for the listed species or critical habitat to address the effects of the storm water discharge(s) and storm water discharge–related activities on listed species and critical habitat. The issuance of a Section 10 permit can pave the way for coverage of a storm water discharge under the EPA construction general permit.

The FWS and NMFS have procedures that must be followed to apply for an ESA Section 10 permit [see 50 CFR section 17.22(b)(1)(FWS) and section 222.22(NMFS)]. More details are available through the nearest FWS or NMFS office, or from the respective Internet Web sites (http://www.fws.gov *or* http://www.nmfs.gov).

Reliance on a previous ESA certification. There is the possibility that several operators may apply for and receive permit coverage for storm water discharges from the same construction project. In this case, the first operator (often the owner, developer, or general contractor) may have performed all the work necessary to allow ESA certification of all the storm water discharges from the entire project, even those storm water discharges that are under the authority of a different operator later in the construction process. If this is the case, then the later operators may achieve ESA certification simply by agreeing to comply with any measures or controls upon which the initial operator's certification was based. However, the initial operator's certification must apply to the later operator's project area and must address the effects from the storm water discharges and storm water discharge–related activities on listed species and critical habitat.

It is very important to note that these later applicants or permittees may be liable for inadequacies or falsehoods in that certification. This potential liability is well described in the certification language of the NOI form, as quoted earlier in this chapter.

Thus, it important for those applicants who choose to rely on another operator's certification that they carefully review that certification and its SWPPP for accuracy and completeness. If the certification appears to be inadequate in any way, applicants should provide an independent basis for their certification in their SWPPP. Relying on a previous ESA certification will usually provide some protection from possible liabilities or penalties, but it does not provide complete protection. Utility service companies that fit the definition of operator that choose to rely on another operator's certification are liable to the same extent as any other operator who relies on another operator's certification.

Protection of historic properties

Congress passed the National Historic Preservation Act (NHPA) in 1966. The law established a national policy for the protection of historic and archaeological sites and outlined responsibilities for federal and state governments to preserve our nation's history. Under the NHPA, identification of historic properties is coordinated by the state historic preservation officers (SHPOs), tribal historic preservation officers (THPOs), or other tribal representatives (in the absence of a THPO).

Section 106 of the NHPA requires federal agency heads to consider the effects of their actions (also known as "Federal undertakings" in the NHPA regulations) on historic and archeological sites that are listed as eligible for the National Register of Historic Places and to seek comments from an independent reviewing agency, the Advisory Council on Historic Preservation (ACHP). Regulations for accomplishing this responsibility have been published in the Federal Register as 36 CFR Part 800: Protection of Historic Properties. They detail the efforts that a federal agency must undertake to protect the historical environment.

The NHPA requires that any activity that obtains a federal permit or license, uses federal funds, or is otherwise assisted or approved by the U.S. government must comply with section 106. Examples of projects that require compliance with this law include

- Federal development activities, such as a new reservoir built by the U.S. Army Corps of Engineers
- Construction of municipal waste water treatment facilities that require a permit from the Environmental Protection Agency
- New highway construction that utilizes federal funds
- Local government improvement projects to rehabilitate or demolish housing using funds from the Housing and Urban Development agency

With regard to NPDES permits, the EPA conducts NHPA consultations on a case-by-case basis as needed. The EPA is working with the Advisory Council on Historic Preservation (ACHP), National Council of State Historic Preservation Officers (NCSHPO), and tribes regarding the development of a comprehensive and efficient approach to ensure that effects to historic properties are given appropriate consideration in the issuance of NPDES storm water discharge permits. Future permits are likely to require a more structured consideration of historic preservation issues.

Storm Water Pollution Prevention Plan

In nonauthorized states, the EPA general permit requires that an SWPPP be completed for each construction project for which an NOI has been submitted. Most authorized states have similar requirements, although some states use different terminology. The SWPPP must be completed and ready to implement at the time that the project begins construction.

Under EPA general permit requirements, the SWPPP is not submitted with the NOI. Instead, the SWPPP is retained on-site by the discharger. Federal, state, and local regulatory agencies have the authority to review the SWPPP at any time. If the reviewing agency finds that the SWPPP is not in compliance with general permit requirements, the discharger has 7 days to revise the SWPPP to achieve compliance. Chapter 11 provides further details on the SWPPP.

Notice of Termination

To terminate coverage, a permittee must submit a Notice of Termination (NOT) form. The NOT must be filed within 30 days after cessation of construction activities and final stabilization of the permittee's portion of the site (or temporary stabilization for residential construction where a homeowner is assuming control of a property). An NOT must also be submitted by a permittee before another operator assumes the previous permittee's liabilities.

The EPA requires the operator(s) to file a Notice of Termination with the EPA within 30 days after project completion. The NOT certifies that specific activities in the SWPPP have ended and either (1) final stabilization is complete, and temporary erosion and sediment controls have been removed from this operator's portion of the site; or (2) the operator has changed, and the new operator is responsible for compliance in all portions of the project area that were the responsibility of this operator. The new operator is responsible for submitting an NOI if activities continue. For example, a developer can pass permit responsibility for lots in a subdivision to the home builder who purchases those lots, provided that the home builder has filed her or his own NOI.

For individual residential construction only, the NOT can be filed as soon as the residence has been transferred to the homeowner. However, the home builder must at least establish temporary stabilization (including perimeter controls) for the lot and inform the homeowner of the need for and benefits of final stabilization.

Final stabilization occurs when all soil-disturbing activities at the site have been completed and a uniform (e.g., evenly distributed, without large bare areas) perennial vegetative cover with a density of 70% of the native background vegetative cover for the area has been established on all unpaved areas and areas not covered by permanent structures. As an alternative, equivalent permanent stabilization measures (such as the use of riprap, gabions, or geotextiles) may be employed. In some parts of the country, background native vegetation will cover less than 100% of the ground (e.g., arid areas). Establishing at least 70% of the natural cover of native vegetation meets the vegetative cover criteria for final stabilization. For example, if the native vegetation covers 50% of the ground, 70% of 50% would require 35% total cover for final stabilization. If the project is located on agricultural land, then the stabilization requirement may be met by returning the site to its preconstruction agricultural use.

The NOT must be signed in accordance with the signatory requirements of 40 CFR 122.22, which were summarized earlier in this chapter with regard to the NOI. Significant parts of the NOT include

- Permittee name and contact information, and site location information
- The permit number that is being terminated
- Permittee certification that he or she understands that submission of the NOT means the permittee no longer will have authorization to discharge storm water associated with construction activity
- Clarification that the authorization to discharge ends at midnight of the day the NOT is postmarked
- The conditions under which an NOT can be submitted

Possible Changes in Construction Permit Requirements

On January 9, 1998, the EPA published proposed regulations for phase II of the National Storm Water Program. The proposed phase II regulations would require the operators of all construction sites that involve the disturbance of at least 1 acre (0.4 ha) to obtain NPDES storm water discharge permit coverage, beginning in about 2002. Projects involving the disturbance of less than 1 acre (0.4 ha) could be required to obtain permit coverage if they are part of a larger common plan of development or sale, as described previously in this chapter.

The EPA currently proposes to provide three waivers, or exemptions, from NPDES storm water permit requirements for small construction sites:

- The permit requirement can be waived if there is a minimal risk of rainfall that might cause serious erosion problems during the period of construction. The allowable risk is defined in terms of an allowable value of 2 for the rainfall erosivity. The rainfall erosivity is discussed later in this chapter, in connection with the revised universal soil loss equation.
- The permit requirement can be waived if there is expected to be minimal erosion during the period of construction. This waiver may be available even if the rainfall waiver (described immediately above) is not available. To be eligible for this waiver, the expected rate of erosion from the construction site cannot exceed 2 tons per acre per year (4.5 metric tons per hectare per year). The rate of erosion is computed using the revised universal soil loss equation (RUSLE) as described below.
- The permit requirement can be waived if the site is covered under an approved state or local plan that adequately protects waters of the United States. This waiver reflects the fact that some state and local governments already have effective sediment and erosion control programs that serve essentially the same purpose as the NPDES storm water discharge permit for construction sites.

Revised Universal Soil Loss Equation (RUSLE)

The most common method for performing soil loss calculations in the United States has been the *universal soil loss equation* (*USLE*). This equation was developed by the USDA Agricultural Research Service (ARS), and it has the following form:

$$A = RKLSCP$$

where A = average soil erosion rate
R = rainfall erosivity
K = soil erodibility
L = slope length
S = slope steepness
C = cover management factor
P = support practice factor

Within the past few years, the ARS has developed a revised USLE (sometimes called the RUSLE) that uses the same equation, but has various refinements in all the factors used in the equation (Agricultural Research Service, 1997).

As noted above, the EPA proposes to waive storm water discharge permit requirements for small construction projects that have a project R value of 2 or less. The project R value is dependent on the duration of the project, the climate of the project area, and the time of year in which the project is scheduled. For example, the total annual R value for San Antonio, Texas, is about 250. Therefore, each day that a project is under way, the accumulated R value is about 250/365, or about 0.7. Therefore, a 3-day project has a total R value of about 2.1. By using this simplified approach, any small construction project with a duration of less than about 3 days in San Antonio, Texas, could be performed without a storm water discharge permit, according to the current EPA proposed regulations.

The simplified approach to the computation of project R values described in the previous paragraph does not account for the large variation in rainfall from one part of the year to another in many locations.

For example, in San Antonio, Texas, the average amount of rainfall for a 15-day period in January is only about 2% of the annual total. Therefore, the average daily R value during the month of January is only about $(0.02/15) \times 250$, or about 0.33. A project undertaken during January could last for about $2/0.33 = 6$ days before the maximum allowable R value of 2 was exceeded. During May, however, the average amount of rainfall for a 15-day period is about 8% of the annual total. Therefore, the average daily R value is about $(0.08/15) \times 250$, or about 1.33 days. The same project undertaken during the month of May would have to be completed in $2/1.33 = 1.5$ days. Some other areas of the United States have even more dramatic differences between the dry and wet periods of an average year.

Even if the project R is greater than 2, the permit requirement could still be waived by showing that the total erosion rate is less than 2 tons/acre per year (4.5 t/ha/per year). This analysis would require the application of the RUSLE method according to USDA guidelines (Agricultural Research Service, 1997). The EPA requires that this analysis be performed assuming no ground cover and no erosion controls in place.

By providing these conditional waivers, the EPA is providing an incentive to keep projects short and small and to get them done during the dry season, when the average daily R values are small. In the long run, this will reduce erosion, even for many sites that do not obtain permit coverage (because they may alter the size or schedule of the project to avoid a permit).

References

Agricultural Research Service, 1997. *Predicting Soil Erosion by Water—A Guide to Conservation Planning with the Revised Universal Soil Loss Equation (RUSLE)*, USDA/HB-703, January.

Environmental Protection Agency, 1992. *NPDES Storm Water Program Question and Answer Document,* vol. 1, EPA 833-F-93-002, March 16.

———, 1993. *NPDES Storm Water Program Question and Answer Document,* vol. 2, EPA 833-F-93-002B, July.

Schueler, T. R., and J. Lugbill, 1990. *Performance of Current Sediment Control Measures at Maryland Construction Sites,* Metropolitan Washington Council of Governments, Washington.

Chapter

11

NPDES Construction Permit Storm Water Pollution Prevention Plan

This is the second of five chapters which deal with the specific requirements for storm water discharges from construction facilities. This chapter describes how to prepare a storm water pollution prevention plan (SWPPP) for a construction site.

Concept of the Storm Water Pollution Prevention Plan

Because of the nature of construction activities and the resulting pollutants, as well as the variable nature of storm events, the EPA has chosen an approach to storm water management for construction sites that uses SWPPPs *designed by the permittee* (*the site operators*). These plans are based on the use of *best management practices* (*BMPs*). For construction sites, there are three main types of BMP:

1. Those that prevent erosion, including the stabilization practices described in Chap. 12
2. Those that trap pollutants before they can be discharged, including the structural practices described in Chap. 13
3. Those that prevent pollutants from the construction materials from mixing with storm water, including the practices described in Chap. 14

Although these three types of BMPs have different functions, the basic principle is the same: These BMPs are designed to prevent, or at least control, the pollution of storm water before it has a chance to affect receiving streams.

Purpose of the SWPPP

The SWPPP provides a description of the site and the project that identifies sources of pollution to storm water discharges. The SWPPP also identifies and describes appropriate measures to reduce pollutants in storm water discharges, and it provides other information necessary for compliance with the storm water discharge permit.

The SWPPP is the centerpiece of the NPDES storm water permit, and the implementation of the SWPPP is the key to controlling pollutants in storm water discharges. Storm water pollution prevention plan requirements were designed to allow maximum flexibility to develop the needed storm water controls based on the specifics of the site. The development of an SWPPP for a particular site should include consideration of local regulations and requirements, the climate and soils of the area, the topography and layout of the site, the project schedule and other operations that will be conducted on the site during the construction project, the sensitivity of nearby water bodies, concerns for public safety, and other site-specific issues.

SWPPP Preparation Date

A SWPPP must be prepared before the NOI is submitted, and the operator(s) must begin implementing the plan at the time that the NOI is submitted (See Chap. 10). The SWPPP generally is not submitted with the NOI; it is filed on-site along with inspection reports (described later in this chapter). Note that some state and local agencies require that the SWPPP be submitted for review and approval.

The planning for pollution prevention measures should be done while the site construction plan is being developed. The best SWPPPs are developed at the same time as the design of the site plan. However, if the site plan design has been completed before one begins to prepare the SWPPP, much of the information needed for the plan should already be included in the design documents. An SWPPP can be prepared for most construction projects by using information from the existing design and modifying the design to accommodate the controls. The plan should be in place prior to project initiation because construction operations pose environmental risks as soon as activity begins. The initial rough-grading activities may contribute a significant amount of pollutants to storm water runoff. However, not all the pollution prevention measures included in the SWPPP must be implemented on the first day of construction; the elements of the plan may be phased in according to a sequence that fits with the construction schedule, seasonal changes, and other factors.

Responsibility for developing an SWPPP typically lies with the operator of the construction project, as defined in Chap. 10. There must be at least one SWPPP for a site which incorporates the required elements for all operators, but there can be separate plans if individual permittees so desire. Operators may be able to achieve cost savings by having a joint SWPPP. For example, the prime developer could assume the inspection responsibilities for the entire site, while each home builder shares in the installation and maintenance of sediment traps serving common areas.

Steps in Developing an SWPPP

The process of developing and implementing an SWPPP for construction activities involves a total of six steps. The first three steps involve the development of the plan:

1. *Site and project description,* which includes collecting information about the site and the proposed construction project
2. *Storm water assessment,* which includes generating quantitative measurements and computations of factors that affect the quantity and rate of storm water and sediment runoff
3. *Control selection and plan design,* which includes selecting pollution prevention measures and documenting their location and operation.

After the SWPPP has been developed, three additional steps must be taken to implement the plan:

4. *Certification and notification,* which includes all the legal and procedural issues involved with preparing to commence construction
5. *Construction and implementation,* which include all the various issues that must be addressed in order to successfully complete the construction and comply with the terms of the EPA general permit
6. *Final stabilization and termination,* which include those tasks necessary to properly terminate permit coverage

The following sections describe the processes involved in each of the steps listed above.

Step 1: Site and Project Description

A good map of the site is probably the most useful document for development of an SWPPP. The map should be drawn to scale, and the scale map should be large enough to easily distinguish important features such as drainage swales and control measures. A map scale of approximately 1:1000 is convenient for small projects. For larger projects, smaller map scales may be necessary. For these projects, swales and other linear features may need to be represented using centerlines only, and point features such as pollution control structures may need to be represented using simplified symbols. The site map should show at least the following:

- The topography of the site if possible, using contour lines. The topography may be based on published sources [such as city or county maps, or U.S. Geological Survey (USGS) maps], or it may be based on topographic survey work performed for the project.
- The existing land use for the site (i.e., wooded area, open grassed area, pavement, building, etc.).
- The location and names (if available) of surface waters that are located on or next to the site (including wetlands, streams, rivers, lakes, and ponds) and the locations of discharge points into these surface waters.
- Other major features and potential pollutant sources, such as locations of impervious structures and soil storage piles.

In addition to the map, a written site description should be prepared and should include at least the following items:

- The general location, size, shape, and other general features of the site, including its existing land use and vegetation.
- The name of receiving waters and a discussion of the size, location, and type of wetlands at the site. If the receiving water is a tributary, include the name of the ultimate body of water if possible. The receiving waters could include a river, lake, stream, creek, run, estuary, wetland, bay, ocean, or any other body of surface water. If the site drains into a municipal separate storm sewer system (MS4), identify the system and indicate the receiving water to which the system discharges. This information usually is available from county, state, or USGS maps.
- The rainfall characteristics of the project area. This information will provide an indication of the climatic factors that may contribute to storm water pollution problems. Figure 13.7 shows the 2-year, 24-h rainfall depths for the eastern United States.
- The type(s) of soils present on the site. This information may be

obtained from a USDA Natural Resources Conservation Service (NRCS) soils map of the area, or other sources as appropriate.

- Information on endangered and threatened species, including whether any endangered species are in proximity to the project. See Chap. 10 for more details.

A written project description should be prepared and should include the following:

- The nature of the construction activity, including the function of the project—residential development, commercial, industrial, institutional, office development, highway projects, roads, streets, or parking lots, recreational areas, underground utility, etc.
- The intended significant activities, presented sequentially, that disturb soil over major portions of the site: clearing and grubbing, excavation and stockpiling, rough-grading, final or finish-grading, preparation for seeding or planting, excavation of trenches, demolition, etc.
- The approximate time it will take to complete the project.
- Off-site locations of equipment storage, material storage, waste storage, and borrow/fill areas.
- Estimates of the total area of the site and the area that is expected to be disturbed by excavation, grading, or other activities, including off-site borrow/fill areas. If the site includes more than one drainage area (as is common), then each drainage area should be described separately. In addition, the drainage area boundaries on the project site may change at different stages of construction. Therefore, it may be necessary to describe the drainage areas, and the disturbed areas in each drainage area, during each major phase of construction.
- A description of any discharge associated with industrial activity other than construction (including storm water discharges from dedicated asphalt plants, concrete plants, etc.) and the location of that activity on the construction site.

Step 2: Storm Water Assessment

After collecting all available information on the site and on the proposed construction project, the next step is to assess the storm water runoff and the potential for storm water pollution.

The assessment should include consideration of anticipated drainage patterns and slopes after major grading activities. The drainage patterns should be indicated on the site map by including drainage basin boundaries and drainage channels or pipes. A *drainage basin* is an area of the site in which water, sediments, and dissolved materials drain to a common outlet (such as a swale or storm drainpipe) from the site. If possible, indicate the revised grades on the same topographic map as the existing grades. Use two separate symbols for existing contours and proposed contour (i.e., dashed and solid lines). If a topographic map of the site cannot be prepared, then examine the proposed plan for the site and indicate on the site map the approximate location, direction, and steepness of slopes.

The assessment should also consider the areas of soil disturbance on the site. Draw the limit of disturbance on the site map so that any soil-disturbing activity such as clearing, stripping, excavation, backfilling, stockpiling (topsoil or other fill material), and paving will be inside the limit. The limit of disturbance should also include roads for construction vehicles unless those roads are paved (or stabilized) and have measures to reduce tracking of sediments.

To provide better estimates of the volume and rate of runoff from each area of the project site, the site's runoff coefficient during and after construction should be assessed. The *runoff coefficient* is defined as the fraction of total precipitation that will appear at a conveyance as runoff (versus infiltrated precipitation). The runoff coefficient is used in the well-known "rational formula" or "rational method" equation

$$Q = CIA$$

where Q = rate of runoff from an area, ft^3/s
C = runoff coefficient, %
I = rainfall intensity, in/h
A = area of drainage basin, acres

The EPA does not require the use of the rational method to design storm water conveyances or management measures, but the method is generally acceptable if applied appropriately. State or local design guidelines may describe appropriate methods to use for estimating design flow rates from each development.

Runoff coefficients can be estimated from site plan maps, which show where impervious surfaces, vegetation, and permeable surfaces will be. Higher C values indicate a higher percentage of impervious cover. For example, the C value of a lawn area is 0.2, which means that only 20% of the rainfall landing on that area will run off—the rest will be absorbed or will evaporate. A paved parking area would have a C value of 0.9, which means that 90% of the rainfall landing on that area will become runoff.

The C value calculated for the SWPPP is the one that represents the final condition of the site after construction is complete. A runoff coefficient should be calculated for each drainage basin on the site. The runoff coefficient or C value for a variety of land uses may be found in Table 11.1.

When a drainage area contains more than one type of surface material with more than one runoff coefficient, a *weighted* C must be calculated. This weighted C will take into account the amount of runoff from all the various parts of the site. A formula used to determine the weighted C is

TABLE 11.1 Typical C Values

Description of area	Runoff coefficients
Business: downtown areas	0.70–0.95
Business: neighborhood areas	0.50–0.70
Residential: single-family areas	0.30–0.50
Residential: multifamily detached	0.40–0.60
Residential: multifamily attached	0.60–0.75
Residential: (suburban)	0.25–0.40
Apartment dwelling areas	0.50–0.70
Industrial: light areas	0.50–0.80
Industrial: heavy areas	0.60–0.90
Parks, cemeteries	0.10–0.25
Playgrounds	0.20–0.35
Railroad yard areas	0.20–0.40
Unimproved areas	0.10–0.30
Streets: asphalt	0.70–0.95
Streets: concrete	0.80–0.95
Streets: brick	0.70–0.85
Drives and walks	0.75–0.85
Roofs	0.75–0.95
Lawns: coarse-textured soil (>85% sand)	
Slope: flat (2%)	0.05–0.10
Slope: average (2–7%)	0.10–0.15
Slope: steep (7%)	0.15–0.20
Lawns: fine-textured soil (>40% clay)	
Slope: flat (2%)	0.13–0.17
Slope: average (2–7%)	0.18–0.22
Slope: steep (7%)	0.25–0.35

SOURCE: American Society of Civil Engineers, 1970.

$$C = \frac{A_1C_1 + A_2C_2 + \cdots + A_xC_x}{\sum_{i=1}^{x} A_i}$$

where A = area in acres and C = runoff coefficient. Therefore, if a drainage area has 15 acres with 5 paved acres ($C = 0.9$), 5 grassed acres ($C = 0.2$), and 5 acres in natural vegetation ($C = 0.1$), a weighted C is calculated as follows:

$$\mathrm{C} = \frac{5 \times 0.9 + 5 \times 0.2 + 5 \times 0.1}{5 + 5 + 5} = \frac{4.5 + 1.0 + 0.5}{15} = \frac{6.0}{15} = 0.4$$

Step 3: Control Selection and Plan Design

After collection of the necessary measurements and other information and an assessment of the storm water and sediment sources on the site, the next step is to design a plan to prevent and control pollution of storm water runoff from the construction site. The SWPPP should address the following types of controls:

- Stabilization practices for erosion and sediment control. These should minimize the amount of disturbed soil, using vegetation or other measures as described in Chap. 12.
- Velocity dissipation measures, which slow down the runoff flowing across the site by using measures such as check dams and surface roughening. These measures are described in Chap. 13.
- Storm water management controls, which prevent runoff from flowing across disturbed areas by using measures such as earth dikes, drainage swales, and pipe slope drains. These measures are described in Chap. 13.
- Sediment capture measures, which remove sediment from on-site runoff before it leaves the site by using measures such as silt fences, straw bale barriers, sediment traps, storm drain inlet protection, and brush barriers. These measures are described in Chap. 13.
- Other pollution prevention measures, which are intended to control spills, minimize airborne dust, reduce off-site vehicle tracking of sediments, provide for proper waste disposal, deal with worker sanitation, and address other issues, as described in Chap. 14.

The SWPPP must describe the intended sequence of major storm water control activities and when, in relation to the construction process, they will be implemented. Specific calendar dates are not required in describing the sequence of implementation, because the

EPA recognizes that many factors can impact the actual construction schedule. All the tasks required for construction of control measures, earth-disturbing construction activities, and maintenance activities for control measures should be listed in the order in which they will occur. Specific timing requirements for installation and maintenance of control measures are dependent upon the measures and/or the construction activities. There are, however, several general principles that will influence the sequence of major activities:

- Install downslope and side-slope perimeter controls before the land-disturbing activity occurs.
- Do not disturb an area until it is necessary for construction to proceed.
- Cover or stabilize as soon as possible.
- Time activities to limit the impact from seasonal climate changes or weather events. Concentrate most of the land disturbance during expected dry seasons, if possible.
- Delay construction of infiltration measures until the end of the construction project when upstream drainage areas have been stabilized.
- Do not remove temporary perimeter controls until all upstream areas have been stabilized.

The site map and descriptive material should identify which parts of the overall project are under the control of each operator. The SWPPP should also identify who will be responsible for implementing each measure contained in the plan. The operators must provide to their contractors and subcontractors necessary information on complying with their SWPPP and the permit. Off-site material storage areas must be addressed in the SWPPP.

The site operator has substantial control over the control measures selected and implemented at a particular location. The EPA construction general permit does not dictate that certain control measures are always required, regardless of the particular project and site conditions. However, the EPA construction general permit does provide guidelines that must be followed in the selection of control measures.

Stabilization measures

The guidelines in the EPA construction general permit state that stabilization measures must be instituted on disturbed areas as soon as practicable, but no more than 14 days after construction activity has temporarily or permanently ceased on any portion of the site, except under the following three conditions:

- When construction activities will resume on a portion of the site within 21 days from suspension of previous construction activities
- When the initiation of stabilization measures is precluded by snow cover or frozen ground, in which case they must be initiated as soon as practicable.
- Where the initiation of stabilization measures is precluded by arid conditions, either seasonal or permanent. The initiation of stabilization measures might be permanently precluded in arid areas (areas with an average annual rainfall of 0 to 10 in) or semiarid areas (10 to 20 in). In other areas that are experiencing droughts, stabilization measures must be initiated as soon as conditions allow.

Chapter 12 provides additional guidance on stabilization measures.

Structural sediment and erosion control measures

The EPA construction general permit guidelines state that structural measures should be implemented according to the size of the disturbed area draining to a common discharge point.

For sites with more than 10 acres (4 ha) disturbed at a time, all of which are served by a common drainage location, a sediment basin providing 3600 ft^3/acre (250 m^3/ha) of drainage area, or equivalent control measures (such as suitably sized dry wells or infiltration structures) must be provided where practicable until final stabilization of the site has been accomplished. In lieu of the default 3600 ft^3/acre (250 m^3/ha), the basin size can be calculated from the expected runoff volume from the local 2-year, 24-h storm event and local runoff coefficient. Flows from off-site or on-site areas that are undisturbed or have undergone final stabilization may be diverted around both the sediment basin and the disturbed area. These diverted flows can be ignored when the sediment basin is designed.

For those drainage locations that serve more than 10 acres (4 ha) disturbed at a time and where a sediment basin designed according to the above guidelines is not feasible, smaller sediment basins or traps should be used. At a minimum, silt fences, vegetative buffer strips, or equivalent sediment controls are required for all downslope and appropriate midslope boundaries of the construction area. Diversion structures should be used on upland boundaries of disturbed areas to prevent run-on from impacting disturbed area. For drainage locations serving 10 acres (4 ha) or less, smaller sediment basins or sediment traps should be used, and at a minimum, silt fences or equivalent sediment controls are required for all downslope and appropriate midslope boundaries of the construction area. Alternatively, a sediment basin providing storage for 3600 ft^3/acre (250 m^3/ha) (or the alternative calculated volume) may be provided. Diversion structures should be installed on upland boundaries of disturbed areas to prevent run-on.

No matter what the size of the disturbed area may be, reliance on a single sediment and erosion control measure is not wise. Most sediment capture measures, even sediment basins with a relatively long holding time for storm water runoff, are seldom able to achieve better than about 80% removal of sediment from storm water. It is much more effective to use a combination of sediment and erosion control measures to achieve maximum pollutant removal. These should include stabilization as well as storm water management. Preventing storm water from coming into contact with disturbed soil is the best sediment control measure. However, this is not possible at all times. Therefore, sediment capture measures can provide a less effective, but necessary, form of backup protection for excessive sediment discharges.

Permanent storm water management controls

The SWPPP should also address permanent storm water management controls constructed to prevent or control pollution of storm water after the construction is completed. Generally, these controls work by controlling the volume and/or velocity of runoff, as well as by reducing the quantity of pollutants discharged after construction has been completed. Options for storm water management measures that should be evaluated in the development of plans include storm water retention/detention structures and wet ponds, on-site infiltration of precipitation, flow attenuation by use of open vegetated swales and natural depressions, velocity dissipation measures, and sequential systems using multiple methods. Additional details are provided in Chap. 13.

The pollution prevention plan must include an explanation of the technical basis used to select control measures, where flows exceed predevelopment levels. This explanation should address how a number of factors were evaluated, including the pollutant removal efficiencies of the measures, costs of the measures, site-specific factors that will affect the utility of the measures, whether the measure is economically achievable at a particular site, and any other relevant factors. In selecting storm water management measures, the impacts of each method on other water resources, such as groundwater, should be considered.

Description of control measures

The SWPPP should include a narrative description of each pollution prevention measure and the location of each control measure on the site map (if possible). It is often convenient to assemble a list of each type of control planned for the site. Include in this list a description of each control, its purpose, and why it is appropriate in this location. The

Figure 11.1 Illustration of sediment accumulation requiring maintenance.

description should include specific information about the control measure such as size, materials, and methods of construction.

The SWPPP should also describe the maintenance, inspection, repair, employee training, and record-keeping procedures that will ensure that control measures remain effective and in working order during the construction activity. There must be a plan for the inspection and maintenance of vegetation, erosion and sediment control measures, and other protective measures. These controls must be in good operating condition until the area they protect has been completely stabilized or the construction project is complete.

An inspection and maintenance checklist should be prepared that addresses each of the control measures proposed for the facility. A blank checklist for the facility should be included in the SWPPP prior to the onset of construction. The inspector should complete a copy of the blank checklist during each inspection. The inspection and maintenance checklist should be prepared based upon the requirements for each individual measure. For example, sediment must be removed from a silt trap when it has been filled to one-third of its depth. Consult state or local manuals for maintenance requirements for control measures. (See Fig. 11.1.)

A good maintenance and inspection plan should include the following:

1. *Areas to be inspected and maintained.* These should include all the disturbed areas and material storage areas of the site.

2. *Measures to be inspected and maintained.* These should include all the erosion and sediment controls identified in the SWPPP.
3. *Inspection schedule.* Provide an inspection schedule for each area and measure, as described in the following section.
4. *Maintenance procedures.* List the typical maintenance procedures for each measure. Chapters 12 and 13 provide maintenance guidelines for various pollution prevention measures. Maintenance requirements should be considered as measures are selected. Some measures require a good deal more maintenance than others, and this may influence the selection of measures.
5. *Responsible parties.* Describe the procedure to be followed if additional repair is required; e.g., who will be responsible or whom to call.
6. *Inspection and maintenance forms.* Provide forms and instructions for record-keeping practices.
7. *Personnel.* List the names of personnel assigned to each task.
8. *Training.* Indicate what training the employees will need to be able to do the job.

The EPA general permit requires that inspections of pollution prevention measures be performed at a minimum once every 14 calendar days plus additional inspections within 24-h after any storm event of greater than 0.5 in (13 mm) of rain per 24-h period. If possible, a "walk-through" inspection of the construction site should be completed before anticipated storm events (or series of events such as intermittent showers over a period of days) that could potentially yield a significant amount of runoff.

For sites that have undergone stabilization (temporary or final), inspections must be conducted at least once per month. Monthly inspections also are adequate for arid or semiarid areas, and during the dry seasons in areas that meet the definition of arid or semiarid areas on a seasonal basis. A *dry season* may be considered as occurring during any month in which the average monthly rainfall is less than 1.67 in (42 mm), based on long-term averages. Where construction activity has been halted due to frozen conditions, inspections are not required until 1 month before snowmelt runoff is expected to commence.

The operator(s) must designate a qualified person or persons to perform the following inspections:

- *Stabilization measures* (*inspected monthly or as needed*). Disturbed areas and areas used for storage of materials that are exposed to precipitation will be inspected for evidence of, or the potential for, pollutants entering the drainage system. The inspection should reveal whether the area was stabilized correctly, whether there has been damage to the area since it was stabilized, and what should be done to correct any problems. Figure 11.2 shows an example of an inspection form for stabilization measures.

- *Structural controls* (*inspected every 14 days or as needed*). Filter-fabric fences and all other erosion and sediment control measures identified in the plan will be inspected regularly for proper positioning, anchoring, and effectiveness in trapping sediments. The inspection should reveal whether the control was installed correctly, whether there has been damage to the control since it was installed, and what should be done to correct any problems. Sediment will be removed from the upstream or upslope side of the filter fabric. Figure 11.3 shows an example of an inspection form for structural controls.
- *Discharge points* (*inspected every 14 days or as needed*). Discharge points or locations will be inspected to determine whether erosion control measures are effective in preventing significant amounts of pollutants from entering receiving waters. This can be done by inspecting the waters for evidence of erosion or sediment introduction. If discharge points are inaccessible, the permit requires that nearby downstream locations be inspected, if practicable.
- *Construction entrances* (*inspected every 14 days or as needed*). Locations where vehicles enter or exit the site will be inspected for evidence of off-site sediment tracking.
- *Areas used for storage of exposed materials* (*inspected every 14 days or as needed*). These are locations where construction materials (including topsoil and related items) are stored, whether on-site or at another location.

If an inspection reveals inadequacies, the site description and pollution prevention measures identified in the SWPPP must be revised. All necessary modifications to the SWPPP must be made within 7 calendar days following the inspection. If existing BMPs need to be modified or if additional BMPs are necessary, implementation must be completed before the next anticipated storm event. If implementation before the next storm event is impracticable, they must be implemented as soon as is practicable.

Inspection

Blank inspection and maintenance forms should be prepared prior to the start of the construction activity. The inspection forms should be specific to the construction project and the SWPPP. The forms should list each of the measures to be inspected on the site. The form should include blanks for the inspector to fill in: her or his name, date of inspection, condition of the measure or area inspected, maintenance or

Inspection Report Form For Stabilization Measures

Inspector ________________ date: ________________________________

Days since last rainfall: ____

amount of last rainfall: ________________________________inches

Area	Date last disturbed	Date of next disturbance	Stabilized?	Stabilized with	Condition

Stabilization required:

____

____

____

____

____

____

____

____

____

____

To be performed by: ________________________________ __________

On or before: ____________________________________ __________

Figure 11.2 Example inspection report form for stabilization measures.

Inspection form for structural controls

Inspector ______________ date: ______________

Days since last rainfall: ______________

amount of last rainfall: ______________ inches

Location of control	In place?	Condition	Sediment depth	Washed out or overtopped?

Maintenance required:

____

____

____

____

____

____

____

____

____

____

To be performed by: ______________

On or before: ______________

Figure 11.3 Example inspection form for structural controls.

repair performed, and any changes which should be made to the SWPPP to control or eliminate unforeseen pollution of storm water. If no incidents of noncompliance were found, the report shall contain a certification that the facility is in compliance with the SWPPP and this permit. Figure 11.4 shows a sample form to be used for certification of the inspection report. The report will be signed by one of the following persons:

- The principal executive officer or ranking elected official of the municipality or agency submitting the SWPPP
- A duly authorized representative of the principal executive officer or ranking elected official

There are no formal requirements for inspectors. Any person authorized and considered qualified by the operator may complete inspections and sign the inspection reports. The EPA recognizes that experience is the best way to develop an understanding of pollution prevention measures, although formal training in sediment and erosion control can be very helpful.

Maintenance of erosion and sediment controls

Erosion and sediment controls can become ineffective if they are damaged or not properly maintained. The SWPPP requires all erosion and sediment control measures to be maintained in effective operating condition. If site inspections identify BMPs that are not operating effectively, maintenance must be performed before the next anticipated storm event. If maintenance before the next anticipated storm event is impracticable, maintenance must be completed as soon as practicable.

Sediment should be removed from the upstream or upslope side of the filter-fabric fences, sediment traps, or other devices when the depth of accumulated sediment reaches about one-third the height of the structure. Sediment should also be removed from storm sewer inlets as needed. After disturbed areas have been stabilized, temporary controls should be removed. Sediment that is removed from structural barriers should be hauled off the site for proper disposal, or used as backfill on the site.

Incorporating state or local requirements

Many states, tribes, municipalities, and counties have developed sediment and erosion control requirements for construction activities. A significant number have also developed complete storm water management requirements. In a situation where there are similar requirements under different programs, a project should comply with the more

Inspection Certification Form

Project: ______________________________

This certification must be completed after each inspection to signify that the inspection has been properly completed and the site has been found to be in compliance with the Storm Water Pollution Prevention Plan.

I certify under penalty of law that this document and all attachments were prepared under my direction or supervision in accordance with a system designed to assure that qualified personnel properly gathered and evaluated the information submitted. Based on my inquiry of the person or persons who manage the system, or those persons directly responsible for gathering the information, the information submitted is, to the best of my knowledge and belief, true, accurate, and complete. I am aware that there are significant penalties for submitting false information, including the possibility of fine and imprisonment for knowing violations.

Signed: ______________________________

Name: ______________________________

Title: ______________________________

Company: ______________________________

Address: ______________________________

Telephone: ______________________________

Date: ______________________________

Figure 11.4 Example inspection certification form.

stringent of the requirements. The EPA requires that all state and local requirements be met. The NPDES storm water discharge is not intended to supersede state or local requirements. Instead, the NPDES storm water permit ensures that a minimum level of pollution prevention is required.

Permittees may use existing plans or local approvals as part of their pollution prevention plans when such use is appropriate, especially if the state or local authority requires review and approval of the sediment and erosion control plan.

Step 4: Certification and Notification

Once the site description and the controls portion of the SWPPP have been prepared, the pollution prevention plan can be certified. An example of a certification form is shown in Fig. 11.5. The Notice of Intent for coverage under the general permit may then be submitted. Chapter 10 provides additional details on the Notice of Intent and certification requirements. The signatory requirements for the SWPPP are exactly the same as those for the NOI.

Regulatory review of SWPPP

Some NPDES storm water permits for construction sites may require that SWPPPs be submitted to the director of the regulatory agency for review. However, the EPA general permit requires only that plans be maintained on-site. Permitting authorities may prefer not to require plans to be submitted, to reduce the administrative burden of reviewing a large number of pollution prevention plans.

Permittees must make SWPPPs available, upon request, to EPA, state, tribal, or local agencies approving sediment and erosion plans, grading plans, or storm water management plans. Plans may also have to be sent to local government officials or to the operator of the municipal separate storm sewer which receives the discharge.

The EPA or the state may notify the permittee at any time that his or her plan does not meet one or more of the requirements. The notification will identify which requirements of the permit are being unmet and which elements of the SWPPP require modification. Within 7 calendar days of receipt of notification from the EPA (or as otherwise requested by the EPA), the required changes to the plan must be made and a certification submitted that the changes have, in fact, been made and implemented.

Public access and notification

The SWPPP and related documents are considered to be reports according to Section 308(b) of the Clean Water Act and, therefore, are

Certification of Storm Water Pollution Prevention Plan

Project: ______________________________

This certification must be completed by an authorized signatory of each operator (generally the Owner and the General Contractor) before the effective date of the Plan.

I certify under penalty of law that this document and all attachments were prepared under my direction or supervision in accordance with a system designed to assure that qualified personnel properly gathered and evaluated the information submitted. Based on my inquiry of the person or persons who manage the system, or those persons directly responsible for gathering the information, the information submitted is, to the best of my knowledge and belief, true, accurate, and complete. I am aware that there are significant penalties for submitting false information, including the possibility of fine and imprisonment for knowing violations.

Signed: ______________________________

Name: ______________________________

Title: ______________________________

Company: ______________________________

Address: ______________________________

Telephone: ______________________________

Date: ______________________________

Figure 11.5 Certification of storm water pollution prevention plan.

available to the public. This is true despite the fact that plans and associated records are not necessarily required to be submitted to the EPA or any other regulatory agency.

A notice about the permit and SWPPP must be conspicuously posted near the main entrance of the site. If displaying near the main entrance is not feasible, the notice can be posted in a local public building such as the town hall or public library. For linear projects, the notice must be posted at a publicly accessible location near the active part of the construction project (e.g., where a pipeline project crosses a public road). The permit notice must include the project's NPDES permit number, the name and phone number of a local contact, a brief project description, and the location of the SWPPP, if not kept on-site.

The permit does not require that the general public have access to the site, nor does it require that operators provide copies of the plan, or mail copies of the plan, to members of the public. However, the EPA will assist members of the public who request assistance in obtaining access to permitting information, including SWPPPs.

Permittees may claim certain portions of their SWPPP as confidential according to the regulations at 40 CFR Part 2. Basically, these regulations state that records which contain trade secret information may be claimed as confidential. These confidential portions may be excluded from public access.

Step 5: Construction and Implementation

The first step in carrying out the SWPPP is to construct or apply the initial controls selected for the plan. This should be done in accordance with state or local standard specifications and/or good engineering practices, as appropriate, and in the order indicated in the sequence of major activities. All workers on the construction site should be made aware of the controls so that workers do not inadvertently disturb or remove them.

Storm water pollution prevention plans are developed based on site-specific features and functions. When there are changes in design, construction, operation, or maintenance, and those changes will have a significant effect on the potential for discharging pollutants in storm water at a site, the SWPPP should be modified by the permittee to reflect the changes and new conditions. For example, a change in the construction schedule or design specifications should be incorporated in the SWPPP. The plan should also be modified when it proves to be ineffective in controlling pollutants. This determination could be based on the results of regular visual inspections. In addition, the plan must be updated to identify any new operator who will implement a portion of the SWPPP.

Report Form For Changes In Pollution Prevention Plan

Inspector: ____________________

Date: ____________________

Summary of required changes:

____

____

____

____

____

Reason(s) for changes:

____

____

____

____

Inspector's signature: ____________________

Date: ____________________

Figure 11.6 Example report form for changes in pollution prevention plan.

Figure 11.6 illustrates a typical form used to request changes in the SWPPP. This type of form should be available to the inspector and to other employees who use the SWPPP.

Step 6: Final Stabilization and Termination

The NPDES permit may remain in effect as long as there is a storm water discharge associated with construction activity, as defined in Chap. 10. The need for a permit may end because the storm water discharge has been eliminated (e.g., through the construction of a large retention basin), because a new operator has taken over the construction project, or because the construction is finished and final stabilization is completed. Final stabilization requirements may vary from permit to permit. However, typically all temporary measures and controls must be removed, all permanent storm water management measures must be in place and functional, and all areas stabilized by vegetation must exhibit at least a 70% viability of the plant cover, as compared with typical natural vegetation in the area.

The EPA general permit requires the permit coverage to be terminated by submitting a Notice of Termination. The NOT is typically the final task required to comply with the requirements of an NPDES storm water permit for a construction activity. Chapter 10 provides some additional details on the NOT.

Recordkeeping is an essential part of compliance with an NPDES storm water discharge permit. The recordkeeping requirements begin with the preparation of the SWPPP and NOI (or permit application) and continue through the NOT and afterward. The SWPPP, NOI, inspection reports, maintenance reports, construction activity reports, NOT, and any other materials relevant to the permit, including spill reports, correspondence with regulatory authorities, and photographs, should be kept. All this information must be kept for least 3 years from the date of final stabilization. This period may be extended by request of the EPA or state.

References

American Society of Civil Engineers, 1967. *Design and Construction of Sanitary and Storm Sewers,* Manual of Practice 37, New York, 1970 (fifth printing, 1982).

Environmental Protection Agency, 1992. *Storm Water Management for Construction Activities. Developing Pollution Prevention Plans and Best Management Practices,* EPA Report EPA/832/R-92/005, September.

National Weather Service, 1961. *Rainfall Frequency Atlas of the United States,* Technical Paper 40, U.S. Department of Commerce, Washington.

———, 1964. *Two- to Ten-Day Precipitation for Return Periods of 2 to 100 Years in the Contiguous United States,* Technical Paper 49, U.S. Department of Commerce, Washington.

———, 1977. *Five- to 60-Minute Precipitation Frequency for the Eastern and Central United States,* Technical Memorandum NWS HYDRO-35, National Oceanographic and Atmospheric Administration, U.S. Department of Commerce, Silver Spring, Md.

Chapter

12

Stabilization Practices for Construction Erosion Control

This is the third of five chapters which deal with the specific requirements for storm water discharges from construction facilities. This chapter provides details on stabilization measures (such as vegetation) for erosion control.

Erosion and Sediment Controls

Erosion controls, as described in this chapter, provide the first line of defense in preventing off-site sediment movement and are designed to prevent erosion by protecting soils. Sediment controls, as described in Chap. 13, are generally designed to remove sediment from runoff before the runoff is discharged from the site. Soil erosion and sediment control are not a new technology. The USDA Soil Conservation Service and a number of state and local agencies have been developing and promoting the use of erosion and sediment control devices for years.

The EPA general permit requires that stabilization practices be implemented within 14 days after activity in a portion of the site ceases. They are not required if construction resumes within 21 days. State permit requirements may vary. Stabilization practices include measures to prevent soil loss from disturbed areas, including temporary or permanent seeding, mulching, geotextiles, sodding, vegetative buffer strips, protection of trees, preservation of mature vegetation, and other appropriate measures.

General Stabilization Guidelines

Stabilization refers to the covering or maintaining of an existing cover over soil. Stabilization of exposed soil is one of the foremost means of minimizing pollutant discharge during construction activities. Stabilization reduces erosion potential by absorbing the kinetic energy of raindrops that would otherwise mobilize unprotected soil; by intercepting water so that it infiltrates the ground instead of running off the surface; and slowing the velocity of runoff, thereby promoting deposition of sediment already being carried. Stabilization provides large reductions in the levels of suspended sediment in discharges and receiving waters.

Several general guidelines pertain to selecting stabilization practices for a construction project:

1. *Minimize impervious surfaces.* Impervious surfaces will increase runoff and may increase erosion and pollutant discharges. Therefore, while soil may be stabilized by covering it with an impervious surface, the runoff from this impervious area may contribute to the destabilization of adjacent areas.
2. *Minimize the disturbed area.* Minimizing the amount of disturbed soil on the construction site will decrease the amount of soil which erodes from the site and can decrease the amount of controls required to remove the sediment from the runoff. If possible, clear only those portions of the site necessary for construction. When a smaller area is disturbed for construction, there is less erosion of soil.
3. *Construct the project in stages.* If the construction project will take place over a widespread area, consider staging or phasing the project so that only a small portion of the site will be disturbed at one time. Phased construction helps to lessen the risk of erosion by minimizing the amount of disturbed soil that is exposed at a given time. There are exceptions to this guideline. For example, if an efficient drainage system for the property is constructed as part of the first phase of construction, then succeeding phases may result in increased sediment downstream because the efficient drainage system will serve to deliver a high proportion of the sediment to the discharge point.
4. *Match existing land contours.* A construction project site should be selected and laid out so that it fits into existing land contours, whenever possible. When a project significantly changes the grades in an area, there is a corresponding increase in the amount of disturbed soil, which in turn increases the amount of erosion.
5. *Use temporary stabilization.* If there are disturbed portions of the site that will not be redisturbed for a long period, then these areas should be stabilized with temporary seeding or mulching. This will

reduce the amount of erosion from these areas until they are disturbed again.

6. *Use final stabilization.* By permanently stabilizing the disturbed areas as soon as possible after construction has been completed in those areas, the amount of sediment which should be trapped before it leaves site can be significantly reduced.

7. *In snow-covered areas, wait for snowmelt.* If snow cover prevents seeding or planting vegetation, then wait until the snow melts before stabilizing the area.

8. *In dry areas, use other methods.* If there is not enough rainfall on the disturbed area to allow vegetation to grow, then seed and irrigate the disturbed area (if allowed by the permit for the facility) or stabilize the disturbed areas by nonvegetative methods, as described later in this chapter.

The remainder of this chapter describes these common stabilization practices:

- *Vegetative stabilization practices:* temporary seeding, sod stabilization, permanent seeding and planting, buffer zones, and preservation of natural vegetation
- *Nonvegetative stabilization practices:* mulching, geotextiles, chemical stabilization, stream bank stabilization, soil-retaining measures, and dust control

Vegetative Stabilization Practices

Preserving existing vegetation or revegetating disturbed soil as soon as possible after construction is the most effective way to control erosion and can provide a sixfold reduction in discharge suspended-sediment levels (Metropolitan Washington Council of Governments, 1990). Vegetation reduces erosion potential through several different processes:

1. *Rainfall protection:* shielding the soil surface from the direct erosive impact (kinetic energy) of raindrops.
2. *Flow protection:* dispersing and decreasing the velocity of surface flow, which reduces erosion and allows sediment to drop out or deposit.
3. *Soil retention:* physically holding the soil in place with plant roots.
4. *Infiltration:* improving the soil's water storage porosity and capacity through the incorporation of roots and plant residues, so that more water can infiltrate the ground.

5. *Transpiration:* conducting moisture into plants, where it is eliminated by evaporation. This decreases the soil moisture content and increases the soil's moisture storage capacity.

Vegetative cover can be grass, trees, vines, or shrubs. Grasses are the most common type of cover used for revegetation because they grow quickly, providing erosion protection within days. Other soil stabilization practices such as mulching may be used during nongrowing seasons to prevent erosion. Newly planted shrubs and trees establish root systems more slowly, so keeping existing ones is a more effective practice.

Vegetative and other site stabilization practices can be either temporary or permanent controls. Temporary controls provide a cover for exposed or disturbed areas for short periods or until permanent erosion controls are put in place. Permanent vegetative practices are used when activities that disturb the soil are completed or when erosion is occurring on a site that is otherwise stabilized.

Temporary seeding

Temporary seeding means growing a short-term vegetative cover (plants) on disturbed site areas that may be in danger of erosion. Temporary seeding provides for temporary stabilization by establishing vegetation of areas of the site which will be disturbed at some time during the construction operation and where work (other than the initial disturbance) is not conducted until some time later in the project. Soils at these areas may be exposed to precipitation for an extended period, even though work is not occurring on these areas.

For temporary seeding, fast-growing grasses are used whose root systems hold down the soil, so that the soil is less apt to be carried off the site by storm water runoff or wind. Temporary seeding also reduces the problems associated with mud and dust from bare soil surfaces during construction. In most climates, temporary seeding typically is appropriate for areas exposed by grading or clearing for more than 7 to 14 days. Temporary seeding practices have been found to be up to 95% effective in reducing erosion (USDA Natural Resources Conservation Service, 1985).

Probably the most common type of grass used for temporary seeding is rye grass. (See Fig. 12.1.) Sterile wheat is also used. A new grass which has received favorable reports is vetiver, which is native to India. Sterile varieties of vetiver are available (National Academy Press, 1992).

Temporary seeding should be performed on areas which have been disturbed by construction and which are likely to be redisturbed, but not for at least 21 days. Typical areas might include denuded areas, soil

Figure 12.1 Construction site seeded with rye grass.

stockpiles, dikes, dams, sides of sediment basins, and temporary road banks. Temporary seeding should be done as soon as practicable following the last land-disturbing activity in an area.

Temporary seeding may not be an effective practice in arid and semiarid regions where the climate prevents fast plant growth, particularly during the dry seasons. In those areas, mulching or chemical stabilization may be better for the short term, as discussed later in this chapter.

Proper seedbed preparation and the use of high-quality seed are needed to grow plants for effective erosion control. Soil that has been compacted by heavy traffic or machinery may need to be loosened. For successful growth, usually the soil must be tilled before the seed is applied. Placement of topsoil is not necessary for temporary seeding; however, it may improve the chances of establishing temporary vegetation in an area. Seedbed preparation may also require the application of fertilizer and/or lime to the soil to make conditions more suitable for plant growth. Proper fertilizer, seeding mixtures, and seeding rates vary depending on the location of the site, soil types, slopes, and season. Local suppliers, state and local regulatory agencies, and the USDA Soil Conservation Service will supply information on the best seed mixes and soil-conditioning methods.

Seeded areas should be covered with mulch to provide protection from the weather. Seeding on slopes of 2:1 or more, in adverse soil conditions, during excessively hot or dry weather, or where heavy rain is expected should be followed by the spreading of mulch. Frequent inspections are needed to check that conditions for growth are good. If the plants do not grow quickly or thickly enough to prevent erosion, the area should be reseeded as soon as possible. Seeded areas should be kept adequately moist. If normal rainfall will not be sufficient, mulching, matting, and controlled watering should be done. If seeded areas are watered, watering rates should be watched so that excessive irrigation (which can cause erosion itself) does not occur.

Temporary seeding costs average about $1 per square yard ($0.84 per square meter), not counting any costs for mulch. This makes it one of the least expensive stabilization measures.

Establishing grass

Where feasible, grading operations should be planned around optimal seeding dates for the particular region. If the time of year is not suitable for seeding a permanent cover (perennial species), a temporary cover crop should be planted. Temporary seeding of annual species (small grains, ryegrasses, or millets) often succeeds during periods of the year that are unsuitable for seeding permanent (perennial) species.

Variations in weather and local site conditions can modify the effects of regional climate on seeding success. For this reason, mixtures including both cool- and warm-season species are often preferred for low-maintenance cover. Such mixtures promote cover which can adapt to a range of conditions. Many of these mixtures are not desirable, however, for high-quality lawns, where variation in texture of the turf is inappropriate.

The best selection of plant species should be considered early in the process of preparing an erosion and sediment control plan, with the goal of achieving a practical, economical stabilization and long-term protection of disturbed sites. Cool-season plants achieve most of their growth during the spring and fall and are relatively inactive or dormant during the hot summer months. Therefore, fall is the most favorable time to plant them. Warm-season plants "green up" late in the spring, grow most actively during the summer, and go dormant at the time of the first frost in fall. Spring and early summer are preferred planting times for warm-season plants.

Even in cases where the long-term goal is to establish a uniform, high-quality turf with only one plant species, the use of more than one species should be considered for erosion control. The addition of a quick-growing annual provides early protection and facilitates establishment of one or two perennials in a mix. More complex mixtures might include a quick-growing annual, one or two legumes, and more than one perennial grass.

The addition of a "nurse" crop (quick-growing annuals added to permanent mixtures) is a sound practice for soil stabilization, particularly on difficult sites—those with steep slopes; poor, rocky, erosive soils; those seeded outside the optimum seeding periods; or in any situation where the development of permanent cover is likely to be slow. The nurse crop germinates and grows rapidly, holding the soil until the slower-growing perennial seedlings become established.

Permanent seeding and planting

Permanent seeding of grass and planting of trees and brush stabilize the soil by holding soil particles in place. Vegetation reduces sediments and runoff to downstream areas by slowing the velocity of runoff and permitting greater infiltration of the runoff. Vegetation also filters sediments, helps the soil absorb water, improves wildlife habitats, and enhances the aesthetics of a site.

Permanent seeding and planting are appropriate for any graded or cleared area where long-lived plant cover is desired. Some areas where permanent seeding is especially important are filter strips, buffer areas, vegetated swales, steep slopes, and stream banks. This practice is effective in areas where soils are unstable because of their texture, structure, a high water table, high winds, or high slope.

For this practice to work, it is important to select appropriate vegetation, prepare a good seedbed, properly time the planting, and condition the soil. Planting local plants during their regular growing season will increase the chances for success and may lessen the need for watering. Check seeded areas frequently for proper watering and growth conditions.

For seeding in cold climates during fall or winter, cover the area with mulch to provide a protective barrier against cold weather. Seeding should also be mulched if the seeded area has a slope of 4:1 or more, if the soil is sandy or clayey, or if the weather is excessively hot or dry. Plant when conditions are most favorable for growth. When possible, use low-maintenance local plant species.

The average cost of permanent seeding can be estimated at about $1 per square yard ($0.84 per square meter), not including any mulch costs. However, there is a wide variation according to the desired appearance of the finished project.

Effective use of topsoil

Topsoil is the surface layer of the soil profile, generally characterized as being darker than the subsoil due to the presence of organic matter. It is the major zone of root development, carrying much of the nutrients available to plants and supplying a large share of the water used by plants. It is sometimes advantageous to strip topsoil from a construction site and stockpile it for later use during final seeding or planting.

Although topsoil provides an excellent growth medium, there are disadvantages to its use. Stripping, stockpiling, and reapplying topsoil, or importing topsoil, may not always be cost-effective. Topsoiling can delay seeding or sodding operations, increasing the exposure time of denuded areas. Most topsoil contains weed seeds, and weeds may compete with desirable species.

Advantages of topsoil include its high organic matter content and friable consistency, water-holding capacity, and nutrient content. In site planning, the option of topsoiling should be compared with that of preparing a seedbed in subsoil. The clay content of subsoils does provide high moisture availability and deter leaching of nutrients. When properly limed and fertilized, subsoils may provide a good growth medium that is generally free of weed seeds.

In many cases topsoiling may not be required for the establishment of less demanding, lower-maintenance plant material. Topsoiling is strongly recommended where ornamental plants or high-maintenance turf will be grown. Topsoiling is a required procedure in establishing vegetation on shallow soils, soils containing potentially toxic materials, and soils of critically low pH (high acid) levels.

If topsoiling is to be done, the following items should be considered:

1. An adequate volume of topsoil must exist on the site or be imported. Topsoil will be spread at a compacted depth of 50 to 100 mm (2 to 4 in). Depths closer to 100 mm (4 in) are preferred. Table 12.1 lists the required volumes of topsoil for various depths of application.
2. The topsoil stockpile must be located so that it meets specifications and does not interfere with work on the site.
3. There must be sufficient time in scheduling for topsoil to be spread and bonded prior to seeding, sodding, or planting.
4. Topsoil should not be applied to a subsoil of contrasting texture. Clayey topsoil over sandy subsoil is a particularly poor combination, as water may creep along the junction between the soil layers, causing the topsoil to slough. Sandy topsoil over a clay subsoil is also likely to fail.
5. If topsoil and subsoil are not properly bonded, water will not infiltrate the soil profile evenly and it will be difficult to establish vegetation. Topsoiling of steep slopes should be discouraged unless good bonding of soils can be achieved.

The site should be explored to determine if there is sufficient surface soil of good quality to justify stripping. Topsoil should be friable and

TABLE 12.1 Required Volumes of Topsoil

Depth		Topsoil volume	
mm	in	m^3/ha	yd^3/ac
25	1	253	134
50	2	506	268
75	3	761	403
100	4	1014	537
125	5	1270	672
150	6	1523	806

SOURCE: U.S. Army Corps of Engineers, 1997.

loamy (loam, sandy loam, silt loam, sandy clay loam, clay loam). It should be relatively free of debris, trash, stumps, rocks, roots, and noxious weeds, and it should give evidence of being able to support healthy vegetation. It should not contain any substance that is potentially toxic to plant growth. All topsoil should be tested by a recognized laboratory for the following criteria:

- Organic matter content should be at least 1.5% by weight.
- The pH range should be 6.0 to 7.5. If pH is less than 6.0, lime should be added in accordance with soil test results or in accordance with the recommendations of the vegetative establishment practice being used.
- Soluble salts should not exceed 500 parts per million.

If additional off-site topsoil is needed, it must meet the standards stated above.

Topsoil operations should not be performed when the soil is wet or frozen. Stripping shall be confined to the immediate construction area. A 100- to 150-mm (4- to 6-in) stripping depth is common, but depth may vary depending on the particular soil. All perimeter dikes, basins, and other sediment controls should be in place prior to stripping.

Topsoil should be stockpiled in such a manner that natural drainage is not obstructed and stockpile erosion is minimized. Stockpiles should be stabilized or protected. Side slopes of the stockpile should not exceed 2:1. Perimeter controls must be placed around the stockpile immediately; seeding of stockpiles should be completed as soon as possible, but no later than 14 days after the formation of the stockpile, unless the stockpile is going to be used again within 21 days after its formation.

Before topsoiling, establish needed erosion and sediment control practices such as diversions, grade stabilization structures, berms, dikes, level spreaders, waterways, and sediment basins. These practices must be maintained during topsoiling. Previously established grades on the areas to be topsoiled should be maintained according to the approved plan.

After the areas to be topsoiled have been brought to grade, and immediately prior to dumping and spreading the topsoil, the subgrade should be loosened by disking or scarifying to a depth of at least 50 mm (2 in) to ensure bonding of the topsoil and subsoil.

Topsoil should not be placed while in a frozen or muddy condition, when topsoil or subgrade is excessively wet, or in a condition that may otherwise be detrimental to proper grading or proposed sodding or seeding. Topsoil should be uniformly distributed to a minimum compacted depth of 25 mm (2 in) on 3:1 or steeper slopes and 100 mm (4 in) on flatter slopes. Any irregularities in the surface, resulting from topsoiling or other operations, should be corrected in order to prevent the formation of depressions or water pockets.

The topsoil should be compacted enough to ensure good contact with the underlying soil and to obtain a level seedbed for the establishment of vegetation. However, undue compaction is to be avoided as it increases runoff velocity and volume and deters seed germination. Special consideration should be given to the types of equipment used to place topsoil in areas to receive fine turf. Avoid unnecessary compaction by heavy machinery whenever possible. In areas that are not going to be mowed, the surface should be left rough.

No sod or seed should be placed on soil that has been treated with soil sterilants until sufficient time has elapsed to permit dissipation of toxic materials.

Sod stabilization

Sod stabilizes an area by immediately covering the surface with vegetation and providing areas where storm water can infiltrate the ground. When installed and maintained properly, sodding can be 99% effective in reducing erosion, making it the most effective vegetation practice available (USDA Natural Resources Conservation Service, 1985). The higher cost of sod stabilization relative to other vegetative controls typically limits its use to exposed soils where a quick vegetative cover is desired and on sites that can be maintained with ground equipment. The average cost of sod is about $4 per square yard ($3.35 per square meter). In addition, sod is sensitive to climate and may require intensive watering and fertilizing.

Sodding is appropriate for any graded or cleared area that might erode and where a permanent, long-lived plant cover is needed immediately. Sodding can be used in buffer zones, stream banks, dikes, swales, slopes, outlets, level spreaders, and filter strips. See Fig. 12.2.

Figure 12.2 Sod being placed behind street curb.

Final grading should be complete before the sod is laid down. Topsoil may be needed in areas where the soil textures are inadequate. Lime and fertilizers should be added to the soil to promote good growth conditions. Sodding can be applied in alternating strips or other patterns, or alternate areas can be seeded to reduce expense. Sod should not be planted during very hot or wet weather.

One advantage of sod is that it can withstand some flow soon after installation, especially if it is installed with staggered joints. The sod blocks may also be pegged into the ground to withstand moderate flow velocities.

Sod should not be placed on slopes greater than 3:1 if they are to be mowed. If placed on steep slopes, sod should be laid with staggered joints and/or pegged. In areas such as steep slopes or next to running waterways, chicken wire, jute, or other netting can be placed over the sod for extra protection against lifting. Roll or compact the sod immediately after installation to ensure firm contact with the underlying topsoil.

Inspect the sod frequently after it is first installed, especially after large storm events, until it has become established as permanent cover. Remove and replace dead sod. Watering may be necessary after planting and during periods of intense heat and/or lack of rain (drought).

Buffer zones

Buffer zones are preserved or planted strips of vegetation at the top and bottom of a slope, outlining property boundaries, or adjacent to receiving waters such as streams or wetlands. Buffer zones can slow runoff flows at critical areas, decreasing erosion and allowing sediment deposition. Buffer zones are different from vegetated filter strips (discussed in Chap. 13) because buffer zone effectiveness is not measured by its ability to improve infiltration. The buffer zone can be an area of vegetation that is left undisturbed during construction, or it can be newly planted.

Buffer zones can be used at almost any site that can support vegetation. Buffer zones are particularly effective on floodplains, next to wetlands, along stream banks, and on steep, unstable slopes. They can be especially useful for very narrow linear construction projects such as underground utilities or pipelines.

If buffer zones are preserved, existing vegetation, good planning, and site management are needed to protect against disturbances such as grade changes, excavation, damage from equipment, and other activities. Establishing new buffer strips requires the establishment of a good dense turf, trees, and shrubs.

Careful maintenance is important to ensure healthy vegetation. The need for routine maintenance such as mowing, fertilizing, liming, irrigating, pruning, and weed and pest control will depend on the species of plants and tress involved, soil types, and climatic conditions. Maintaining planted areas may require debris removal and protection against unintended uses or traffic.

Many state and local storm water program or zoning agencies have regulations which define required or allowable buffer zones, especially near sensitive areas such as wetlands. Contact the appropriate state and local agencies for their requirements.

If buffer zones must be established by planting, the costs may be estimated at about $1 per square yard ($0.84 per square meter), plus any costs for mulch. If buffer zones consist of existing native vegetation, the costs vary according to the effort required to preserve and maintain the vegetation.

Preservation of natural vegetation

The preservation of natural vegetation (existing trees, vines, brushes, and grasses) provides natural buffer zones. Preserving stabilized areas minimizes erosion potential, protects water quality, and provides aesthetic benefits. Preservation of natural vegetation qualifies as a permanent storm water control measure, as described in Chap. 13.

Mature trees have extensive canopy and root systems which help to hold soil in place. Shade trees also keep soil from drying rapidly and becoming susceptible to erosion. Measures taken to protect trees can vary significantly, from simple measures such as installing tree fencing around the drip line and installing tree armoring, to more complex measures such as building retaining walls and tree wells.

Areas where preserving vegetation can be particularly beneficial include floodplains, wetlands, stream banks, steep slopes, and other areas where erosion controls would be difficult to establish, install, or maintain.

Preservation of vegetation on a site should be planned before any site disturbance begins. Preservation requires good site management to minimize the impact of construction activities on existing vegetation. Clearly mark the trees to be preserved, and protect them from ground disturbances around the base of the tree.

Proper maintenance is important to ensure healthy vegetation that can control erosion, especially during construction. Different species, soil types, and climatic conditions will require different maintenance activities such as mowing, fertilizing, liming, irrigation, pruning, and weed and pest control.

The EPA originally considered the requirement that 15% of the original vegetation be left in place on all sites. However, the EPA eventually rejected this approach because of the need to allow flexibility. However, some state and local regulations require natural vegetation to be preserved in sensitive areas; consult the appropriate state and local agencies for more information on their regulations.

The cost to preserve existing vegetation can vary widely. To protect existing mature trees, costs may be estimated at $30 to $200 per tree, but can be much higher for isolated large trees.

Nonvegetative Stabilization Practices

Nonvegetative covers such as mulches and stone aggregates protect soils from erosion. Like vegetative covers, these ground covers can shield the soil surface from the impact of failing rain, reduce flow velocity, and disperse flow. Each of these types of cover provides a rough surface that slows the runoff velocity and promotes infiltration and deposition of sediment. The condition as well as the type of ground cover influences the rate and volume of runoff.

Note that although impervious surfaces (such as parking lots) protect the covered area, they prevent infiltration and consequently increase the peak flow rate. This, along with the possible increase in flow velocities due to the hydraulic efficiency of flow over these surfaces, may increase the potential for erosion at the discharge.

Mulching

Mulching is a temporary soil stabilization or erosion control practice in which materials such as grass, hay, wood chips, wood fibers, straw, or gravel are placed on the soil surface. In addition to stabilizing soils, mulching can reduce the speed of storm water runoff over an area. When used together with seeding or planting, mulching can aid in plant growth by holding the seeds, fertilizers, and topsoil in place; by helping to retain moisture; and by insulating against extreme temperatures.

Mulching often is used alone in areas where temporary seeding cannot be used because of the season or climate. Where temporary seeding and permanent seeding are not feasible, exposed soils can be stabilized by applying plant residues or other suitable materials to the soil surface. Although generally not as effective as seeding practices, mulching, by itself, can provide immediate and inexpensive erosion control. On steep slopes and critical areas such as waterways, mulch matting is used with netting or anchoring to hold it in place.

Mulching is also typically used as part of permanent and temporary seeding practices. Mulching in conjunction with seeding practices provides erosion protection prior to the onset of vegetation growth. In addition, mulching protects seeding practices, providing a higher likelihood of their success. Seeded and planted areas should be mulched wherever slopes are steeper than 2:1, where runoff is flowing across the area, or when seedlings need protection from bad weather. Table 12.2 provides details on mulch application.

Mulch is more effective when it is secured by using a binder or netting or by tacking the mulch to the ground. Mulch binders should be applied according to the manufacturer's recommended rates and methods.

TABLE 12.2 Application Rates and Quality Standards for Various Mulch Materials

Mulch material	Quality standards	Application rates
Straw	Air-dried; free from undesirable seed and coarse material	2–3 in thick; 2–3 bales per 1000 ft^2 or 2–3 tons/acre
Wood-fiber cellulose	No growth-inhibiting factors	Approx. 25–30 lb/1000 ft^2 or 1000–1500 lb/acre
Compost	No visible water or dust during handling. Must be purchased from supplier with solid waste handling permit	2-in-thick minimum; approx. 100 tons/acre (approx. 800 lb/yd)
Chipped site vegetation	Average size shall be several inches	2-in minimum thickness

SOURCE: U.S. Army Corps of Engineers, 1997.

Final grading is not necessary before mulching. Mulched areas should be inspected often to find where mulched material has been loosened or removed. Such areas should be reseeded (if necessary) and the mulch cover replaced immediately.

Mulch costs may be estimated at about \$1.25/yd^2 (\$1.05/m^2)

Geotextiles

Geotextiles are porous fabrics used for erosion control and other construction purposes. Geotextiles are manufactured by weaving or bonding fibers made from synthetic materials such as polypropylene, polyester, polyethylene, nylon, polyvinyl chloride, glass, and various mixtures of these. Geotextiles are typically used as mulch netting, matting, and separators. Geotextiles are also used as separators. For example, a geotextile may be placed between riprap and soil to maintain the base of the riprap and to prevent the soil from being eroded.

Many types of geotextiles are available for different purposes. State or local requirements, design procedures, and any other applicable requirements should also be consulted. In the field, important concerns include regular inspections to determine whether cracks, tears, or breaches are present in the fabric and appropriate repairs should be made. Effective netting and matting require firm, continuous contact between the materials and the soil. If there is no contact, the material will not hold the soil and erosion will occur underneath the material.

Degradable soil stabilization blankets

A degradable soil stabilization blanket may be used to help establish vegetation on previously disturbed slopes—normally slopes of 3:1 or greater. These blankets may include a photodegradable plastic netting that covers and is intertwined with a natural organic or artificial mulch. The mulching material may consist of wood fibers, wood excelsior, straw, coconut fiber, or artifical fibers, or a combination of these. The mulching material and fibers must interlock or entwine to form a dense layer which not only resists raindrop impact, but also will allow vegetation to penetrate the blanket. The blanket should be nontoxic to vegetation and to the germination of seed and should not be injurious to the unprotected skin of humans. At a minimum, the plastic netting must cover the top side of the blanket and possess a high web strength. The netting should be entwined with the mulching material or fiber to maximize strength and provide for ease of handling. Instead of a plastic netting, the blanket may include a jute mesh that can be used without the mulch in certain applications.

Since the materials in the soil stabilization blankets will deteriorate over time, they should be used in permanent conveyance channels, with the realization that the system's resistance to erosion is based on the type of vegetation planted and the existing soil characteristics. During the establishment of vegetation, degradable soil stabilization blankets should not be subjected to shallow or deep concentrated flows moving at greater than 1 m/s (3 ft/s).

Degradable soil stabilization blankets provide the following benefits in the achievement of vegetative stabilization, when properly applied:

1. Protection of the seed and soil from raindrop impact and subsequent displacement
2. Thermal consistency and moisture retention for seedbed area
3. Stronger and faster germination of grasses and legumes
4. Capacity for conveying excess storm water runoff
5. Prevention of sloughing of topsoil added to steeper slopes

Soil stabilization blankets are anchored to the ground with staples. Typical applications require staples made of No. 11 gauge wire or heavier. The typical staple length is 150 mm (6 in). A larger staple with a length of up to 300 mm (12 in) should be used on loose, sandy, or unstable soils.

Nondegradable soil stabilization mats

A soil stabilization mat may consist of a nondegradable, three-dimensional structure of nylon, polyethylene, or randomly oriented monofilament which can be filled with soil prior to planting. This configuration provides a matrix for root growth where the matting becomes entangled and penetrated by roots, forming continuous anchorage for surface growth and promoting enhanced energy dissipation. These products contain ultraviolet (uv) inhibiting stabilizers, added to the compounds to ensure endurance and provide "permanent root reinforcement."

Soil stabilization mats can be used on problem slopes (normally 3:1 or greater) and in storm water conveyance channels. In addition to those benefits noted above for degradable soil stabilization blankets, soil stabilization mats provide the following benefits in the achievement of vegetative stabilization and in the replacement of more traditional channel linings such as concrete and riprap:

1. They cause soil to drop out of storm water and fill the matrix with fine soils which become the growth medium for the development of roots.
2. When embedded in the soil within storm water channels, they act with the vegetative root system to form an erosion-resistant cover which resists hydraulic lift and shear forces.

Since soil stabilization mats are nondegradable, they can be used in permanent conveyance channels and can withstand higher velocities of flow than the vegetation and soil would normally allow. However, a flow velocity of 3 m/s (10 ft/s) should be the maximum allowed in a conveyance system that utilizes soil stabilization mats.

Selection of the appropriate matting materials and proper installation are critical factors in the success of this practice. Consultation with the supplier or the manufacturer, and thorough evaluation of performance data to ensure proper selection of a soil stabilization matting, are also essential. Although many manufacturers claim their products may inhibit erosion associated with channel velocities of up to 6 m/s (20 ft/s), it is recommended that any velocities that exceed 3 m/s (10 ft/s) be properly protected with some form of structural lining.

Chemical stabilization

Chemical stabilization is a temporary erosion control practice in which materials made of vinyl, asphalt, or rubber are sprayed onto the surface of the soil to hold the soil in place and to protect against erosion from storm water runoff and wind. These materials are often referred to as *chemical mulch, soil binder,* or *soil palliative.*

Asphalt emulsions, latex emulsions, or resin in water can be sprayed onto mineral soil to serve as soil adhesives. Calcium chloride may be applied by mechanical spreader as loose, dry granules or flakes.

Chemical stabilization can be used as an alternative in areas where temporary seeding practices cannot be used because of the season or climate. It can provide immediate, effective, and inexpensive erosion control anywhere that erosion is occurring on a site.

The application rates and procedures recommended by the manufacturer of a chemical stabilization product should be followed as closely as possible to prevent the products from forming ponds and from creating large areas where moisture cannot get through.

Stream bank stabilization

Many erosion problems, and a high proportion of sediment volume, result from various types of stream bank erosion and bank failures, rather than erosion from the land surface itself. Stream bank stabilization is used to prevent stream bank erosion due to high velocities and quantities of storm water runoff. Typical methods include riprap, gabions, slope paving and bulkheads, log cribbing, grid pavers, and asphalt paving.

Riprap is large angular stones placed along the bank line. It protects soil from erosion and is often used on steep slopes built with fill materials that are subject to harsh weather or seepage. Riprap can also be used for flow channel liners, inlet and outlet protection at culverts, stream bank protection, and protection of shorelines subject to wave action. It is used where water is turbulent and fast-flowing and where soil may erode under the design flow conditions. Riprap is often placed over a filter blanket (i.e., a gravel layer or filter cloth). Riprap is either a uniform size or graded (different sizes) and usually is applied in an even layer throughout the stream. Figure 12.3 illustrates a typical riprap installation. Riprap costs may be estimated at about $45 per square yard ($38 per square meter).

Gabions are rock-filled wire-mesh cages that are used to create a new stream bank. Gabions should be installed according to the manufacturer's recommendations. Gabions are sometimes placed over a filter blanket, such as a gravel layer or filter cloth. Figure 12.4 illustrates a typical gabion installation.

Slope paving and bulkheads replace natural stream banks and create a nonerosive surface. Reinforced-concrete structures may require positive drainage behind the bulkhead or retaining wall to prevent erosion around the structure. Concrete slope paving typically costs about $65 per square yard ($55 per square meter).

Log cribbing consists of retaining walls built of logs to anchor the soils against erosive forces. These are usually built on the outside of stream bends. Only pressure-treated logs should be used. Figure 12.5 illustrates a typical log cribbing installation.

Grid pavers are precast or poured-in-place concrete units that are placed along stream banks to stabilize them and to create open spaces where vegetation can be established. Grid pavers should be installed according to the manufacturer's recommendations. Figure 12.6 illustrates a typical installation of grid pavers.

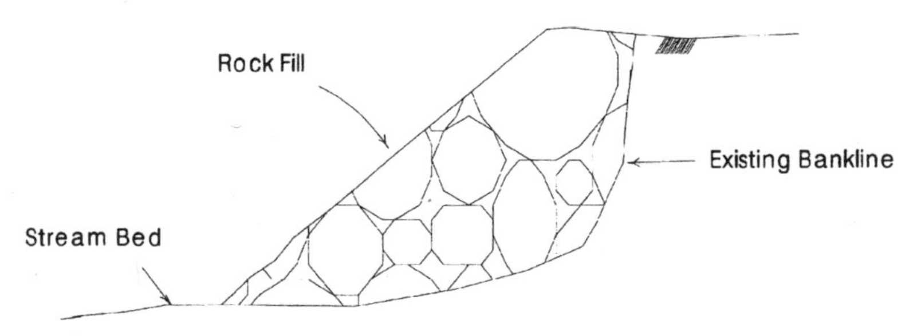

Figure 12.3 Stream bank stabilization by riprap (Environmental Protection Agency, 1992).

Figure 12.4 Stream bank stabilization by gabions.

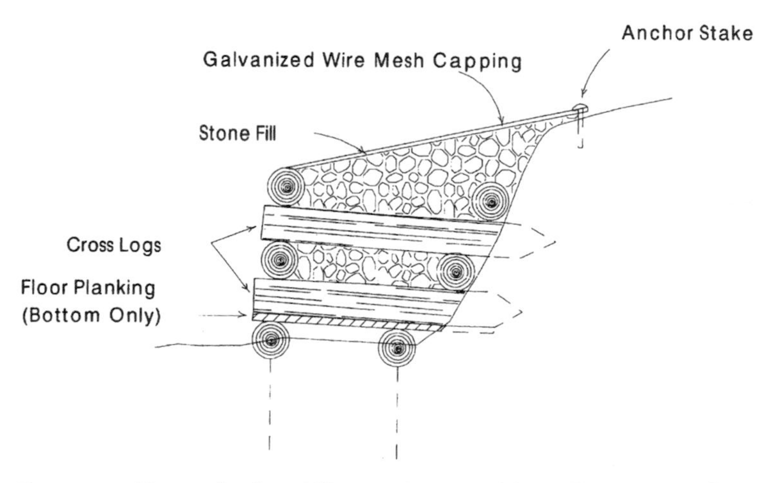

Figure 12.5 Stream bank stabilization by log cribbing (Environmental Protection Agency, 1992).

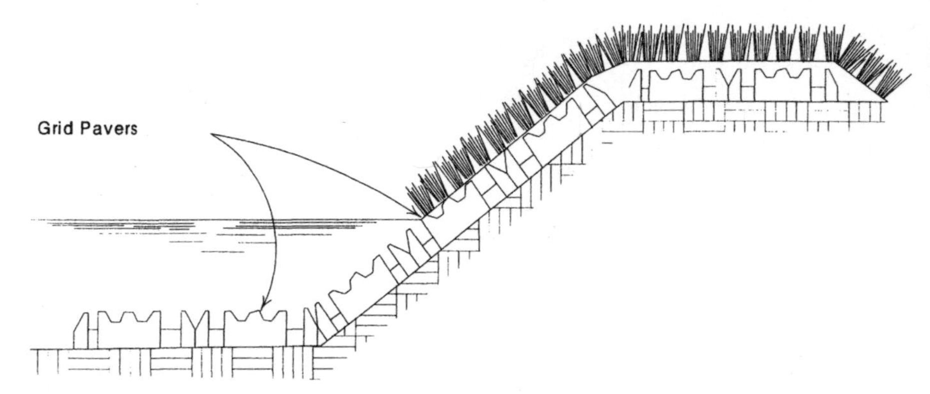

Figure 12.6 Stream bank stabilization by grid pavers (Environmental Protection Agency, 1992).

Asphalt paving is placed along the natural stream bank to create a nonerosive surface. Asphalt paving costs may be estimated at about $35 per square yard ($30 per square meter).

Stream bank stabilization is used where vegetative stabilization practices are not practical and where the stream banks are subject to heavy erosion due to increased flows or disturbance during construction. Stabilization should be achieved prior to any land development in the watershed area. Stabilization can also be retrofitted when erosion of a stream bank occurs, although it is much better to anticipate erosion damage whenever possible.

Stream bank stabilization structures should be planned and designed by a professional engineer licensed in the state where the site is located. Applicable federal, state, and local requirements should be followed. Many types of stream bank stabilization will require permits from the Army Corps of Engineers, under Section 404 of the Clean Water Act.

An important design feature of stream bank stabilization methods is the foundation of the structure; the potential for the stream to erode the sides and bottom of the channel should be considered, to ensure that the stabilization measure will be supported properly.

Structures can be designed to protect and improve natural wildlife habitats; e.g., log structures and grid pavers can be designed to keep vegetation. Permanent structures should be designed to handle expected flood conditions.

Stream bank stabilization structures should be inspected regularly and after each large storm event. Structures should be maintained as installed. Structural damage should be repaired as soon as possible to prevent further damage or erosion to the stream bank.

Soil-retaining measures

Soil-retaining measures refer to structures or vegetative stabilization practices used to hold the soil firmly to its original place or to confine as much as possible within the site boundary. There are many different methods for retaining soil; some are used to control erosion while others are used to protect the safety of the workers (i.e., during excavations). Examples of soil-retaining measures include reinforced soil-retaining systems, wind breaks, and stream bank protection by shrubs and reeds.

Reinforced soil-retaining measures refer to the use of structural measures to hold in place loose or unstable soil. During excavation, for example, soil tiebacks and retaining walls are used to prevent cave-ins and accidents. These same methods can be used to retain soils and prevent them from moving. The following types of soil-retaining measures are commonly used:

- *Skeleton sheeting*. Skeleton sheeting, the least expensive soil-bracing system, requires cohesive soils, such as clay soils. Construction-grade lumber is used to brace the excavated face of the slope.
- *Continuous sheeting*. Continuous sheeting involves use of a material that covers the face of the slope in a continuous manner. Struts and boards are placed along the slope which provide continuous support to the slope face. The material can be steel, concrete, or wood.
- *Permanent retaining walls*. Permanent construction walls may be needed to provide support to the slope well after the construction is completed. In this instance, concrete masonry or wood (railroad tie) retaining walls can be constructed and left in place.

Some sites may have slopes or soils that do not lend themselves to ordinary practices of soil retention, such as vegetation. In these instances, a reinforced soil-retaining measure should be considered.

As emphasized earlier, the use of reinforced soil-retaining practices serves both safety and erosion control purposes. Since safety is the first concern, the design should be performed by qualified and certified engineers. Such design normally requires an understanding of the nature of soil, location of the groundwater table, the expected loads, and other important design considerations.

Dust control

Construction activities inevitably result in the exposure and disturbance of soil. Fugitive dust is emitted both during the activities (i.e., excavation, demolition, vehicle traffic, human activity) and as a result of wind erosion over the exposed earth surfaces. Large quantities of dust typically are generated in "heavy" construction activities, such as road and street construction and subdivision, commercial, or industrial development which involve disturbance of significant areas of the soil surface. Research of construction sites has established an average dust emission rate of 2700 kg/ha per month (1.2 tons/acre per month) for active construction. Earthmoving activities comprise the major source of construction dust emissions, but traffic and general disturbance of the soil also generate significant dust emissions.

In planning for dust control, limiting the amount of soil disturbance at any one time should be a key objective. Therefore, phased clearing and grading operations and the utilization of temporary stabilization can significantly reduce dust emissions. Undisturbed vegetative buffers, with 15-m (50-ft) minimum widths, left between graded areas and protected areas can also be very helpful in dust control. The following measures should be considered for controlling dust during construction:

- *Vegetative cover.* In areas subject to little or no construction traffic, a vegetatively stabilized surface will reduce dust emissions.
- *Mulch.* When properly applied, mulch offers a fast, effective means of controlling dust. However, it is not recommended for areas within heavy traffic pathways. Binders or tackifiers should be used to tack organic mulches.
- *Tillage.* This practice is designed to roughen and bring clods to the surface. It is an emergency measure that should be used before wind erosion starts. Begin plowing on windward side of site. Chisel-type plows spaced about 300 mm (12 in) apart, spring-toothed harrows, and similar plows are examples of equipment which may produce the desired effect.
- *Irrigation.* This is the most commonly used dust control practice. The site is sprinkled with water until the surface is wet. Repeat as needed. It offers fast protection for haul roads and other heavy traffic routes.
- *Spray-on adhesives.* Tremendous progress has been made in recent years in the development of products of this type. Most are effective on "mineral" soils and are ineffective on "muck" soils. These coherics are derived from a variety of compounds, both organic and synthetic-based. Many of the adhesives will withstand heavy traffic loads. The organics include derivatives from pine tar and vegetable gum; synthetics may be acrylic- or petroleum-based. Table 12.3 lists various adhesives and provides corresponding information on mixing and application.

- *Stone.* Stone can be used to stabilize roads or other areas during construction using crushed stone or coarse gravel, as described in Chap. 14.
- *Barriers.* A board fence, wind fence, sediment fence, or similar barrier can help to control air currents and blowing soil. Place barriers perpendicular to prevailing air currents at intervals of about 15 times the barrier height. Where dust is a known problem, existing windbreak vegetation should be preserved. Constructed wind barriers typically cost about $2.50 per lineal foot ($0.75 per meter).
- *Calcium chloride.* This chemical may be applied by mechanical spreader as loose, dry granules or as flakes at a rate that keeps the surface moist but not so high as to cause water pollution or plant damage. Application rates should be strictly in accordance with suppliers' specified rates.

Contouring and protection of sensitive areas

Contouring refers to the practice of building in harmony with the natural flow and contour of the land. By minimizing changes in the natural contour of the land, existing drainage patterns are preserved as much as possible, thereby reducing erosion. Minimizing the amount of regrading done will also reduce the amount of soil being disturbed.

The preservation of sensitive areas at a site, such as steep slopes and wetlands, should also be a priority. Disturbance of soil on steep slopes should be avoided due to vulnerability to erosion. Wetlands should be protected because they provide flood protection, pollution mitigation, and an essential aquatic habitat.

TABLE 12.3 Adhesives Used for Dust Control

Adhesive	Water dilution adhesive:water	Type of nozzle	Application rate, L/ha	Application rate, gal/acre
Anionic asphalt emulsion	7:1	Coarse spray	1000	1200
Latex emulsion	12.5:1	Fine spray	220	235
Resin in water	4:1	Fine spray	280	300
Acrylic emulsion (nontraffic)	7:1	Coarse spray	420	450
Acrylic emulsion (traffic)	3.5:1	Coarse spray	328	350

SOURCE: U.S. Army Corps of Engineers, 1997.

References

Environmental Protection Agency, 1992. *Storm Water Management for Construction Activities. Developing Pollution Prevention Plans and Best Management Practices,* EPA Report 832/R-92/005, September.

Metropolitan Washington Council of Governments, 1998. *Performance of Current Sediment Control Measures at Maryland Construction Sites,* Washington, January.

National Academy Press, 1992. *Vetiver: A Thin Green Line Against Erosion,* Washington.

U.S. Army Corps of Engineers, 1997. *Handbook for the Preparation of Storm Water Pollution Prevention Plans for Construction Activities,* EP 1110-1-16, Washington, DC, February 28.

USDA Natural Resources Conservation Service, 1985. *Guides for Erosion and Sediment Control in California,* Davis, Calif.

Chapter

13

Structural Practices for Construction Sediment Control

This is the fourth of five chapters which deal with the specific requirements for storm water discharges from construction facilities. This chapter provides details on structural measures (such as sediment basins) for erosion and sediment control.

Erosion and Sediment Controls

Erosion controls, as described in Chap. 12, provide the first line of defense in preventing off-site sediment movement and are designed to prevent erosion by protecting soils. Sediment controls, as described in this chapter, are generally designed to remove sediment before the runoff is discharged from the site. Structural controls are necessary because vegetative controls cannot be employed at areas of the site which are continually disturbed and because a finite time is required before vegetative practices are fully effective.

Structural practices selected for incorporation into a construction storm water pollution prevention plan (SWPPP) are to be based on what is attainable at a given site. Structural practices involve the installation of devices to divert flow, store flow, or limit runoff. Options for such controls include straw bale dikes, silt fences, earth dikes, brush barriers, drainage swales, check dams, subsurface drain, pipe slope drain, level spreaders, storm drain inlet protection, outlet protection, sediment traps, and temporary sediment basins.

Structural Practices

Structural practices involve the installation of devices to divert, store, or limit runoff. Structural practices have several objectives. First, structural practices can be designed to prevent water from flowing onto disturbed areas where erosion may occur. This involves diverting runoff from undisturbed, upslope areas through the use of earth dikes, temporary swales, perimeter dikes, or other diversions to stable areas. A second objective of structural practices may be to remove sediment before the runoff leaves the site. Methods for removing sediment from runoff include diverting flows to a trapping or storage device and filtering diffuse flows through on-site silt fences. All structural practices require proper maintenance (e.g., removal of collected sediment) to remain functional, and should be designed to avoid presenting a safety hazard, especially in areas frequented by children.

Structural practices generally fall into the following categories:

- *Velocity dissipation.* This includes measures which reduce the erosive forces of runoff waters, including outlet protection, check dams, surface roughening, and gradient terraces.
- *Sediment capture.* This includes measures which remove sediment runoff before it is carried off the site, including silt fences, straw bale dikes, brush barriers, gravel or stone filter berms, storm drain inlet protection, sediment traps, and temporary sediment basins.
- *Temporary storm water management.* This includes measures which divert flows away from exposed areas or divert sediment-laden flows into controlled areas, including earth dikes, drainage swales, interceptor dikes and swales, temporary storm drain diversions, pipe slope drains, and subsurface drains.
- *Permanent storm water management.* This includes measures which remain in place after the construction has been completed, including on-site infiltration, outfall velocity dissipation devices, storm water retention structures and artificial wetlands, and storm water detention structures. These measures are intended to provide long-term improvements in the quality of runoff from the project site.

General Permit Requirements

The EPA general permit has several specific requirements for structural erosion and sediment control measures:

1. *Velocity dissipation measures* must be implemented for all construction projects.

2. A *sediment basin* is required if the disturbed area contributing to a common drainage basin is equal to or greater than 10 acres (4 ha).
3. A *sediment basin and/or sediment traps* must be used as needed if the disturbed area contributing to a common drainage basin is less than 10 acres (4 ha).
4. *Silt fabric fences* are required for all side slope and downslope site boundaries. However, if a sediment basin with a volume of 3600 ft^3/acre (25 m^3/ha) of drainage area is provided, then no other structural practices are required for drainage basins of less than 10 acres (4 ha).
5. *Permanent storm water management measures* are required for all projects.

The remainder of this chapter describes each of these measures in greater detail.

Velocity Dissipation Measures

Velocity dissipation measures reduce the erosive forces of runoff waters by decreasing the flow velocity. The quantity and size of the soil particles that are loosened and removed increase with the velocity of the runoff. This is because high runoff velocities reduce infiltration of the soil (and therefore also increase runoff volume) and exert greater forces on the soil particles, causing them to detach. High flow velocities are associated with severe rill and gully erosion. There are several ways in which velocities can be reduced:

1. *Reduce slopes.* When you prepare the grading plan, try to make grades as gradual as possible without modifying the existing site conditions significantly. Steeper slopes result in faster-moving runoff, which results in greater erosion.

2. *Protect steep slopes.* Steeply sloped areas can be protected from erosion in a number of ways. Flow can be diverted away from the face of the slope. On the slope itself, gradient terraces should be used to break the slope and slow the speed of the runoff flowing down the hillside. Surface roughening can also be used on sloped areas to slow down overland flow on a steep slope.

3. *Provide vegetative cover.* In addition to holding soil in place and shielding it from the impact of raindrops, vegetative cover increases the surface roughness, which decreases the flow velocity.

4. *Decrease channel flow velocities.* Concentrated runoff can be more erosive than overland flow. Runoff concentrated into swales or channels can be slowed by reducing the slope and increasing the width of a

channel, or by using check dams. Runoff can also be slowed in channels by establishing a vegetative cover. Sod can provide a quick method of establishing vegetative cover. Geotextiles are often used to hold the channel soil in place while the grass is growing.

The following sections provide additional details on various velocity dissipation measures.

Velocity dissipation by level spreaders

Level spreaders are outlets for dikes and diversions consisting of an excavated depression constructed at zero grade across a slope. Level spreaders convert concentrated runoff to diffuse runoff and release it onto areas stabilized by existing vegetation. Figure 13.1 illustrates a typical level spreader.

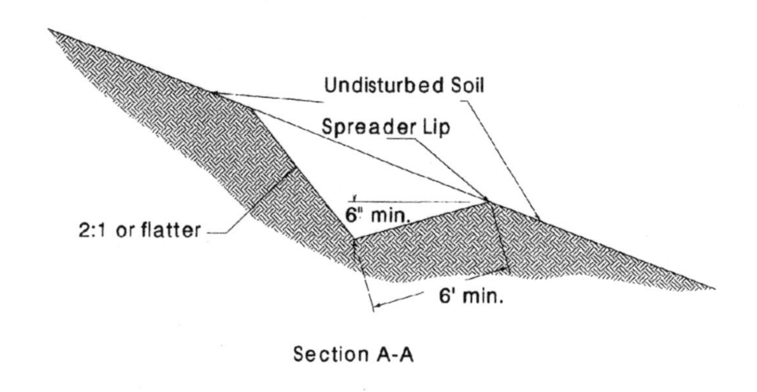

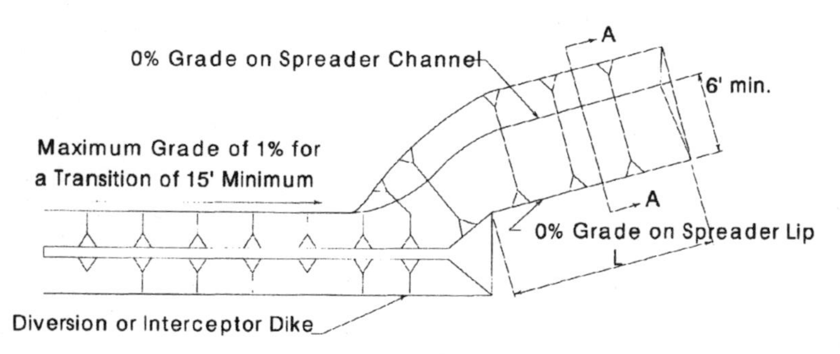

Figure 13.1 Velocity dissipation by level spreader (*Harris County, 1992*).

Velocity dissipation by outlet protection

Outlet protection reduces the speed of concentrated storm water flows, and therefore it reduces erosion or scouring at the outlet end of culverts or paved channel sections. In addition, outlet protection lowers the potential for downstream erosion. This type of protection can be achieved through a variety of techniques, including stone or riprap, concrete aprons, paved sections, and settling basins installed below the storm drain outlet. Figure 13.2 illustrates culvert outlet protection using a riprap apron.

Outlet protection should be installed at all pipe, interceptor dike, swale, or channel section outlets where the velocity of flow may cause erosion at the pipe outlet and in the receiving channel. Outlet protection should also be used at outlets where the velocity of flow at design capacity may result in plunge pools (small permanent pools formed by erosion). Outlet protection should be installed early during construction activities, but may be added at any time, as necessary.

The exit velocity of the runoff as it leaves the outlet protection structure should be reduced to levels that minimize erosion. Outlet protection should be inspected on a regular schedule to check for erosion and scouring. Repairs should be made promptly.

Outlet velocity protection measures which are properly designed and installed may later be converted to use as a permanent storm water

Figure 13.2 Culvert outlet protection.

management structure. This may satisfy the requirements of the EPA general permit for permanent storm water control measures, as described later in this chapter.

Velocity dissipation by check dams

A *check dam* is a small, temporary or permanent dam constructed across a drainage ditch, swale, or channel to lower the speed of concentrated flows. Reduced runoff speed reduces channel erosion and allows sediments to settle out.

A check dam should be installed in steeply sloped swales or in swales where adequate vegetation has not been established. A check dam may be built from logs, stone, covered straw bales, or sandbags filled with pea gravel.

Check dams should be used only in small open channels which will not be overtopped by flow once the dams are constructed. The dams should not be placed in existing natural streams (unless approved by state authorities). Use the following design guidelines for check dams:

- *Drainage area:* The drainage area above the check dam should be between 2 and 10 acres (1.5 and 4 ha).
- *Spacing:* The dams must be spaced so that the top of the upstream dam is never any higher than the top of the downstream dam.
- *Height:* The center of the dam must be 6 to 9 in (150 to 225 mm) lower than either edge, and the maximum height of the dam should be 24 in (0.6 m).
- *Width:* The check dam should be as much as 18 in wider than the banks of the channel to prevent undercutting as overflow water reenters the channel.
- *Sediment sump:* Excavating a sediment sump immediately upstream from the check dam improves its effectiveness.
- *Erosion protection:* Provide outlet stabilization below the lowest check dam where the risk of erosion is greatest.
- *Channel linings:* Consider the use of channel linings or protection such as plastic sheeting or riprap where there may be significant erosion or prolonged submergence.

For *rock check dams,* use stone 2 to 15 in (50 to 375 mm) in diameter. Place the stones on the filter fabric either by hand or by using appropriate machinery; do not simply dump them in place. Extend the stone 18 in (0.5 m) beyond the banks, and keep the side slopes 2:1 or flatter. As a suggested option, line the upstream side of the dam with 0.75- to 1.25-in (20 to 30 mm) gravel to a depth of 1 ft (0.3 m). Figure 13.3 illustrates a typical rock check dam.

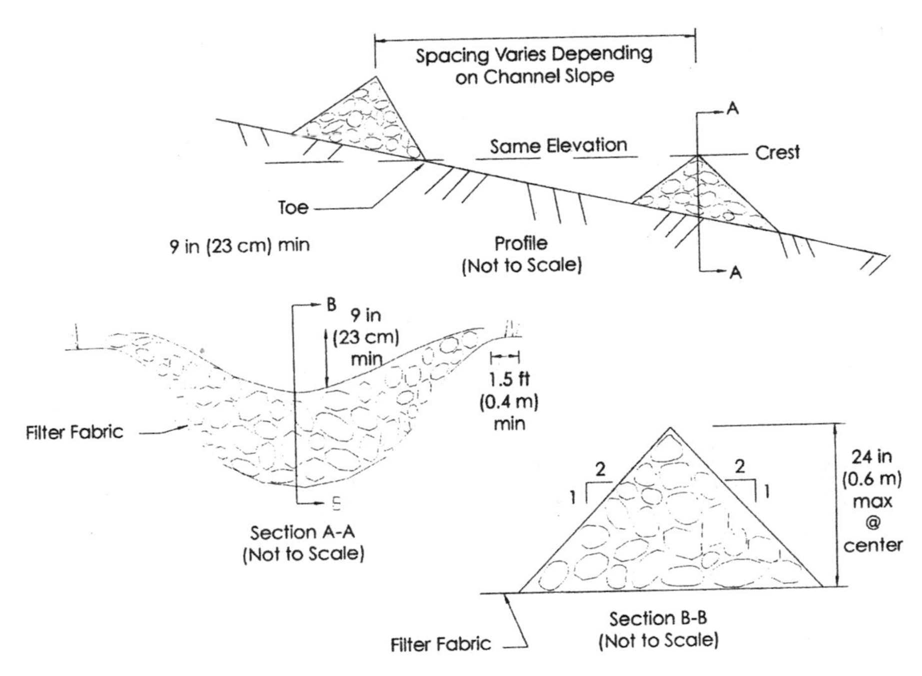

Figure 13.3 Rock check dam (*Environmental Protection Agency, 1992*).

For *log check dams,* use logs 6 to 8 in (150 to 200 mm) in diameter. Logs must be firmly embedded in the ground; 18 in (0.5 m) is the recommended minimum depth. Figure 13.4 illustrates a typical log check dam.

For *sandbag check dams,* use sandbags filled with pea gravel. Be sure that bags are all securely sealed. Place bags by hand or use appropriate machinery.

The cost for the construction of check dams varies with the material used. Rock costs about $100 per dam. Log check dams are usually slightly less expensive than rock check dams. All costs vary depending on the width of channel.

After each significant rainfall, check dams should be inspected for sediment and debris accumulation. Sediment should be removed when it reaches one half the original dam height. Check for erosion at edges, and repair promptly as required. If sandbags are used, the fabric of the bags should be inspected for signs of deterioration.

Check dams should remain in place and operational until the drainage area and channel are completely stabilized or up to 30 days after the permanent site stabilization is achieved. Restore the channel lining or establish vegetation when each check dam is removed.

It will be important to know the expected erosion rates and runoff flow rate for the swale in which this measure is to be installed. Contact

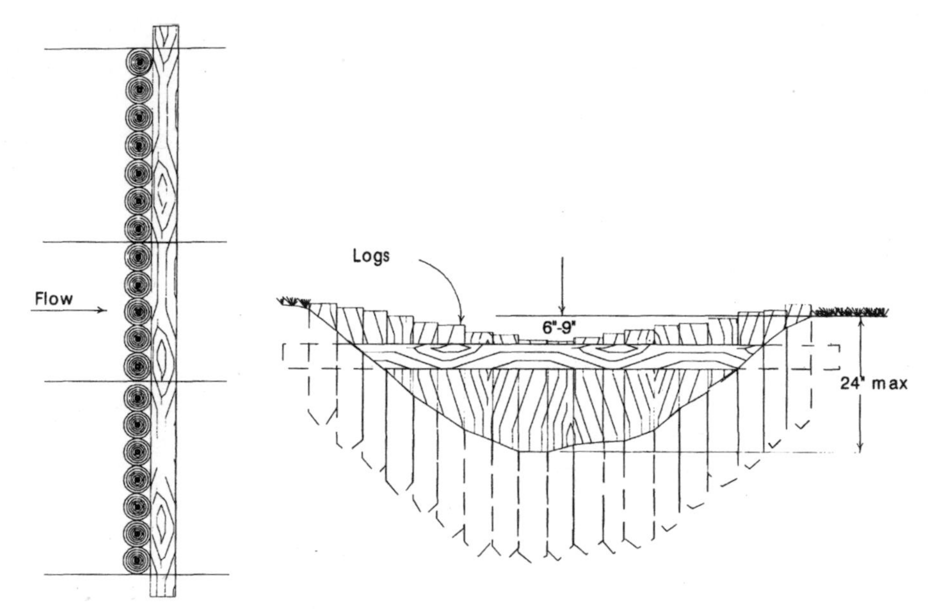

Figure 13.4 Log check dam (*Environmental Protection Agency, 1992*).

the state or local storm water program agency or a licensed engineer for assistance in designing this measure.

Velocity dissipation by surface roughening

Surface roughening is a temporary erosion control practice. The soil surface is roughened by the creation of horizontal grooves, depressions, or steps that run parallel to the contour of the land. Slopes that are not fine-graded and that are left in a roughened condition can also control erosion. Surface roughening reduces the speed of runoff, increases infiltration, and traps sediment. Surface roughening also helps establish vegetative cover by reducing runoff velocity and giving seed an opportunity to take hold and grow.

To slow erosion, surface roughening should be done as soon as possible after the vegetation has been removed from the slope. Roughening can be used with both seeding and planting and temporary mulching to stabilize an area. For steeper slopes and slopes that will be left roughened for longer periods, a combination of surface roughening and vegetation is appropriate. Surface roughening should be performed immediately after grading activities have ceased in an area, even if grading activities will be resumed later.

Different methods can be used to roughen the soil surface on slopes. They include stairstep grading, grooving with an implement, and tracking (driving a crawler tractor up and down a slope, leaving the cleat imprints parallel to the slope contour). The selection of an appropriate method depends on the grade of the slope, mowing requirements after vegetative cover is established, whether the slope was formed by cutting or filling, and the type of equipment available.

Any gradient with a slope greater than 2:1 should be stairstep-graded. Each step catches material discarded from above and provides a level site where vegetation can grow. Stairs should be wide enough to work with standard earthmoving equipment.

Cut slopes with a gradient steeper than 3:1 but less than 2:1 should be stairstep-graded or groove-cut. Stairstep grading works well with soils with large amounts of small rock. Grooving can be done by any implement that can be safely operated on the slope, including disks, spring harrows, or teeth on a front-end loader. Grooves should not be less than 3 in (75 mm) deep or more than 15 in (375 mm) apart. Medium slopes should be compacted every 9 in (225 mm) of depth. The face of the slope should consist of loose, uncompacted fill 4 to 6 in deep.

Any cut or filled slope that will be mowed should have a gradient less than 3:1. Such a slope can be roughened with shallow grooves parallel to the slope contour by normal tilling. Grooves should be close together [less than 10 in (250 mm)] and not less than 1 in (25 mm) deep.

It is important to avoid excessive compacting of the soil surface, especially during tracking, because soil compaction inhibits vegetation growth and causes higher runoff speed. Therefore, it is best to limit roughening with tracked machinery to sandy soils that do not compact easily and to avoid tracking on clay soils. Surface-roughened areas should be seeded as quickly as possible. Also, regular inspections should be made of all surface roughened areas, especially after storms. If rills (small watercourses that have steep sides and are usually only a few inches deep) appear, they should be filled, graded again, and reseeded immediately. Proper dust control procedures should be followed during surface roughening.

Velocity dissipation by gradient terraces

Gradient terraces are earth embankments or ridge and channels constructed along the face of a slope at regular intervals. Gradient terraces are constructed at a positive grade. They reduce erosion damage by capturing surface runoff and directing it to a stable outlet at a speed that minimizes erosion. Figure 13.5 illustrates a typical gradient terrace.

Gradient terraces are usually limited to use on long, steep slopes with a water erosion problem, or where it is anticipated that water erosion will be a problem. Gradient terraces should not be constructed on slopes with sandy or rocky soils. They will be effective only where suitable runoff outlets are or will be made available.

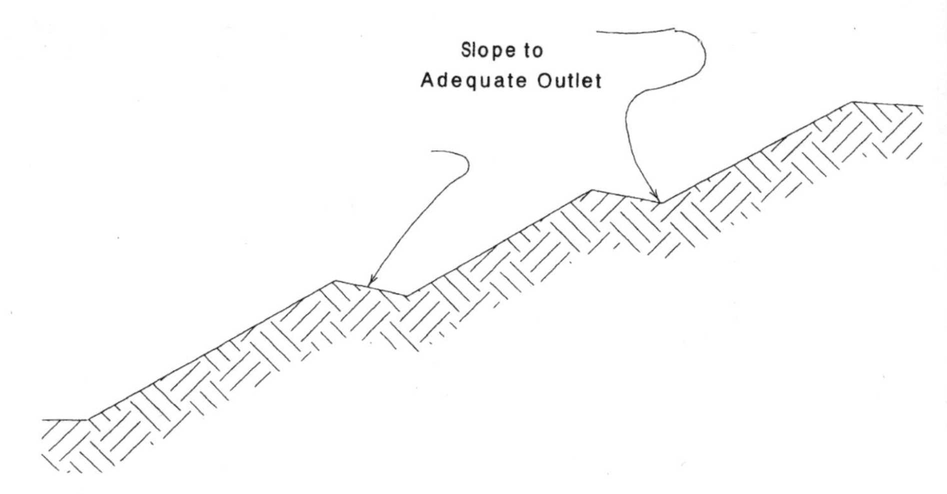

Figure 13.5 Gradient terraces (*Environmental Protection Agency, 1992*).

Gradient terraces should be designed and installed according to a plan determined by an engineering survey and layout. It is important that gradient terraces be designed with adequate outlets, such as a grassed waterway, vegetated area, or tile outlet. In all cases, the outlet should direct the runoff from the terrace system to a point where the outflow will not cause erosion or other damage. Vegetative cover should be used in the outlet where possible. The design elevation of the water surface of the terrace should not be lower than the design elevation of the water surface in the outlet at their junction, when both are operating at design flow. Terraces should be inspected regularly, at least once a year and after major storms. Proper stabilization practices should be followed in the construction of these features.

Sediment Capture Measures

It is necessary to have some disturbed portions of a construction site exposed to possible rainfall, at least briefly. Therefore, it is necessary to install measures which can remove sediment from runoff before it flows off the construction site.

Sediment capture measures prevent sediment from moving off the site by physically interposing a barrier to the movement of the sediment. Such measures include silt fences, straw bale dikes, brush barriers, gravel or stone filter berms, storm drain inlet protection, sediment traps, and temporary sediment basins.

The sediment capture device which is most suitable for large disturbed areas is the *sediment basin.* The EPA general permit for construction storm water discharges requires that a sediment basin be installed at all locations where there is an upstream disturbed area of

10 acres (4 ha) or more, wherever attainable. Authorized states may have different requirements for sediment basins. In addition, many local drainage regulatory agencies and soil conservation districts have regulations affecting construction practices.

Disturbed areas of less than 10 acres (4 ha) have greater variety in the measures which are suitable for sediment capture. Several types of measures can be used for sediment control, including sediment basins, sediment traps, silt fences, and gravel filter berms. The selection among these measures depends upon a number of situations, including the following:

- *Overland flow.* Runoff which passes over disturbed soil should pass through sediment controls before it can be allowed to flow off the construction site. Therefore the entire downslope and side-slope borders of the disturbed area should be lined with filtration devices, such as silt fences or gravel filter berms. These methods have limitations regarding the specific conditions in which they are effective. As an alternative, overland flow runoff from a disturbed area can also be directed to a sediment trap or a temporary sediment basin by using diversion devices such as an earth dike or an interceptor dike and swale.
- *Concentrated flow.* Sediment should be removed from concentrated runoff by either a sediment trap or a temporary sediment basin, depending on the disturbed area upstream. Filtration measures are generally not effective when used in concentrated flow because flow will overtop the filtering device.
- *Flow into storm drain inlets.* If there is a yard drain or curb inlet which receives flow from a disturbed area, then a sediment basin, sediment trap, or inlet protection should be constructed to remove the sediment from the runoff before it flows into the inlet.

Sediment capture by temporary sediment basins

A *temporary sediment basin* is a settling pond with a controlled storm water release structure used to collect and store sediment produced by construction activities. A sediment basin can be constructed by excavation and/or by placing an earthen embankment across a low area or drainage swale. Sediment basins can be designed to maintain a permanent pool or to drain completely dry. The basin detains sediment-laden runoff from larger drainage areas long enough to allow most of the sediment to settle out. Figure 13.6 illustrates a temporary sediment basin.

To meet the requirements of the EPA general permit, sediment basins designed to serve disturbed areas of at least 10 acres (4 ha) must have a volume adequate to contain the runoff calculated using the local

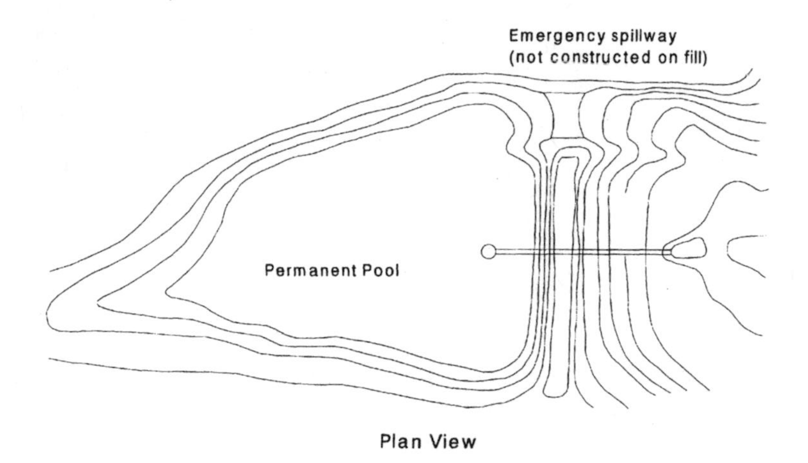

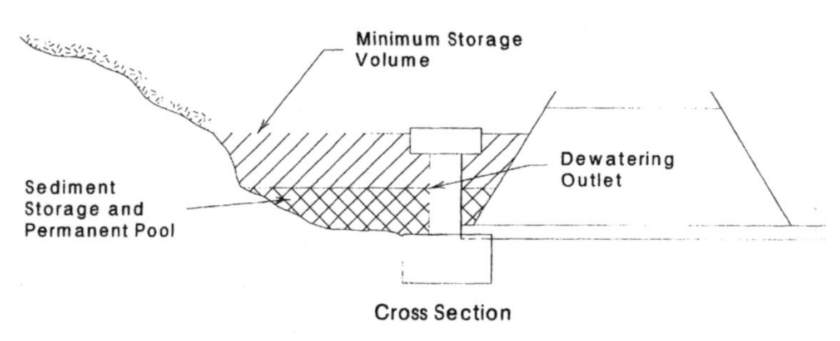

Figure 13.6 Temporary sediment basin (*Environmental Protection Agency, 1992*).

2-year, 24-h storm and runoff coefficient from each disturbed acre drained. Figure 13.7 shows the 2-year, 24-h rainfall amounts for the eastern United States.

To use this figure, do the following:

1. Locate your project site on the map, and interpolate the 2-year, 24-h rainfall amount. For example, the 2-year, 24-h rainfall for Chicago is approximately 2.8 in (71 mm), because Chicago is located between the 2.5-in and 3.0-in contours on the exhibit.
2. Determine the drainage area for the sediment basin, as described in Chap. 11.
3. Calculate the runoff coefficient for the drainage area. This is also described in Chap. 11.
4. Multiply the drainage area, the rainfall depth, and the runoff coefficient together to compute the required detention basin volume. The appropriate unit conversions must be made.

Figure 13.7 Two-year, 24-hour rainfall for eastern United States (*National Weather Service, 1961*).

As an example, assume that we want to design a sedimentation basin in the Chicago area. As noted above, the 2-year 24-h rainfall is approximately 2.8 in (71 mm) in this area. Assume that the drainage area has been determined to be 12.4 acres (5 ha) and the runoff coefficient is 0.35. Therefore, the required volume of the sedimentation basin is

$$\frac{2.8\text{ in}}{12} \times (12.4 \times 43{,}560) \times 0.35 = 0.23 \times 479{,}160 \times 0.35 = 44{,}000\text{ ft}^3$$

$$\frac{71\text{ mm}}{1000} \times 5 \times 10{,}000 \times 0.35 = 0.071 \times 5000 \times 0.35 = 1250\text{ m}^3$$

Where no such calculation has been performed, a temporary (or permanent) sediment basin providing 3600 ft^3 of storage per acre (250 m^3/ha) of total drainage area, or equivalent control measures, must be provided where attainable until final stabilization of the site. For the example above, the required volume computed by using the default method is

$$12.4 \text{ acres} \times 3600 \text{ ft}^3/\text{acre} = 45{,}000 \text{ ft}^3$$

$$5 \text{ ha} \times 250 \text{ m}^3/\text{ha} = 1250 \text{ m}^3$$

Obviously, for this example, the two methods give almost the same results.

A sediment basin may not be economically attainable at a particular location because of shallow bedrock that prevents excavation of a basin, topographic difficulties that prohibit the construction of a basin of adequate storage volume; lack of space available at the common drainage location to construct a basin, due to the presence of existing structures, pavement, or utilities which cannot be relocated; lack of property rights available at the only common drainage location (the only feasible location may be beyond the property line or right-of-way of the construction activity and a temporary construction easement cannot be obtained); or legal restrictions as a result of state, local, or other federal regulations that may prohibit a basin or the construction of a basin in the common drainage locations.

If a sediment basin is not economically feasible under the criteria listed above, then sediment traps, silt fences, or other equivalent sediment control measures such as gravel filter berms should be installed instead.

The sediment basin has a riser and pipe outlet with a gravel outlet or spillway to slow the release of runoff and provide some sediment filtration. By removing sediment, the basin helps prevent the clogging of off-site conveyance systems and the sediment loading of receiving waterways. In this way, the basin helps prevent destruction of waterway habitats.

A temporary sediment basin should be installed before clearing and grading are undertaken. It should not be built in an active (existing natural) stream. The creation of a dam in such a site may result in the destruction of aquatic habitats. The risk of flooding damages downstream in case of dam failure should also be considered in the design of the temporary sediment basin.

A temporary sediment basin used in combination with other control measures, such as seeding or mulching, is especially effective for removing sediments. The sediment trapping efficiency is improved by providing the maximum surface area possible. Because finer silts may not settle out completely, additional erosion control measures should be used to minimize the release of fine silt. Runoff should enter the basin as far from the outlet as possible, to provide maximum retention time. To avoid "short-circuit" flows, baffles made of plywood may be installed in the sediment basin to force the flow to follow a longer path from the inlet to the outlet.

The useful life of a temporary sediment basin is dependent upon adequate maintenance. Sediment basins should be readily accessible for maintenance and sediment removal. They should be inspected after each rainfall and cleaned out when about one-half the volume has been filled with sediment. The sediment basin should remain in operation and be properly maintained until the site area is permanently stabilized by vegetation and/or when permanent structures are in place. The embankment forming the sedimentation pool should be well compacted and stabilized with vegetation. If the pond is located near a residential area, a sign should be posted and the area should be secured by a fence to increase public safety and to reduce the possibility of vandalism.

The cost of sediment basins varies widely according to size and site conditions. Costs typically range from about $5000 to about $50,000 or more.

A well-built temporary sediment basin that is large enough to handle the postconstruction runoff volume may be converted later to use as a permanent storm water management structure. This may satisfy the requirements of the EPA general permit for permanent storm water control measures, as described later in this chapter.

The sediment basin outlet pipe and spillway should be designed by an engineer based upon an analysis of the expected runoff flow rates from the site.

Sediment capture by sediment traps

A *sediment trap* is formed by excavating a pond or by placing an earthen embankment across a low area or drainage swale. An outlet or spillway is constructed by using large stones or aggregate to slow the release of runoff. The trap retains the runoff long enough to allow most of the silt to settle out. Figure 13.8 illustrates a typical sediment trap.

A temporary sediment trap may be used in conjunction with other temporary measures, such as gravel construction entrances, vehicle wash areas, slope drains, diversion dikes and swales, or diversion channels. Temporary sediment traps are appropriate in the following locations:

- At the outlet of the perimeter controls installed during the first stage of construction
- At the outlet of any structure which concentrates sediment-laden runoff, e.g., at the discharge
- Above a storm water inlet that is in line to receive sediment-laden runoff

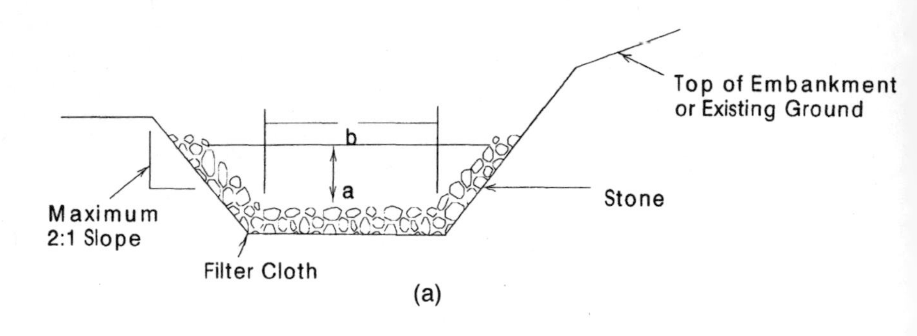

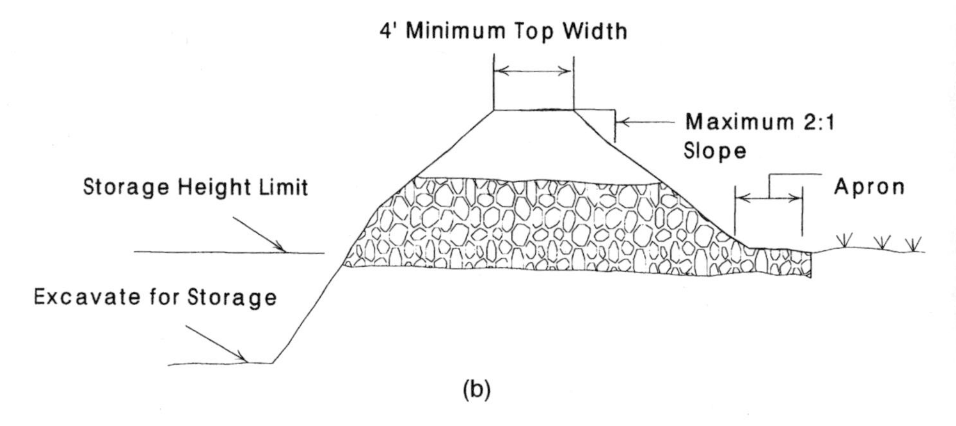

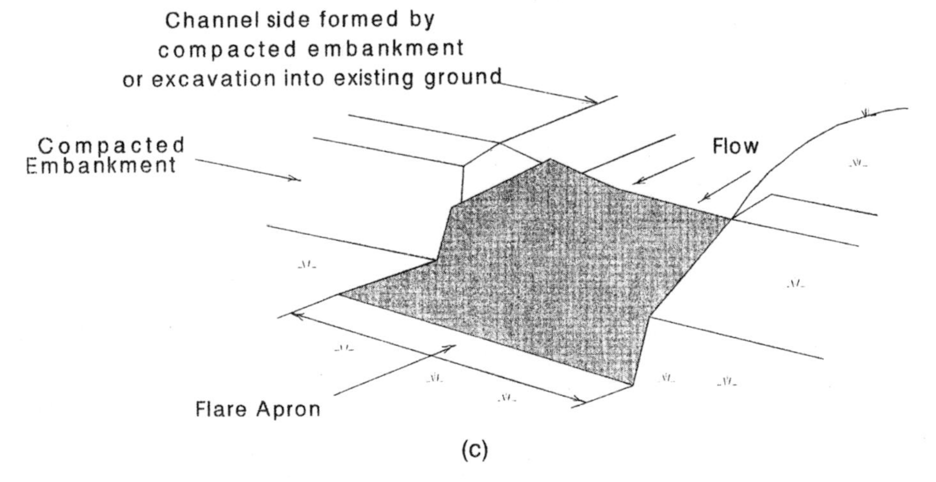

Figure 13.8 Sediment trap (*Environmental Protection Agency, 1992*).

TABLE 13.1 Recommended Dimensions of Sediment Trap

Embankment height, ft (m)	Outlet height, ft	Width, ft (m)
1.5 (0.5)	0.5 (15 cm)	2.0 (0.6)
2.0 (0.6)	1.0 (0.3 m)	2.0 (0.6)
2.5 (0.8)	1.5 (0.5 m)	2.5 (0.8)
3.0 (0.9)	2.0 (0.6 m)	2.5 (0.8)
3.5 (1.1)	2.5 (0.8 m)	3.0 (0.9)
4.0 (1.2)	3.0 (0.9 m)	3.0 (0.9)
4.5 (1.4)	3.5 (1.1 m)	4.0 (1.2)
5.0 (1.5)	4.0 (1.2 m)	4.5 (1.4)

Temporary sediment traps may be constructed by excavation alone or by excavation in combination with an embankment. Temporary sediment traps are often used in conjunction with a diversion dike or swale. Sediment traps should not be planned to remain in place longer than about 18 to 24 months.

Sediment traps are suitable for small drainage areas, usually no more than 10 acres (4 ha) total, or no more than 5 disturbed acres (2 ha). The trap should be large enough to allow the sediments to settle, and should have a capacity to store the collected sediment until it is removed. The capacity of the sedimentation pool should provide storage volume for at least 3600 ft^3/acre (250 m^3/ha) of drainage area, and more if conditions require.

The outlet should be designed to provide a 2-ft (0.6-m) settling depth and an additional sediment storage area 1.5 ft (0.5 m) deep at the bottom of the trap. The embankment height should not exceed 5 ft (1.5 m). The recommended minimum width at the top of the embankment is between 2 and 5 ft (0.6 and 1.5 m). The minimum recommended length of the weir is between 3 and 4 ft (1.0 and 1.3 m), and the maximum length is 12 ft (3.7 m). Table 13.1 illustrates the typical relationship between the embankment height, the height of the outlet, and the width at the top of the embankment.

The following steps are involved in constructing a temporary sediment trap:

1. *Clearing.* Clear the area of all trees, brush, stumps, or other obstructions.

2. *Embankment.* Construct the embankment in 8-in (20-cm) lifts, compacting each lift with the appropriate earthmoving equipment. Fill material must be free of woody vegetation, roots, and large stones. Keep cut-and-fill slopes between 3:1 and 2:1 or flatter.

3. *Filter fabric.* Line the outlet area with filter fabric prior to placing stone or gravel.
4. *Gravel outlet.* Construct the gravel outlet using coarse aggregate between 2 and 14 in (5 and 36 cm) in diameter, and face the upstream side with a 12-in (0.3-m) layer of 0.75- to 1.5-in (2- to 4-cm) washed gravel on the upstream side.
5. *Stabilization.* Seed and mulch the embankment as soon as possible to ensure stabilization.

Costs for a sediment trap vary widely based upon the size and the amount of excavation and stone required. It usually can be installed for $500 to $7000.

The effective life of a sediment trap depends upon adequate maintenance. The trap should be readily accessible for periodic maintenance and sediment removal. Traps should be inspected regularly and after each rainfall. Make any repairs necessary to ensure the measure is in good working order. Check the embankment regularly to make sure it is structurally sound. At a minimum, sediment should be removed and the trap restored to its original volume when sediment reaches 50% of the original volume. Sediment removed from the trap must be properly disposed of.

The trap should remain in operation and be properly maintained until the site area is permanently stabilized by vegetation and/or when permanent structures are in place.

Sediment capture by silt fences

A *silt fence,* also called a *filter fence,* is a temporary measure for sedimentation control. The silt fence is used to intercept sediment in diffuse (shallow overland) flow. It usually consists of posts with geotextile fabric (filter cloth) stretched across them and sometimes with a wire support fence. The lower edge of the fence is vertically trenched and covered by backfill. See Fig. 13.9.

A silt fence is used in small drainage areas to detain sediment. These fences are most effective where there is overland flow (runoff that flows over the surface of the ground as a thin, even layer) or in minor swales or drainageways. They prevent sediment from entering receiving waters. Silt fences are also used to catch wind-blown sand and to create an anchor for sand dune creation.

A silt fence is not appropriate for controlling runoff from a large area. For slopes between 50:1 and 5:1, the maximum allowable upstream flow path length to the fence is 100 ft; for slopes of 2:1 and steeper, the maximum is 20 ft. The maximum upslope grade perpendicular to the fence line should not exceed 1:1.

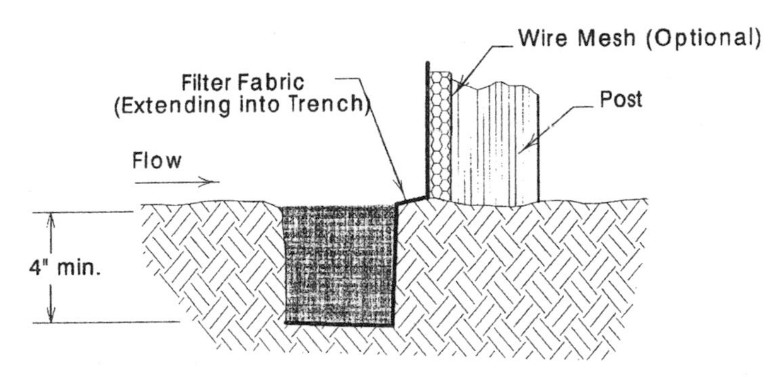

Figure 13.9 Silt fence construction detail (*Environmental Protection Agency, 1992*).

A silt fence should be installed prior to major soil disturbance in the drainage area. The fence should be placed across the bottom of a slope along a contour line (a line of uniform elevation). This will place the fence perpendicular to the direction of flow. It can be used at the outer boundary of the work area. However, the fence does not have to surround the work area completely. In addition, a silt fence is effective where sheet and rill erosion may be a problem. Silt fences should not be constructed in streams or large swales.

Silt fences can be more effective than a straw bale barrier if properly installed and maintained. They may be used in combination with other erosion and sediment practices. Ponding should not be allowed behind silt fences since they will collapse under high pressure; the design should provide sufficient outlets to prevent overtopping.

Aside from the traditional wooden post and filter fabric method, several variations of silt fence installation including silt fence fabric which can be purchased with pockets presewn to fit steel fenceposts.

The maximum height of the filter fence should range between 18 and 36 in (0.5 and 1 m) above the ground surface (depending on the amount of upslope ponding expected). The following elements must be considered:

1. *Filter fabric.* Synthetic filter fabric should be a pervious sheet of polypropylene, nylon, polyester, or polyethylene yarn. Synthetic filter fabric should contain ultraviolet ray inhibitors and stabilizers to provide a minimum of 6 months of expected usable construction life at a temperature range of 0 to 120°F. Burlap fabric can also be used, but is only acceptable for periods of up to 60 days. The filter fabric should be purchased in a continuous roll to avoid joints. Where joints in the fabric are required, the filter cloth should be spliced together only at a support post, with a minimum 6-in (15-cm) overlap, and securely sealed. See Table 13.2.

TABLE 13.2 Synthetic Filter-Fabric Requirements

Physical property	Requirements
Filtering efficiency	75%–85% (minimum)
Tensile strength at 20% (maximum elongation)	Standard strength: 30 lb/lin in (minimum); extra strength: 50 lb/lin in (minimum)
Slurry flow rate	0.3 gal/(ft^2/min) (minimum)

2. *Reinforcing wire.* While not required, wire fencing may be used as a backing to reinforce standard-strength filter fabric. The wire fence (14 gauge minimum) should be 24 to 48 in (60 to 120 cm) wide and should have a maximum mesh spacing of 6 in (15 cm). If standard-strength filter fabric is to be used, the optional wire-mesh support fence may be fastened to the upslope side of the posts by using 1-in heavy-duty wire staples, tie wires, or hog rings. Extend the wire-mesh support to the bottom of the trench. The filter fabric should then be stapled or wired to the fence, and 8 to 20 in (20 to 50 cm) of fabric should extend into the trench. Extra-strength filter fabric does not require a wire-mesh support fence. Staple or wire the filter fabric directly to the posts, and extend 8 to 20 in (20 to 50 cm) of the fabric into the trench.

3. *Posts.* Posts should be 2 to 4 ft (0.6 to 1.5 cm) long and should be composed of either 2-in by 2-in (5-cm by 5-cm) or 2-in by 4-in (5-cm by 10-cm) pine (or equivalent) or steel. Steel posts should have projections for fastening wire and fabric to them. Posts should be spaced 8 to 10 ft (2.5 to 3 m) apart when a wire-mesh support fence is used and no more than 6 ft (2 m) apart when extra-strength filter fabric (without a wire fence) is used. The posts should extend 12 to 30 in (0.3 to 0.8 m) into the ground. Trees should not be used as substitutes for posts.

4. *Trench.* A trench should be excavated 4 to 8 in (10 to 20 cm) wide and 4 to 12 in (10 to 30 cm) deep along the upslope side of the line of posts. Backfill the trench with compacted soil or 0.75-in (2-cm) minimum-diameter gravel placed over the filter fabric.

Silt fence installation costs approximately $6 per linear ft ($2 per meter). See Fig. 13.10.

The fence requires frequent inspection and prompt maintenance to maintain its effectiveness. Inspect filter fences daily during periods of prolonged rainfall, immediately after each rainfall event, and weekly during periods of no rainfall. Check for areas where runoff eroded a channel beneath the fence, or where the fence was caused to sag or collapse by runoff flowing over the top. Make any required repairs immediately. Remove and properly dispose of sediment when it is one-third

Figure 13.10 Typical silt fence installation along stream bank.

to one-half the height of the fence or after each storm. Take care to avoid damaging the fence during cleanout. See Fig. 13.11.

Filter fences should not be removed until the upslope area has been permanently stabilized. Any sediment deposits remaining in place after the filter fence has been removed should be dressed to conform with the existing grade, prepared, and seeded.

Sediment capture by straw bale dikes or brush barriers

Straw bales are temporary barriers of straw or similar material used to intercept sediment in runoff from small drainage areas of disturbed soil. When installed and maintained properly, straw bale dikes can remove approximately two-thirds of the sediment in runoff, but this can be achieved only through careful maintenance with special attention to replacing rotted or broken bales.

Brush barriers are composed of tree limbs, weeds, vines, root mat, soil, rock, and other cleared materials placed at the toe of a slope.

Many state and local agencies allow straw bales or brush barriers as an alternative to silt fences. However, straw bales and brush barriers are not listed in the EPA general permit because of questions regard-

Figure 13.11 Silt fabric fence with sediment accumulated during storm event.

ing their effectiveness, primarily because of improper installation and inadequate maintenance. Figure 13.12 illustrates a straw bale installation requiring maintenance.

Sediment capture by gravel or stone filter berms

A *gravel* or *stone filter berm* is a temporary ridge constructed of loose gravel, stone, or crushed rock. It slows and filters flow, diverting it from an exposed traffic area. Diversions constructed of compacted soil may be used where there will be little or no construction traffic within the right-of way. They are also used for directing runoff from the right-of-way to a stabilized outlet. Figure 13.13 illustrates a typical gravel filter berm.

Gravel or stone filter berms are appropriate where roads and other rights-of-way under construction should accommodate vehicular traffic. Berms are meant for use in areas with gentle slopes. They may also be used at traffic areas within the construction site.

Berm material should be well-graded gravel or crushed rock. The spacing of the berms will depend on the steepness of the slope: Berms should be placed closer together as the slope increases.

Figure 3.12 Straw bales used for curb inlet protection.

The diversion should be inspected regularly after each rainfall or if breached by construction or other vehicles. All needed repairs should be performed immediately. Accumulated sediment should be removed and properly disposed of and the filter material replaced, as necessary.

Sediment capture by storm drain inlet protection

Storm drain inlet protection is a filtering measure or excavated impounding area placed around any inlet or drain to trap sediment. This prevents the sediment from entering inlet structures. Additionally, it serves to prevent the silting of inlets, storm drainage systems, or receiving channels.

Storm drain inlet protection is appropriate for small drainage areas where storm drain inlets will be ready for use before final stabilization. Storm drain inlet protection is also used where a permanent storm drain structure is being constructed on the site. Storm drain inlet pro-

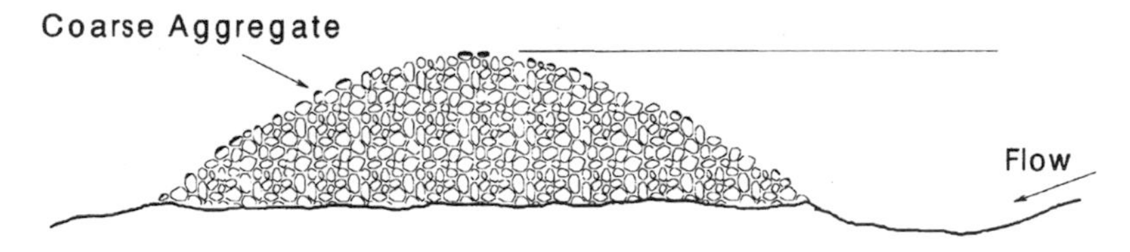

Figure 13.13 Gravel filter berm (*Environmental Protection Agency, 1992*).

tection may be constructed of filter fabric, excavated gravel, concrete block and gravel, or other materials. Sod may be used where sediments in the storm water runoff are low. Gravel and mesh filters can be used where flows are higher and subject to disturbance by site traffic. Straw bales are used as storm drain inlet protection in many locations. However, the EPA does not recommend straw bales for this purpose.

The cost of storm drain inlet protection varies depending upon the size and type of inlet to be protected but generally is about $300 per inlet.

Storm drain inlet protection is not meant for use in drainage areas exceeding 1 acre (0.4 ha) or for large concentrated storm water flows. Installation of this measure should take place prior to any soil disturbance in the drainage area. The type of material used will depend on the site conditions and the size of the drainage area. Inlet protection should be used with other measures, such as small impoundments or sediment traps, to provide more effective sediment removal.

Inlet protection structures should be inspected regularly, especially after a rainstorm. Repairs and silt removal should be performed as necessary. Storm drain inlet protection structures should be removed only after the disturbed areas are completely stabilized.

Filter-fabric inlet protection. *Filter fabric* is used for inlet protection when storm water flows are relatively small with low velocities. This practice cannot be used where inlets are paved because the filter fabric should be staked. The drainage area should be 1 acre (0.4 ha) or less, with slopes of 5% or less. The area immediately surrounding the inlet should not exceed a slope of 1%. Overland flow to the inlet should be no greater than 0.5 ft^3/s (0.01 m^3/s). Figure 13.14 illustrates a typical filter-fabric inlet protection measure.

To avoid failure caused by pressure against the fabric when overtopping occurs, the height of the filter fabric should be limited to 1.5 ft (0.5 m) above the crest of the drop inlet. A sediment-trapping sump of 1- to 2-ft (0.3- to 0.6-m) depth with side slopes of 2:1 is recommended. The following elements must be considered:

1. *Filter fabric.* This is the same as that used for silt fences.
2. *Posts.* Use wooden 2-in by 2-in (5-cm by 5-cm) or 2-in by 4-in (5-cm by 10-cm) with a minimum length of 3 ft (1 m). Place a stake at each corner of the inlet and around the edges no more than 3 ft (1 m) apart. Stakes should be driven into the ground 18 in (0.5 m) or at a minimum 8 in (200 cm).
3. *Framework.* For stability, a framework of wood strips should be installed around the stakes at the crest of the overflow area 1.5 ft (0.5 m) above the crest of the drop inlet. Staple the filter fabric to the wooden stakes with heavyduty staples at least ½ in (12 mm) long, overlap-

ping the joints to the next stake. Ensure that 12 to 32 in (30 to 80 cm) of filter fabric extends at the bottom so it can be formed into the trench.

4. *Trench.* Excavate a trench of 8 to 12 in (20 to 30 cm) deep around the outside perimeter of the stakes. If a sediment-trapping sump is being provided, then the excavation may be as deep as 2 ft (0.6 m). Place the bottom of the fabric in the trench, and backfill the trench all the way around, using washed gravel 0.75 in (2 cm) in diameter to a minimum depth of 4 in (10 cm).

Inspect regularly and after every storm. Make sure that the stakes are firmly seated in the ground and that the filter fabric continues to be securely anchored. Make any repairs necessary to ensure that the measure is in good working order.

Sediment should be removed and the trap restored to its original dimensions when sediment has accumulated to 50% of the design depth of the trap. All removed sediments should be properly disposed of. If the filter fabric becomes clogged, it should be replaced immediately.

Inlet protection should remain in place and operational until the drainage area has been completely stabilized or up to 30 days after the permanent site stabilization is achieved.

Excavated gravel inlet protection. Excavated gravel inlet protection may be constructed by excavating around a storm drain inlet located in an unpaved area. The excavation provides storage for sediment. Figure 13.15 illustrates a typical excavated gravel inlet protection measure.

Figure 13.14 Filter fabric storm drain inlet protection.

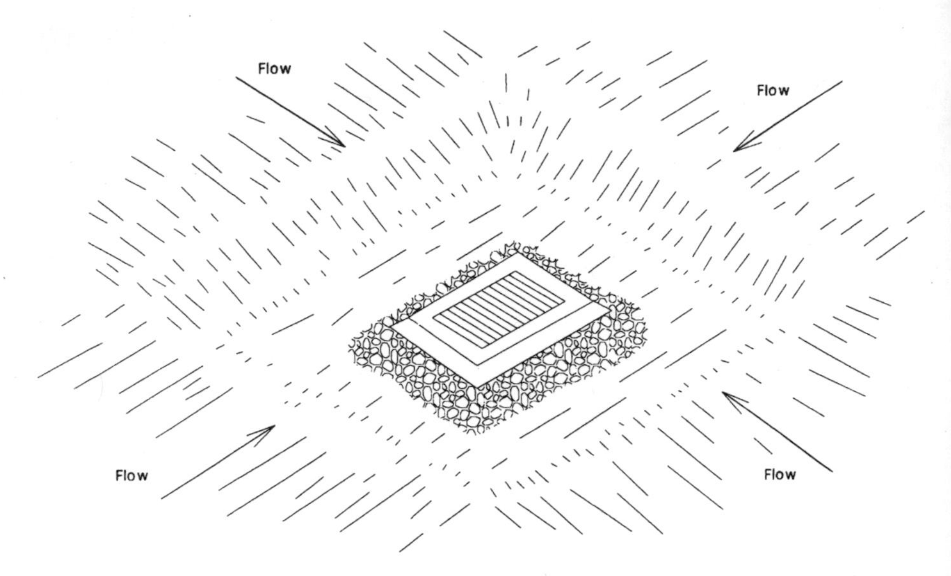

Figure 13.15 Excavated gravel inlet protection (*Environmental Protection Agency, 1992*).

Excavated gravel and mesh inlet protection may be used with most inlets where overflow capability is needed and in areas of heavy flows, 0.5 ft^3/s (0.01 m^3/s) or greater. The drainage area should be fairly flat with slopes of 5% or less. The trap should have a sediment-trapping sump of 1 to 2 ft (0.3 to 0.6 m) measured from the crest of the inlet. Side slopes should be 2:1. The recommended volume of excavation is 35 yd^3/acre (2.4 m^3/ha) disturbed. To achieve maximum trapping efficiency, the longest dimension of the basin should be oriented toward the longest inflow area. The following steps are followed in constructing excavated gravel inlet protection:

1. *Excavation.* Remove any obstructions to excavating and grading. Excavate the sump area, grade the slopes, and properly dispose of the excavated soil.
2. *Inlet grate.* Secure the inlet grate to prevent seepage of sediment-laden water.
3. *Wire mesh.* Place hardware cloth or wire mesh with ½-in (12-mm) openings over the drop inlet so that the wire extends a minimum of 1 ft (30 cm) beyond each side of the inlet structure. Overlap the strips of mesh if more than one is necessary.
4. *Filter fabric.* Place filter fabric (as specified for silt fences) over the mesh, extending it at least 18 in (0.5 m) beyond the inlet opening on

all sides. Ensure that weep holes in the inlet structure are protected by filter fabric and gravel.

5. *Gravel*. Place washed gravel ¾ to 4 in (2 to 10 cm) in diameter over the fabric and wire mesh to a depth of at least 1 ft (30 cm).

Inspect regularly and after every storm. Make any repairs necessary to ensure that the measure is in good working order. Clean or remove and replace the stone filter or filter fabric if it becomes clogged. Sediment should be removed and the trap restored to its original dimensions when sediment has accumulated to one-half the design depth of the trap.

Inlet protection should remain in place and operational until the drainage area has completely stabilized or up to 30 days after the permanent site stabilization has been achieved.

Block-and-gravel inlet protection. Concrete block-and-gravel filters can be used where velocities are higher than those allowed for excavated gravel inlet protection. Figure 13.16 illustrates a typical block-and-gravel inlet protection measure.

Block-and-gravel inlet protection may be used with most types of inlets where overflow capability is needed and in areas of heavy flows, 0.5 ft^3/s (0.01 m^3/s) or greater. The drainage area should not exceed 1 acre (0.4 ha). The drainage area should be fairly flat with slopes of 5% or less. To achieve maximum trapping efficiency, the longest dimension of the basin should be oriented toward the longest inflow area. Where possible, the trap should have a sediment-trapping sump 1 to 2 ft (0.3 m to 0.6 m) deep with side slopes of 2:1. The following steps are used in constructing this type of inlet protection:

1. *Inlet grate*. Secure the inlet grate to prevent seepage of sediment-laden water.
2. *Wire mesh*. Place hardware cloth or wire mesh with ½-in (12-mm) openings over the drop inlet so that the wire extends a minimum of 12 to 18 in (0.3 m to 0.5 m) beyond each side of the inlet structure. Overlap the strips of mesh if more than one is necessary.
3. *Filter fabric* (*optional*). Place filter fabric (as specified for silt fences) over the mesh, and extend it at least 18 in (0.5 m) beyond the inlet structure.
4. *Concrete blocks*. Place concrete blocks—4 to 12 in (10 cm to 30 cm) wide—over the filter fabric in a single row lengthwise on their sides along the sides of the inlet. The foundation should be excavated a minimum of 2 in (5 cm) below the crest of the inlet, and the bottom row of blocks should be against the edge of the structure for lateral support. The open ends of the block should face outward, not upward, and the ends of adjacent blocks should abut. Lay one block on each side of the

structure on its side to allow for dewatering of the pool. The block barrier should be at least 12 in (0.3 m) high, and it may be up to a maximum of 24 in (0.6 m) high and may be from 4 to 12 in (10 to 30 cm) deep, depending on the size of block used. Prior to backfilling, place wire mesh over the outside vertical end of the blocks so that stone does not wash down the inlet.

5. *Gravel.* Place washed gravel, ¾ to 4 in (2 to 10 cm) in diameter against the wire mesh to the top of the blocks.

For curb inlets, follow these installation procedures:

1. *Spacer blocks.* Place two concrete blocks on their sides perpendicular to the curb at either end of the inlet opening.
2. *Front blocks.* Place concrete blocks on their sides across the front of the opening and abutting the spacer blocks. The openings in the blocks should face outward, not upward.
3. *Retaining stud.* Cut a 2-in by 4-in (5-cm by 10-cm) wooden stud the length of the curb inlet plus the width of the spacer blocks. Place the stud through the outer hole of each spacer block to keep the front blocks in place.
4. *Wire mesh.* Place wire mesh over the outside vertical face (open ends) of the front concrete blocks to prevent stone from being washed through the openings in the blocks. Use chicken wire or hardware cloth with 0.5-in (12-mm) openings.
5. *Gravel.* Place 2- to 3-in (5- to 7-cm) gravel against the wire to the top of the barrier.

Inspect regularly and after every storm. Make any repairs necessary to ensure that the measure is in good working order. Sediment should be removed and the trap restored to its original dimensions when sediment has accumulated to one-half the design depth of the trap. All sediments removed should be properly disposed of.

Inlet protection should remain in place and operational until the drainage area has completely stabilized or up to 30 days after the permanent site stabilization has been achieved.

Temporary storm water management measures

Diverting off-site runoff around a disturbed area reduces the amount of storm water which comes into contact with the exposed soils. If there is less runoff coming in contact with exposed soil, then there will be less erosion of the soil and less storm water which has to be treated to remove sediment. The following conditions may be considered:

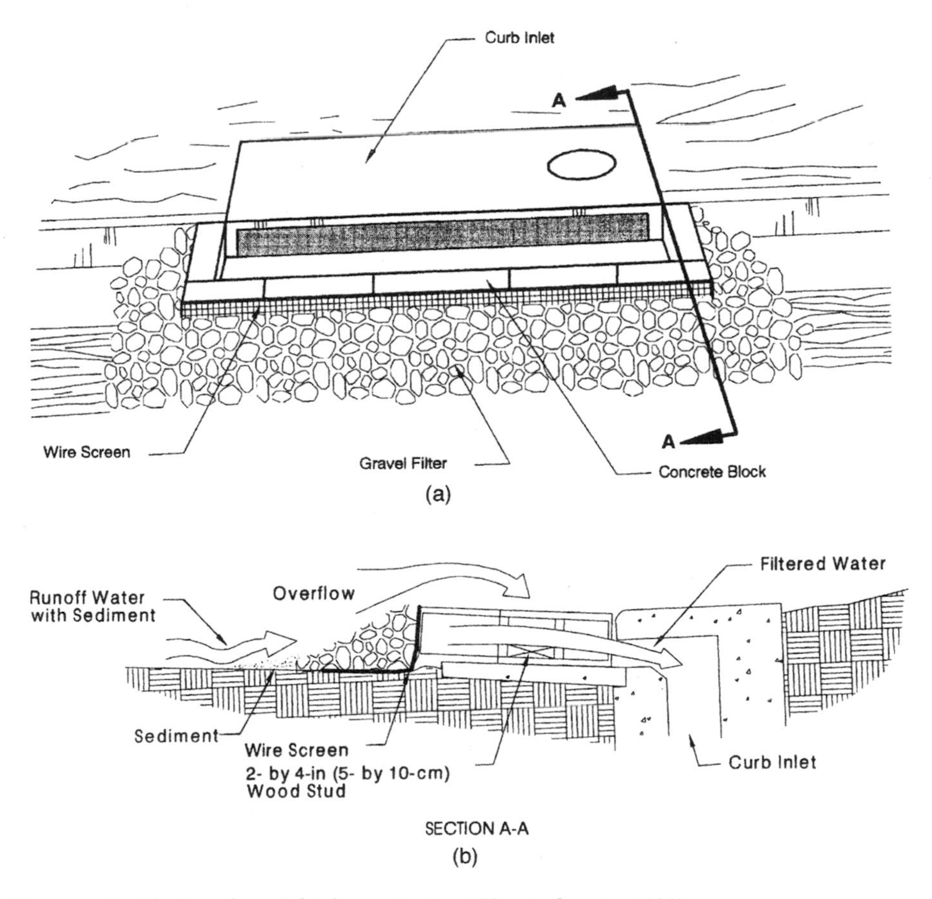

Figure 13.16 Block-and-gravel inlet protection (*Harris County, 1992*).

- *Off-site flows entering site.* Overland flow can be diverted around a construction site by installing an earth dike, an interceptor dike and swale, or a drainage swale. The choice of diversion methods depends upon the size of the uphill area and the steepness of the slope that the diversion must go down. Interceptor dikes and swales are effective in diverting overland flows from smaller areas of 3 acres (1.2 ha) or less down gentle slopes (10% or less). A temporary swale is most effective in diverting runoff from concentrated channels, and an earth dike is capable of diverting both sheet and concentrated flows from larger areas down steeper slopes. These devices should be installed from the uphill side of the site down to a point where they can discharge to an undisturbed area on the downhill side of the site.

- *Steep slopes.* Steeply sloped areas are especially susceptible to erosion. If there are steep areas on your site which will be disturbed, then an earth dike or interceptor dike and swale may be used to divert the runoff from the top of the slope to the inlet of a pipe slope drain or to a less steeply sloped area. These measures will minimize the amount of runoff flowing across the face of a slope and will decrease the erosion of that slope.
- *On-site swales or streams.* Swales and streams which run through construction sites must be protected from erosion and sediment because they can be significantly damaged. Streams and other water bodies should be protected by preservation of natural vegetation or buffer zones, as described in Chap. 12. Where possible, these techniques should also be used to protect swales or intermittent streams. Where construction requires that the stream or swale be disturbed, the amount of area and time of disturbance should be minimized. All stream and channel crossings should be made at right angles to the stream, preferably at the narrowest portion of the channel. Once a stream or swale has been disturbed, construction should proceed as quickly as possible in that area. Once completed, the stream banks should be stabilized as described in Chap. 12. Swales and intermittent streams disturbed by construction should be seeded and stabilized as soon as possible.
- *Construction crossings of on-site streams.* If it is necessary to cross a swale or stream to get to all or parts of your construction site, then before you begin working on the opposite side of the stream, you should construct a temporary stream crossing. Stream crossings can be either permanent or temporary, depending upon the need to cross the stream after construction is completed.

The following sections describe storm water management measures for erosion and sediment control.

Earth dikes

An *earth dike* is a temporary ridge or berm of compacted soil used to protect work areas from upslope runoff or to divert sediment-laden water to appropriate traps or stable outlets. The dike consists of compacted soil and stone, riprap, or vegetation to stabilize the channel. Figure 13.17 illustrates a typical earth dike.

Earth dikes can be used in the following situations:

- Above disturbed existing slopes and above cut-or-fill slopes to prevent runoff over the slope
- Across unprotected slopes, as slope breaks, to reduce slope length

- Below slopes to divert excess runoff to stabilized outlets
- To divert sediment-laden water to sediment traps
- At or near the perimeter of the construction area to keep sediment from leaving the site
- Above disturbed areas before stabilization to prevent erosion and maintain acceptable working conditions
- To serve as sediment traps when the site has been overexcavated on a flat grade or in conjunction with a sediment fence

Despite an earth dike's simplicity, improper design can limit its effectiveness; therefore, the state or local requirements should be consulted. Some general considerations include proper compaction of the earth dike, appropriate location to divert the intercepted runoff, and properly designed ridge height and thicknesses. Earth dikes should be constructed along a positive grade. There should be no dips or low points in an earth dike where the storm water will collect (other than the discharge point). Also the intercepted runoff from disturbed areas should be diverted to a sediment-trapping device. Runoff from undisturbed areas can be channeled to an existing swale or to a level spreader. Stabilization for the dike and flow channel of the drainage swales should be accomplished as soon as possible. Stabilization materials can include vegetation or stone or riprap.

When the drainage area to the earth dike is greater than 10 acres (4 ha), the USDA Natural Resources Conservation Service (NRCS) standards and specifications for diversions should be consulted. Table 13.3 contains suggested dike design criteria.

If the dike is constructed from coarse aggregate, the side slopes should be 3:1 or flatter. The channel formed behind the dike should have a positive grade to a stabilized outlet. The channel should be sta-

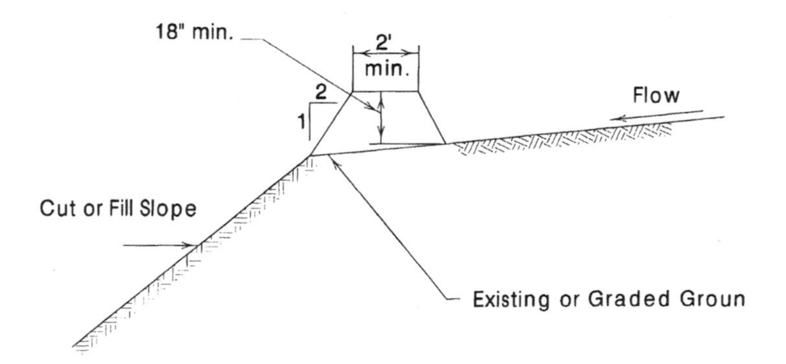

Figure 13.17 Earth dike (*Harris County, 1992*).

TABLE 13.3 Recommended Design Criteria for Earth Dikes

Drainage area	Under 5 acres (2 ha)	5–10 acres (2–4 ha)
Dike height (upstream side)	18 in (0.5 m)	30 in (0.8 m)
Dike top width	24 in (0.6 m)	36 in (0.9 m)
Dike base width	6 ft (2 m)	8 ft (2.5 m)
Dike side slopes	2:1 or less	2:1 or less
Flow width	4 ft (1.2 m)	6 ft (2 m)
Flow depth	12 in (0.3 m)	24 in (0.6 m)
Grade	0.5%–10%	0.5%–10%

bilized with vegetative or other stabilization measures. Grades over 10% may require an engineering design.

Construct the dike where it will not interfere with major areas of construction traffic so that vehicle damage to the dike will be kept to a minimum. Diversion dikes should be installed prior to the majority of soil-disturbing activity and may be removed when stabilization of the drainage area and outlet is completed.

Clear the area of all trees, brush, stumps, or other obstructions before constructing the dike. Construct the dike to the designed cross section, line, and grade, making sure that there are no irregularities or bank projections to impede flow. The dike should be compacted by using earthmoving equipment to prevent failure of the dike. The dike must be stabilized as soon as possible after installation.

The cost associated with earth dike construction is roughly $4.50 per linear foot ($14.00 per linear meter), which covers the earthwork involved in preparing the dike. Also added to this cost is approximately $1 per linear foot ($3 per linear meter) for stabilization practices. For many construction projects, the cost of earth dike construction is insignificant when compared to the overall earthwork project costs.

Inspect the dike, flow channel, and outlet regularly and after every storm, and make any repairs necessary to ensure that the measure is in good working order. If material must be added to the dike, be sure it is properly compacted. Reseed or stabilize the dike as needed to maintain its stability regardless of whether there has been a storm event.

Drainage swales

A *drainage swale* is a channel with a lining of vegetation, riprap, asphalt, concrete, or other material. Drainage swales are installed to convey runoff without causing erosion. They are constructed by excavating a channel and applying the appropriate stabilization.

Drainage swales can be used to convey runoff from the bottom or top of a slope. Temporary drainage swales are appropriate in the following situations:

- To divert upslope flows away from disturbed areas such as cut or fill slopes and to divert runoff to a stabilized outlet
- To reduce the length of the slope that runoff will cross
- At the perimeter of the construction site to prevent sedimentladen runoff from leaving the site
- To direct sediment-laden runoff to a sediment-trapping device

Since design flows, channel linings, and appropriate outlet devices will need to be considered, consult your state's requirements on such erosion control measures prior to constructing a drainage swale. General considerations include these:

- Divert the intercepted runoff to an appropriate outlet.
- The swale should be lined with geotextiles, grass, sod, riprap, asphalt, or concrete. The selection of the liner is dependent upon the volume and velocity of the anticipated runoff.
- The swale should have a positive grade. There should be no dips or low points in the swale where storm water can collect.

When the drainage area is greater than 10 acres (4 ha), the USDA Natural Resources Conservation Service standards and specifications for diversions should be consulted.

Swales may have side slopes ranging from 3:1 to 2:1. The minimum channel depth should be between 12 and 18 in (0.3 and 0.5 m). The minimum width at the bottom of the channel should be 24 in (0.6 m), and the bottom should be level. The channel should have a uniform positive grade between 2% and 5%, with no sudden decreases where sediments may accumulate and cause overtopping. Grades greater than 10% may require an engineering design. The channel should be stabilized by means of temporary or permanent stabilization measures. Runoff must discharge to a stabilized outlet.

Construct the swale away from areas of major construction traffic. Clear the area of all trees, brush, stumps, or other obstructions. Construct the swale to the designed cross section, line, and grade, making sure that there are no irregularities or bank projections to impede flow. The lining should be well compacted by using earthmoving equipment, and stabilization should be initiated as soon as possible. Stabilize lining with grass seed, sod, or riprap. Surplus material should be properly distributed or disposed of so that it does not interfere with the functioning of the swale. Outlet dissipation measures should be used to avoid the risk of erosion.

Inspect the flow channel and outlet for deficiencies or signs of erosion. Inspect regularly, and after every storm, make any repairs necessary to ensure that the measure is in good working order. If the surface of the channel requires material to be added, ensure that it is properly compacted. Reseed or stabilize the channel as needed to prevent erosion during a storm event.

Drainage swales can vary widely depending on the geometry of the swale and the type of lining material: Grass costs $3 per square yard ($3.60 per square meter); sod, $4 per square yard ($4.80 per square meter); riprap, $45 per square yard ($54 per square meter). No matter which liner type is used, the entire swale must be seeded and mulched at a cost of about $1.25 per square yard ($1.50 per square meter).

Interceptor dikes and swales

Interceptor dikes (ridges of compacted soil) and swales (excavated depressions) are used to keep upslope runoff from crossing areas where there is a high risk of erosion. The dikes reduce the amount and speed of flow and then guide it to a stabilized outfall (point of discharge) or sediment-trapping area (see sections on sediment traps and sediment basins). Figure 13.18 illustrates a typical interceptor dike and swale.

Interceptor dikes and swales divert runoff by using a combination of earth dike and vegetated swale. Runoff is channeled away from locations where there is a high risk of erosion by placing a diversion dike or swale at the top of a sloping disturbed area. Dikes and swales also collect overland flow, changing it to concentrated flows. Interceptor dikes and swales can be either temporary or permanent storm water control structures.

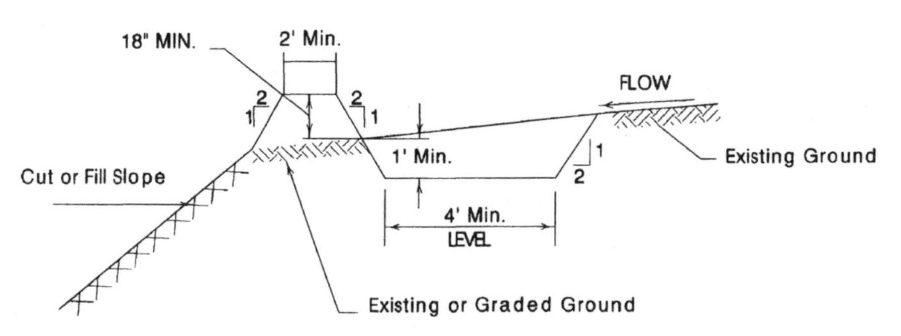

Figure 13.18 Interceptor dike with swale (*Harris County, 1992*).

Interceptor dikes and swales are generally built around the perimeter of a construction site before any major soil-disturbing activity takes place. Temporary dikes or swales may also be used to protect existing buildings, areas such as stockpiles, or other small areas that have not yet been fully stabilized.

When constructed along the upslope perimeter of a disturbed or high-risk area (though not necessarily all the way around it), dikes or swales prevent runoff from uphill areas from crossing the unprotected slope. For short slopes, a dike or swale at the top of the slope reduces the amount of runoff reaching the disturbed area. For longer slopes, several dikes or swales are placed across the slope at intervals. This practice reduces the amount of runoff that accumulates on the face of the slope and carries the runoff safely down the slope. Temporary dikes or swales constructed on the downslope side of the disturbed or high-risk area will prevent runoff that contains sediment from leaving the site before sediment is removed. In all cases, runoff is guided to a sediment-trapping area or a stabilized outfall prior to release.

Temporary dikes and swales are used in areas of overland flow. If they remain in place longer than 15 days, they should be stabilized. Runoff channeled by a dike or swale should be directed to an adequate sediment-trapping area or stabilized outfall. Care should be taken to provide enough slope for drainage but not so much slope as to cause erosion due to the high runoff flow speed. Temporary interceptor dikes and swales may remain in place as long as 12 to 18 months (with proper stabilization) or may be rebuilt at the end of each day's activities. Dikes or swales should remain in place until the area they were built to protect has permanently stabilized.

Interceptor dikes and swales which are properly designed and installed may later be converted to use as permanent storm water management structures. This may satisfy the requirements of the EPA general permit for permanent storm water control measures, described later in this chapter.

Temporary storm water control measures should be inspected once per week on a regular schedule and after every storm. Repairs necessary to the dike and flow channel should be made promptly.

The cost associated with earth dike construction is roughly $4.50 per linear feet ($15 per linear meter); this covers the earthwork involved in preparing the dike. Drainage swales can vary widely depending on the geometry of the swale and the type of lining material: Grass costs $3 per square yard ($3.60 per square meter); sod, $4 per square yard ($4.80 per square meter); riprap, $45 per square yard ($54 per square meter). No matter which liner type is used, the entire dike and swale must be seeded and mulched at a cost of about $1.25 per square yard ($1.50 per square meter).

Temporary stream crossings

A *temporary stream crossing* is a bridge or culvert across a stream or watercourse for short-term use by construction vehicles or heavy equipment. Vehicles moving over unprotected stream banks will damage the bank, thereby releasing sediments and degrading the stream banks. A stream crossing provides a means for construction vehicles to cross streams or watercourses without moving sediment to streams, damaging the streambed or channel, or causing flooding.

A temporary stream crossing is used when heavy equipment should be moved from one side of a stream channel to another or when light-duty construction vehicles have to cross the stream channel frequently for a short time. Temporary stream crossings should be constructed only when it is necessary to cross a stream and a permanent crossing has not yet been constructed.

Where available materials and designs are adequate to bear the expected loadings, *bridges* are preferred as a temporary stream crossing.

Culverts are the most common type of stream crossings and are relatively easy to construct. A pipe, to carry the flow, is laid into the channel and covered by gravel.

When feasible, always attempt to minimize or eliminate the need to cross streams. Temporary stream crossings are a direct source of pollution; therefore, every effort should be made to use an alternate method (e.g., longer detour), when feasible. When it becomes necessary to cross a stream, a well-planned approach will minimize the damage to the stream bank and will reduce erosion. The design of temporary stream crossings requires knowledge of the design flows and other information; therefore, a professional engineer and specific state and local requirements should be consulted. State and local jurisdictions may require a separate permit for temporary stream crossings; contact them directly to learn about their exact requirements.

The specific loads and the stream conditions will dictate what type of stream crossing to employ. Bridges are the preferred method to cross a stream because they provide the least obstruction to flows and fish migration.

The cost of temporary stream crossings will vary widely according to the size of the stream, the type of equipment using the stream crossing, and other factors. Typically, costs range from $500 to $1500, but they may be much higher.

Temporary storm drain diversions

A *temporary storm drain* is a pipe which redirects an existing storm drain system or outfall channel to discharge into a sediment trap or basin.

Use storm drain diversions to temporarily divert flow going to a permanent outfall. This diverted flow should be directed to a sediment-trapping device. A temporary storm drain diversion should remain in place as long as the area draining to the storm sewer remains disturbed. Another method is to delay completion of the permanent outfall and instead use temporary diversions to a sediment-trapping device before discharge. Finally, a sediment trap or basin can be constructed below a permanent storm drain outfall. The basin is designed to trap any sediment before final discharge.

Since the existing storm drain systems will be modified, careful consideration should be given to the piping configuration and the resulting impact of installing a temporary storm drain diversion. The temporary diversions will also need to be moved, once the construction has ceased and it is necessary to restore the original storm drainage systems. Therefore, appropriate restoration measures should be taken such as flushing the storm drain prior to removal of the sediment trap or basin, stabilizing the outfall, and restoration of grade areas. Finally, the state or local requirements should be consulted.

Slope drains

Slope drains reduce the risk of erosion by discharging runoff to stabilized areas. Made of flexible or rigid pipe or paved chutes, they carry concentrated runoff from the top to the bottom of a slope that has already been damaged by erosion or is at high risk for erosion. Slope drains are also used to drain saturated slopes that have the potential for soil slides. Slope drains can be either temporary or permanent depending on the method of installation and the material used.

Pipe slope drains are appropriate in the following general locations:

- On cut or fill slopes before permanent storm water drainage structures have been installed
- Where earth dikes or other diversion measures have been used to concentrate flows
- On any slope where concentrated runoff crossing the face of the slope may cause gullies, channel erosion, or saturation of slide-prone soils
- As an outlet for a natural drainage-way

Temporary pipe slope drains, usually made of flexible tubing or conduit, may be installed prior to the construction of permanent drainage structures. Permanent slope drains may be placed on or beneath the ground surface. Slope drains may be used with other devices, including diversion dikes or swales, sediment traps, and level spreaders (used to spread storm water runoff uniformly over the surface of the ground).

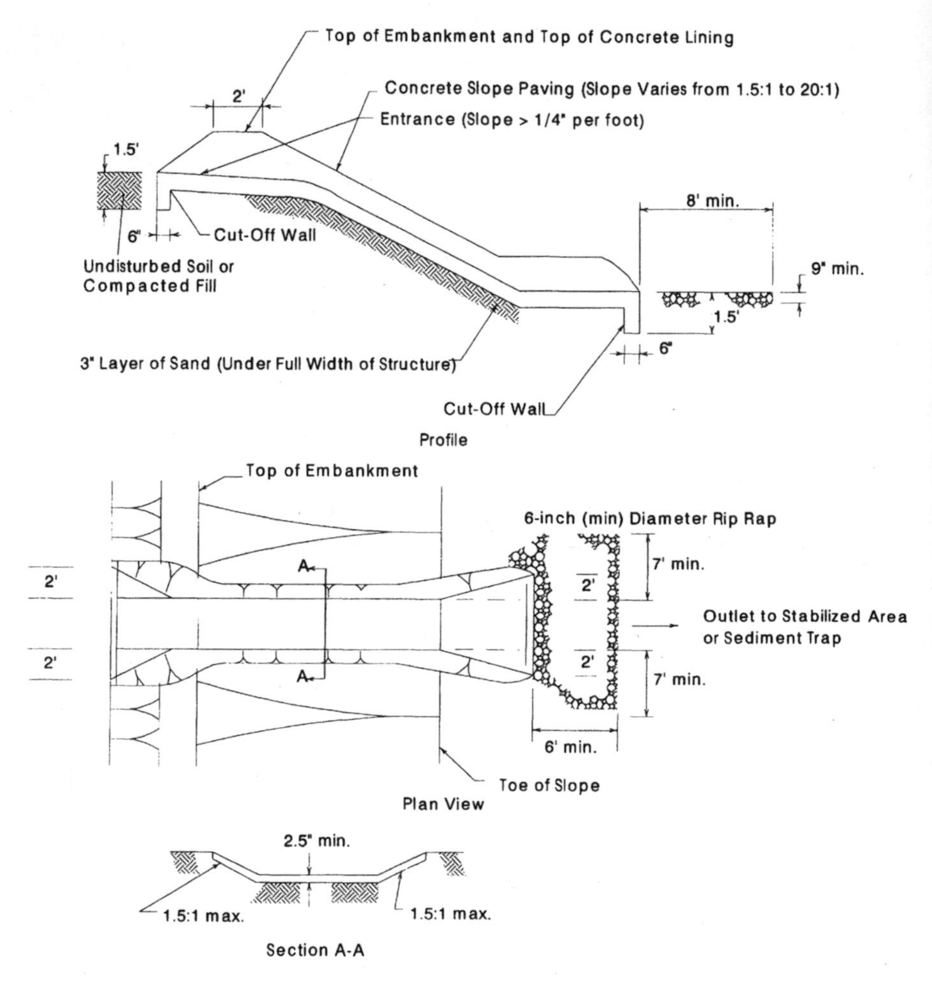

Figure 13.19 Slope drain structure (*Harris County, 1992*).

A slope drain constructed on the surface of the slope is called a *paved chute*; it may be covered with a surface of concrete or other impenetrable material. Figure 13.19 illustrates a typical paved chute.

Subsurface slope drains can be constructed of concrete, polyvinyl chloride, clay tile, corrugated metal, or other permanent material. Figure 13.20 illustrates a typical pipe slope drain.

The drainage area for the slope drain may be up to 10 acres (4 ha); however, in many jurisdictions 5 acres (2 ha) is the recommended maximum. Some guidelines recommend that the slope drain design handle the peak runoff for the 10-year storm. Typical relationships between area and pipe diameter are shown in Table 13.4, although these will vary with local conditions.

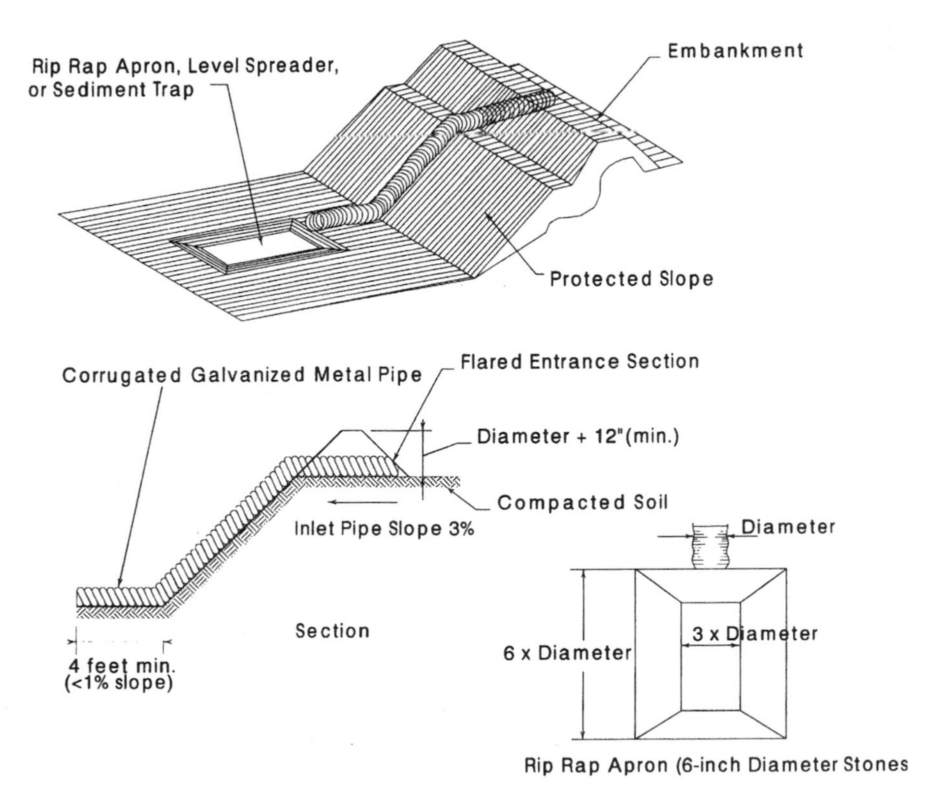

Figure 13.20 Pipe slope drain (*Environmental Protection Agency, 1992*).

TABLE 13.4 Recommended Design of Pipe Slope Drains

Maximum drainage area	Pipe diameter
0.50 acres (0.2 ha)	12 in (30 cm)
0.75 acres (0.3 ha)	15 in (38 cm)
1.00 acres (0.4 ha)	18 in (45 cm)

The following additional design guidelines may be applied to slope drains:

Dike. A dike is used to direct water into the pipe inlet and to prevent overtopping the slope. The height at the centerline of the earth dike should range from a minimum of 1.0 ft (0.3 m) over the pipe to twice the diameter of the pipe measured from the invert of the pipe. It should also be at least 6 in (15 cm) higher than the adjoining ridge on either side. At no point along the dike will the elevation of the top of the dike be less than 6 in (15 cm) higher than the top of the pipe.

Inlet. The inlet of a pipe slope drain should be stabilized. A standard flared end section secured with a watertight fitting should be used for the inlet. A standard T-section fitting may also be used. Soil around and under the entrance section must be hand-tamped in 4- to 8-in (10- to 20-cm) lifts to the top of the dike to prevent piping failure around the inlet. Place filter cloth under the inlet, extend 5 ft (1.5 m) in front of the inlet, and key in 6 in (15-cm) on all sides to prevent erosion. A 6-in (15-cm) metal toe plate may also be used for this purpose.

Base. Place the pipe slope drain on undisturbed or wellcompacted soil. Ensure firm contact between the pipe and the soil at all points by backfilling around and under the pipe with stable soil material hand-compacted in lifts of 4 to 8 in (10 to 20 cm). Securely stake the pipe slope drain to the slope, using grommets provided for this purpose at intervals of 10 ft (3 m) or less.

Pipe material. Pipe may be heavy-duty flexible tubing designed for this purpose, e.g., nonperforated, corrugated plastic pipe, corrugated metal pipe, bituminous fiber pipe, or specially designed flexible tubing. Extension collars should be 12-in- (30-cm-) long sections of corrugated pipe. Ensure that all slope drain sections are securely fastened together and have watertight fittings.

Discharge. Extend the pipe beyond the toe of the slope, and discharge at a nonerosive velocity into a stabilized area or to a sedimentation trap or pond. The soil at the discharge end of the pipe should be stabilized with riprap (a combination of large stones, cobbles, and boulders). The riprap should be placed along the bottom of a swale which leads to a sediment-trapping structure or another stabilized area.

The slope drain should have a slope of 3% or more. Immediately stabilize all areas disturbed by installation or removal of the slope drain. Do not allow construction traffic to cross the slope drain, and do not place any material on it.

Pipe slope drain costs are generally based upon the type and size of pipe, which is generally flexible polyvinyl chloride at $5 per linear foot ($16 per linear meter). Also added to this cost are any expenses associated with inlet and outlet structures.

Inspect the slope drain regularly and after every storm. Be sure that the inlet from the pipe is properly installed to prevent bypassing the inlet and undercutting the structure. If necessary, install a headwall, riprap, or sandbags around the inlet. Check the outlet point for erosion, and check the pipe for breaks or clogs. Install outlet protection if needed, and promptly clear breaks and clogs. Make any other necessary repairs. If a sediment trap has been provided, clean it out when the sediment level reaches one-third to one-half the design volume.

The slope drain should remain in place until the slope has been completely stabilized or up to 30 days following permanent slope stabilization.

Subsurface drains

A *subsurface drain* is a perforated pipe or conduit placed beneath the surface of the ground at a designed depth and grade. It is used to drain an area by lowering the water table. A high water table can saturate soils and prevent the growth of certain types of vegetation. Saturated soils on slopes will sometimes "slip" down the hill. Installing subsurface drains can help prevent these problems. Figure 13.21 illustrates the possible effects of subsurface drains on the water table.

Drains can be made of tile, pipe, or tubing. There are two types of subsurface drains: relief drains and interceptor drains.

Relief drains are used to dewater an area where the water table is high. They may be placed in a grid-iron, herringbone, or random pattern. Figure 13.22 illustrates several typical patterns.

Interceptor drains are used to remove water where sloping soils are excessively wet or subject to slippage. The drains are usually placed as single pipes instead of in patterns. Generally, subsurface drains are suitable only in areas where the soil is deep enough for proper installation. They are not recommended where they pass under heavy vehicle crossings.

Drains should be placed so that tree roots will not interfere with drainage pipes. The drain design should be adequate to handle the volume of flow. Areas disturbed by the installation of a drain should be stabilized, or else they, too, will be subject to erosion. The soil layer must be deep enough to allow proper installation.

Backfill immediately after the pipe is placed. The material used for backfill should be open granular soil that is highly permeable. The out-

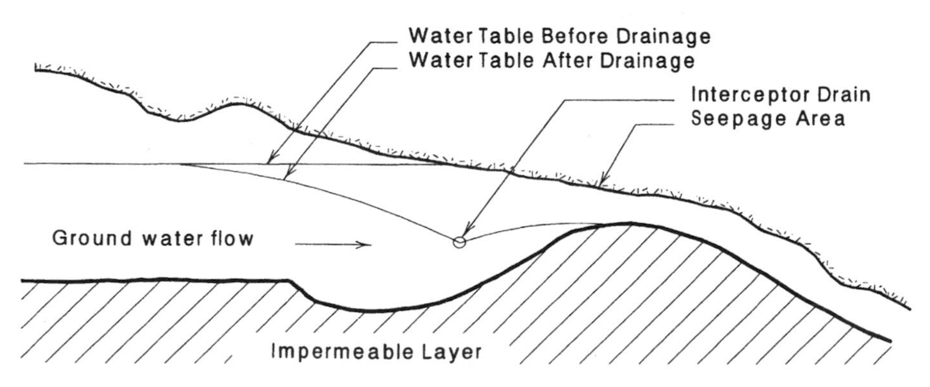

Figure 13.21 Effects of subsurface drains on water table (*Environmental Protection Agency, 1992*).

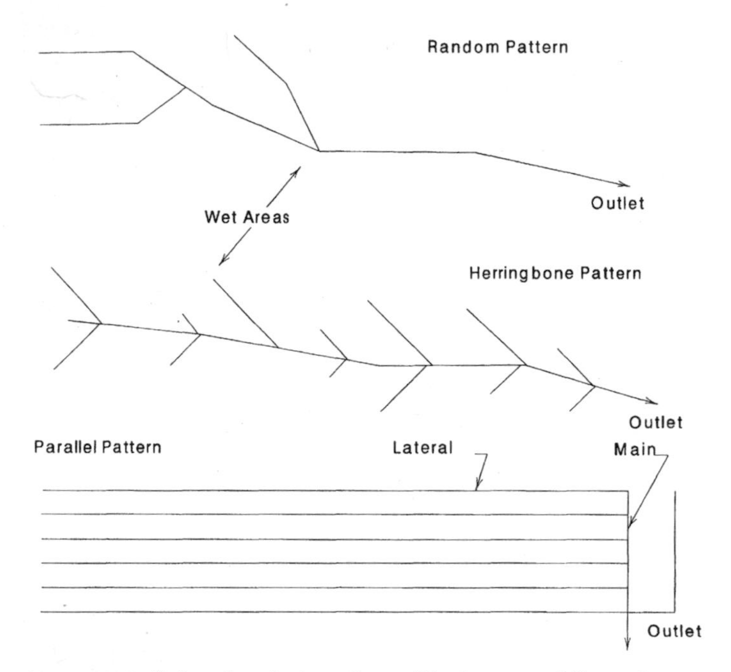

Figure 13.22 Subsurface drain patterns (*Environmental Protection Agency, 1992*).

let should be stabilized and should direct sediment-laden storm water runoff to a sediment-trapping structure or another stabilized area.

Subsurface drain costs are generally based upon the pipe type and size. Generally, drain pipe is about $2.25 per linear foot ($7.50 per linear meter). Added to this cost are installation and outlet costs.

Inspect subsurface drains on a regular schedule, and check for evidence of pipe breaks or clogging by sediment, debris, or tree roots. Remove any blockage immediately, replace any broken sections, and restabilize the surface. If the blockage is from tree roots, it may be necessary to relocate the drain. Check inlets and outlets for sediment or debris. Remove and dispose of these materials properly.

Permanent Storm Water Management Measures

Permanent storm water management controls are generally those controls which are installed during the construction process, but primarily result in reductions of pollutants in storm water discharged from the site after the construction has been completed.

Construction activities often result in a significant change in land use. These changes in land use typically involve an increase in the overall imperviousness of the site, which can result in dramatic changes in the runoff patterns of a site. As the amount of runoff from a site increases, the amount of pollutants carried by the runoff increases. Storm water management controls generally attempt to limit increases in the amount of runoff and pollution discharged from land impacted by construction.

Major classes of storm water management controls include infiltration of runoff on the site; flow attenuation by vegetation or natural depressions; outfall velocity dissipation devices; storm water retention structures and artificial wetlands; and storm water detention structures. For many sites, a combination of these controls may be appropriate. In designing storm water controls, features that would pose a safety hazard—especially for children—should be avoided and/or have limited public access.

Infiltration of runoff on site

A variety of infiltration technologies can be used to reduce the volume and pollutant loadings of storm water discharges from a site, including infiltration trenches and infiltration basins. Infiltration devices tend to mitigate changes to an area's natural hydrologic conditions. Properly designed and installed infiltration devices can reduce peak discharges, provide groundwater recharge, augment low-flow conditions of receiving streams, reduce storm water discharge volumes and pollutant loads, and protect downstream channels from erosion.

Infiltration devices are a feasible option where soils are permeable and the water table and bedrock are well below the surface. Infiltration basins can also be used as sediment basins during construction. Infiltration trenches can be more easily placed into less active areas of a development and can be used for small sites and infill developments. However, trenches may require regular maintenance to prevent clogs, particularly where grass inlets or other pollutant-removing inlets are not used. In some situations, such as low-density areas of parking lots, porous pavement can provide for infiltration.

Flow attenuation by vegetation or natural depressions

Flow attenuation provided by vegetation or natural depressions can provide pollutant removal and infiltration and can lower the erosive potential of flows. Vegetative flow attenuation devices include grass swales and filter strips as well as trees that are either preserved or planted during construction.

Typically the costs of vegetative controls are small compared to those of other storm water practices. The use of check dams incorporated into flow paths can provide additional infiltration and flow attenuation. Given the limited capacity to accept large volumes of runoff and the potential erosion problems associated with large concentrated flows, vegetative controls should typically be used in combination with other storm water devices.

Grass swales are typically used in low or medium residential development and highway medians as an alternative to curb and gutter drainage systems.

Outfall velocity dissipation devices

Outfall velocity dissipation devices include riprap and stone or concrete flow spreaders, described earlier in this chapter. Outfall velocity dissipation devices slow the flow of water discharged from a site to lessen the amount of erosion caused by the discharge.

Storm water retention structures

Properly designed and maintained storm water retention structures, also referred to as *wet ponds,* can achieve a high removal rate of sediment, biochemical oxygen demands, organic nutrients, and metals. Retention basins are most cost-effective in larger, more intensively developed sites. See Chap. 4 for more details on wet ponds.

Retention structures and artificial wetlands

Retention structures include ponds and artificial wetlands that are designed to maintain a permanent pool of water. Properly installed and maintained retention structures and artificial wetlands can achieve a high removal rate of sediment, biochemical oxygen demand (BOD), organic nutrients, and metals, and are most cost-effective when used to control runoff from larger, intensively developed sites. These devices rely on settling and biological processes to remove pollutants.

Retention ponds and artificial wetlands can also become wildlife habitats, recreation, and landscape amenities, and increase local property values. Artificial wetlands can be one of the most effective long-term storm water management measures, but may contribute to problems at certain sites. This could be the case at airports where bird populations drawn to wetlands proximate to runways and taxiways may endanger moving aircraft. Structures which maintain continuous habitat for wildlife should not be constructed within 10,000 ft (3000 m) of a public-use airport serving turbine-powered aircraft, or within 5000 ft (1500 m) of a public-use airport serving piston-powered air-

Figure 13.23 Storm water detention basin.

craft [63 FR 07835 (Feb. 17, 1998)]. See Chap. 4 for more information on these types of facilities.

Water quality detention structures

Storm water detention structures include extended detention ponds, which control the rate at which the pond drains after a storm event. Extended detention ponds are usually designed to completely drain in about 24 to 40 h, and they remain dry at other times. They can provide pollutant removal efficiencies that are similar to those of retention ponds. Extended detention systems are typically designed to provide both water quality and water quantity (flood control) benefits. See Fig. 13.23. Chapter 4 provides more information on water quality detention structures.

References

Environmental Protection Agency, 1992. *Storm Water Management for Construction Activities. Developing Pollution Prevention Plans and Best Management Practices,* EPA report 832/R-92/005, September.

Harris County/Harris County Flood Control District/City of Houston, 1992. *Storm Water Management Handbook for Construction Activities,* September 17.

National Weather Service, 1961. *Rainfall Frequency Atlas of the United States,* Technical Paper 40, U.S. Department of Commerce, Washington.

Chapter

14

Other Storm Water Pollution Prevention Measures

This is the last of five chapters which deal with the specific requirements for storm water discharges from construction facilities. This chapter describes pollution prevention measures for construction sites other than erosion and sediment control.

Best Management Practices for Construction Sites

Erosion and sediment are not the only potential sources of pollution from construction activity. Other pollutants, including toxic chemicals, may be present during construction. The controls and practices which limit the discharge of pollutants in storm water are called *best management practices* (BMPs). BMPs are an important part of site-specific controls in a storm water pollution prevention plan (SWPPP). The BMPs in this chapter deal with prevention—limiting contact between storm water and potential pollutants. There are no specific BMPs that are applicable to all construction sites. Only the controls which best address site-specific conditions should be implemented to control or eliminate contamination of storm water.

There are several areas of control (in addition to erosion and sedimentation controls and storm water management) that should be addressed in each SWPPP:

1. *Materials management.* Provide protected storage areas for chemicals, paints, solvents, fertilizers, and other potentially toxic materials.

2. *Waste disposal.* Provide waste receptacles at convenient locations and provide regular collection of wastes, including building material wastes.
3. *Off-site tracking.* Minimize off-site tracking of sediments.
4. *Spill prevention and response.* Make adequate preparations, including training and equipment, to contain spills of oil and hazardous materials.
5. *Sanitation.* Comply with applicable state or local waste disposal, sanitary sewer, or septic system regulations.
6. *Non–storm water discharges.* Use appropriate pollution prevention measures for allowable non–storm water components of discharge.

Most of these measures involve the day-to-day operations of the construction site. These operations are usually under the control of the general contractor. Therefore, the measures described in this chapter commonly are carried out or directly supervised by the general contractor. However, depending upon the language of the NPDES permit for the construction project, the owner of the project may still be fully responsible for these measures.

Many BMPs could be described as what EPA terms *good housekeeping. Good housekeeping involves keeping a clean, orderly construction site.* One of the first steps toward preventing storm water contamination is to improve housekeeping practices and use common sense. Good housekeeping practices reduce the possibility of accidental spills, improve the response time if there is a spill, and reduce safety hazards as well. Good housekeeping practices generally are inexpensive, relatively easy to implement, and often effective in preventing storm water contamination.

Pollutants that may enter water from construction sites due to poor housekeeping include oils; grease; paints; gasoline; concrete truck washdown; raw materials used in the manufacture of concrete, including sand, aggregate, and cement; solvents; litter; debris; and sanitary wastes. Construction site SWPPPs should address the following to prevent the discharge of pollutants:

- Designate and control areas for equipment maintenance and repair.
- Provide waste receptacles at convenient locations and regular collection of wastes.
- Locate equipment washdown areas on-site, and provide appropriate control of wash water to prevent unauthorized dry-weather discharges and avoid mixing with storm water.

- Provide protected storage areas for chemicals, paints, solvents, fertilizers, and other potentially toxic materials.
- Provide adequately maintained sanitary facilities.

Materials Management

On a construction site, *the material storage area can become a major source of risk* due to possible mishandling of materials or accidental spills. An inventory should be made of the material storage area and of the site. Special care should be taken to identify any materials that have the potential to come in contact with storm water. This will help to raise employee awareness and plan effective controls. See Fig. 14.1.

Some of the materials commonly found on a construction site include pesticides, petroleum products, fertilizers and detergents (nutrients), construction chemicals, other pollutants, and hazardous products. (See Fig. 14.2.) The materials inventory list should include these for risk assessment. These questions should be addressed when risks are identified:

Figure 14.1 Construction material (topsoil) stored under highway overpass to reduce storm water contact.

Figure 14.2 Weatherproof material and equipment storage container at construction site.

- What types of materials are stored on the site?
- How long will the materials be stored before use?
- Are all the materials really needed?
- Can a smaller quantity of each material be stored on the site?
- How are the materials stored and distributed?
- How can potential storm water contact be avoided?

Storage of pesticides

Pesticides include insecticides, rodenticides, and herbicides that are often used on construction sites. The management practices used to reduce the amounts of pesticides that could contact storm water include the following:

- Handle pesticides as infrequently as possible.
- Leave pesticides in their original shipping containers whenever possible, because those containers often have special handling instructions printed on them.
- Observe all applicable federal, state, and local regulations when using, handling, or disposing of pesticides.

- Store pesticides in a dry, covered area.
- Provide curbs or dikes to contain the pesticide if it should spill.
- Have measures on-site to contain and clean up spills of pesticides.
- Strictly follow recommended pesticide application rates and methods.

Storage of petroleum products

Oil, gasoline, lubricants, and asphaltic substances such as paving materials are petroleum products. These materials should be handled carefully to minimize their exposure to storm water. Petroleum products usually are found in two site areas.

1. Areas where road construction of some type is occurring
2. Vehicle storage areas or areas of on-site fueling or equipment maintenance

These management practices will help reduce the risks of using petroleum products:

- Have equipment to contain and clean up petroleum spills in fuel storage areas or on board maintenance and fueling vehicles.
- Where possible, store petroleum products and fuel vehicles in covered areas and construct dikes to contain any spills. (See Fig. 14.3.)
- Contain and clean up petroleum spills immediately.
- Perform preventive maintenance for on-site equipment to prevent leakage (e.g., check for and fix gas or oil leaks in construction vehicles on a regular basis).
- Apply asphaltic substances properly, according to the manufacturer's instructions.

Storage of fertilizers and detergents (nutrients)

Nutrients such as phosphorus and nitrogen are found on construction sites in both fertilizers and detergents. Fertilizers are needed on construction sites to provide the nutrients for plant growth; however, when excess quantities of fertilizers are used or when fertilizers are washed away by storm water runoff, they may be a major source of pollution. An excess of nutrients reaching a body of water can cause an overgrowth of water plants, which then use up the oxygen in the water, creating an unfavorable environment. These steps can be taken to reduce the risks of nutrient pollution:

Figure 14.3 Petroleum storage area without containment dikes or curbs.

- Limit the application of fertilizers to the minimum area and the minimum recommended amounts.
- Reduce exposure of nutrients to storm water runoff by working the fertilizer into the soil to a depth of 4 to 6 in (10 to 15 cm) instead of letting it remain on the surface.
- Apply fertilizer more frequently, but at lower application rates.
- Limit hydroseeding, in which lime and fertilizers are applied to the ground surface in one application.
- Implement good erosion and sediment control to help reduce the amount of fertilizer lost as a result of erosion.
- Limit the use of detergents on the site. Wash water containing detergents should not be discharged in the storm water system. (*Note:* The EPA general permit for storm water discharges from construction sites prohibits the discharge of wash waters containing detergent residues, as discussed later in this chapter.)
- Apply fertilizer and use detergents only in the recommended manner and amounts.

Figure 14.4 Uncovered drum storage area at construction site.

Storage of hazardous products

Hazardous materials include (but are not limited to) paints, acids for cleaning masonry surfaces, cleaning solvents, chemical additives used for soil stabilization, and concrete curing compounds. Most problem situations involving hazardous materials and other pollutants are the result of carelessness or not using common sense. These practices will help to avoid pollution of storm water by these materials:

- Keep equipment to contain and clean up spills of hazardous materials in the areas where these materials are stored or used.
- Contain and clean up spills immediately after they occur.
- Keep materials in a dry, covered area. Store materials in the original manufacturer's containers whenever possible, because special handling instructions are often printed on these containers. (See Fig. 14.4.)

Waste Disposal

Proper management and disposal of building materials and other construction site wastes are an important part of pollution prevention. The EPA general permit for storm water discharges from construction sites

Figure 14.5 "Floatable" debris discharged from construction site.

prohibits the discharge of *floatables* (floating objects) in storm water. Waste materials are generally the largest sources of floatables. (See Fig. 14.5)

Construction wastes include surplus or refuse building materials as well as hazardous wastes. Possible management practices for these construction wastes include trash disposal, recycling, material handling, and spill prevention and cleanup measures. Controls and practices should also meet other federal, state, and local requirements.

Construction projects tend to generate a great deal of solid waste material which is unique to this activity. Construction wastes may include but are not limited to the following:

- Trees and shrubs removed during clearing and grubbing or other phases of construction
- Packaging materials (including wood, paper, plastic, etc.)
- Scrap or surplus building materials, e.g., scrap metals, rubber, plastic and glass pieces, masonry products, and other solid waste materials
- Paints and paint thinners
- Materials resulting from the demolition of structures (rubble)

The following steps will help ensure proper disposal of construction wastes:

- *Collection area.* Select a designated waste collection area on the site.
- *Covered containers.* Provide an adequate number of containers with lids or covers that can be placed over the container prior to rainfall. Whenever possible, locate containers in a covered area. (See Fig. 14.6.)
- *Collection.* Arrange for waste collection before containers overflow.
- *Spill response.* If a container spills or overflows, clean up immediately.
- *Adjustment of collection activity.* Plan for additional containers and more frequent pickups during the demolition phase of construction.
- *Disposal.* Ensure that construction waste is collected, removed, and disposed of only at authorized disposal areas.
- *Subcontractor participation.* Since painting and many other waste-generating construction activities are often performed by subcontractors, make sure that the subcontractors are aware of the requirements of the NPDES permit for the construction project.

Check with the local solid waste management agency for specific guidance.

Figure 14.6 Open waste disposal container at construction site.

Disposal of hazardous products

Many of the materials found at a construction site may be hazardous to the environment or to personnel. It is always important to *read the labels of the materials or products kept on the site.* At a minimum, consider paints, acids for cleaning masonry surfaces, cleaning solvents, chemical additives used for soil stabilization (e.g., palliative such as calcium chloride), and concrete curing compounds and additives to be hazardous substances.

The following management practices will help you to avoid problems associated with the disposal of hazardous materials:

- Check with local waste management authorities to determine the requirements for disposing of hazardous materials.
- Use all of the product before disposing of the container.
- Do not remove the original product label from the container; it often contains important information.
- Do not mix products unless it is specifically recommended by the manufacturer.
- Follow the manufacturer's recommended method for disposal, which is often found on the label.

Disposal of contaminated soils

Contaminated soils are soils which have been exposed to and still contain hazardous substances. Contaminated soils may be encountered on the site during earthmoving activities or during the cleanup of a spill or of a leak of a hazardous product. Material storage areas may also have been contaminated by undetected spills. The nature of the contaminants may or may not be known.

State or local solid waste regulatory agencies should be contacted concerning information and procedures necessary to treat or dispose of contaminated soils. Some landfills may accept contaminated soil; however, laboratory tests may be required before a final decision is made. Private firms can also be consulted concerning disposal options.

Disposal of waste concrete

Most construction projects include some sort of concrete work. Usually, concrete is mixed off-site and delivered to the project by mixer truck. The concrete is poured, and there is a residual amount of concrete remaining in the truck. Sometimes excess concrete is delivered, or the concrete is found to be unacceptable and is rejected by the construction inspector or foreman. The truck must be cleaned and the residual concrete dumped before it "sets up" (hardens) in the truck.

Emptying or washout of excess concrete may be allowed on the site. *Excess concrete and wash water should be disposed of in a manner that prevents contact between these materials and storm water which will be discharged from the site.* For example, dikes could be constructed around the area to contain these materials until they harden, at which time they may be properly disposed of.

If possible, a beneficial use of waste concrete should be found. For example, an area may be set aside where waste concrete can be poured into simple forms to make riprap for channel or outlet protection. Concrete mixer truck drivers are often under pressure to eliminate waste concrete; providing a beneficial use can reduce or eliminate illicit dumping.

Disposal of sandblasting grits

Sandblasting is a commonly used technique to remove paint, dirt, etc., from surfaces. Sand is sprayed on the surface to be cleaned. Sandblasting grits consist of both the spent sand and the particles of paint and dirt removed from the surface. *Sandblasting grits are hazardous waste if they were used to clean old structures where lead-, cadmium-, or chrome-based paints were used.*

Sandblasting grits containing hazardous materials should not be washed into the storm or sanitary sewer. A licensed waste management or transport and disposal firm should be contacted to dispose of this type of used grit.

Recycling of construction and demolition debris

Many building materials and the majority of asphalt and concrete in the waste stream can be recycled. This can result in a significant reduction of disposed waste due to the large quantities of this material generated on site. Waste generation and disposal are usually handled by contractors, but alternate diversion methods can be specified in the contract. Recycling building materials, such as rebar, asphalt, and concrete, and selling salvaged lumber, building materials, and fixtures are more economical than disposal and can result in a lower bid price if a tipping fee is charged. Recycling asphalt and concrete on-site provides a source of crushed stone product for base applications and reduces the cost of new material.

Current practices of recycling asphalt and concrete include recycling the material with a local recycling center or recycling the material on site. Local recyclers of asphalt and concrete may include construction material companies, sand and gravel producers, and asphalt plants. Information on local recyclers with stationary asphalt and concrete recycling equipment should be available from the local government solid waste authority. Asphalt and concrete rubble can be recycled on-site without further processing for use as general rubble fill in low-lying areas or quarries. However, a more beneficial use of the material is to crush the asphalt and concrete with a portable crusher. The crushed product can be used as a substitute for crushed stone products in road construction (pavement and road base) and as a fill material (footing and foundation backfill and general fill). The use of crushed concrete for landfill roadways or landfill cover should also be investigated as a potential market. Large slabs of concrete can be used as riprap to prevent shoreline and stream erosion, or used in various projects such as artificial reef development and underground cavern stabilization.

Recycling asphalt pavements on site can be beneficial, as these materials are immediately reused and do not enter the waste stream. The two types of on-site recycling include surface recycling and road base recycling. Surface recycling involves breaking up the top layer of a pavement structure to a depth of about 1 in (2.5 cm) by milling or crushing the material, and then recompacting the loose material. Performance specifications for the pavement surface should be considered prior to using the recycled asphalt. A new surface is placed on top of the recycled asphalt, and the pavement is compacted again. Road base recycling involves recycling both the asphalt surface and the base material to produce a new road base. Equipment is available that uses a downward rotation of a cutter drum to raise asphalt and base material by cutting a path 10 ft (3 m) wide and as much as 24 in (0.6 m) deep. The base aggregate and asphalt top coat are blended and crushed by the machine to produce a well-graded base course, and then deposited back on the roadway, all in one pass. A new surface is next placed on top of the recycled road base. In some applications, the base is never hauled away from the project site. Specifications for pavement base course are not as stringent, and can generally contain 100% recycled material.

Alternatively, recycled asphalt can be transported to an asphalt batch plant and processed. The quality of the recycled pavement should be documented to help determine the percentage of *reclaimed asphalt pavement* (RAP) that can be combined with virgin asphalt to produce new asphalt pavement. The percentage of RAP that can be used in the intermediate or surface asphalt concrete course depends primarily on the performance specifications of the road being constructed. Governing specifications should be consulted to determine the percentages of recycled material or RAP allowed in asphalt pavements.

The cost for a contractor to crush asphalt and concrete on-site will vary depending on the volume of material collected, the size requirements of the finished product, and the distance to the facility. Average costs for crushing are approximately \$5 per ton (\$5.50 per metric ton). Alternatively, costs for hauling and disposing of the material with a local recycling center can range from $5 to $13 per ton ($5.50 to $14.30 per metric ton). Cost savings may be obtained by minimizing crushed stone product purchases and reducing landfill disposal fees (Cosper et al., 1993; von Stein and Savage, 1994).

Off-site Vehicle Tracking

The EPA general permit for storm water discharges from construction sites states that off-site vehicle tracking of sediments shall be minimized. Stabilized construction entrances and construction access road stabilization can be used to reduce off-site vehicle tracking. In addition to these measures, paved streets adjacent to the site should be swept to remove any excess mud, dirt, or rock tracked from the site. Off-site sediment removal should be conducted at a frequency necessary to minimize impacts. Deliveries or other traffic may be scheduled at a time when site personnel are available to provide cleanup if necessary. Vehicle wash racks could be provided to clean vehicles before they leave the site, if needed. All these measures should be considered, but not all are required to be implemented for a particular construction project.

Stabilized construction entrance

A *stabilized construction entrance* is a portion of the construction road which is constructed with filter fabric and large stone. The primary purpose of a stabilized construction entrance is to reduce the amount of soil tracked off the construction site by vehicles leaving the site. The rough surface of the stone will shake and pull the soil off the vehicles' tires as they drive over the entrance. The stone will also reduce erosion and rutting on the portion of the road where it is installed, by protecting the soil below.

The filter fabric separates the stone from the soil below, preventing the large stone from being ground into the soil. The fabric also reduces the amount of rutting caused by the vehicle tires by spreading the weight of the vehicles over a larger soil area than just the tire width.

A stabilized construction entrance should be installed (before construction begins on site) *at every point where traffic leaves or enters a disturbed area*. This measure is appropriate in the following locations:

- Wherever vehicles are leaving a construction site and enter onto a public road. If the number of vehicles is less than about 24 per day, then a stabilized construction entrance may not be necessary.
- At any unpaved entrance or exit location where there is risk of transporting mud or sediment onto paved roads.

A stabilized construction entrance should not be installed over an existing pavement (except for a slight overlap). Where the construction will require a permanent access road or driveway, a stabilized construction entrance should be installed before the permanent pavement. See Fig. 14.7.

Figure 14.8 illustrates a typical stabilized construction entrance. Stabilized construction entrances should be wide enough and long enough that the largest construction vehicle will fit in the entrance with room to spare. The width should be at least 10 to 12 ft (3 to 3.7 m) or across the entire width of the access. At sites where traffic volume is high, the entrance should be wide enough for two vehicles to pass safely. The length should be 50 to 75 ft (15 to 23 m). The entrance should be flared where it meets the existing road to provide a turning radius.

If vehicles will be turning onto the paved road or drive from the stabilized construction entrance, an apron should be provided so that vehicles do not go off the stabilized construction entrance before they leave the site.

If the stabilized construction entrance has to cross a swale or stream, then a stream crossing should be provided.

Stabilized construction entrances cost from $1500 to $5000. The following steps are required in their construction:

1. *Site preparation.* Clear all vegetation, roots, and other obstructions in preparation for grading. Prior to placing geotextiles, make sure that the entrance is properly graded and compacted.
2. *Geotextile.* To reduce maintenance and loss of aggregate, place geotextile over the existing ground before placing the stone for the entrance. Table 14.1 lists the required properties of the geotextile.
3. *Stone.* Crushed stone 2 to 4 in (5 to 10 cm) in diameter should be placed to a depth of 6 in (15 cm) or greater for the entire width and length of the entrance. Stone used for the construction entrance should be large enough that it does not get picked up and tracked off the site by the vehicle traffic. Sharp-edged stone should not be used, so as to avoid puncturing tires.
4. *Drainage.* Runoff from a stabilized construction entrance should drain to a sediment trap or sediment basin. If a culvert is placed under the entrance to handle runoff, fill should be placed along the entire length of the culvert to protect it from damage.
5. *Vehicle wash rack.* The construction entrance may be provided with a vehicle wash rack which drains to a temporary sediment trap or other sediment-removing measure. This will allow vehicle tires to be washed prior to leaving the site and will ensure that wash water sediments are removed and can be properly disposed of. Vehicle wash racks, if required, cost about $2000 each.

Figure 14.7 Stabilized construction entrance.

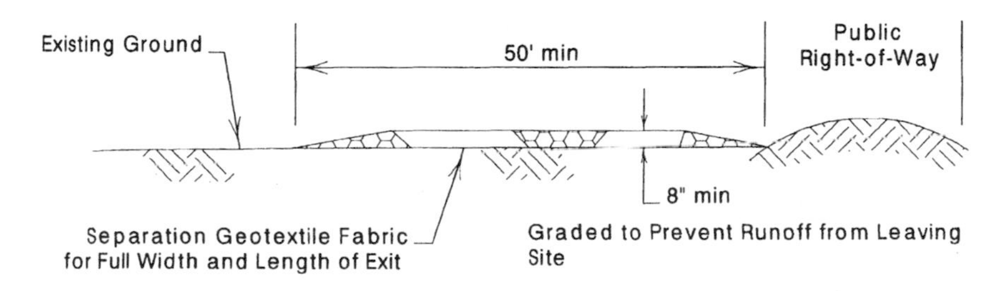

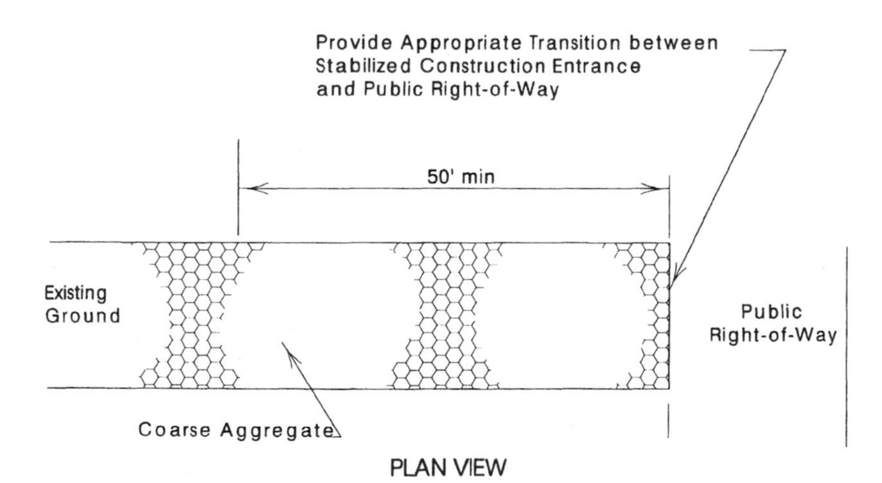

Figure 14.8 Stabilized construction entrance details (*Harris County, 1992*).

TABLE 14.1 Geotextile Requirements for Stabilized Construction Entrance

Physical property	Requirements
Grab tensile strength	220 lb (ASTM D1682)
Elongation failure	60% (ASTM D1682)
Mullen burst strength	430 lb (ASTM D3768)
Puncture strength	125 lb (ASTM D751) (modified)
Equivalent opening size	40–80 (U.S. standard sieve) (CW02215)

Stabilized construction entrances should be inspected on a regular basis and after there has been a high volume of traffic or a storm event. Additional stone must be applied periodically and when repair is required. Sediments or any other materials tracked onto the public roadway should be removed immediately. Ensure that associated sediment control measures are in good working condition.

Construction road stabilization

A *stabilized construction road* is a road built to provide a means for construction vehicles to move around the site without causing significant erosion. A stabilized construction road is designed to be well drained so that water does not puddle or flood the road during wet weather. It typically will have a swale along one side or both sides of the road, to collect and carry away runoff. Stabilized construction roads should have a layer of crushed stone or gravel which will cover and protect the soil below from erosion.

A stabilized construction road should be installed in a disturbed area where a high volume of construction traffic is expected. A construction road should be stabilized at the beginning of construction and maintained throughout construction. Construction parking areas should be stabilized as well as the roads. A stabilized construction road should not be located in a cut or fill area until after grading has been performed.

Stabilized construction roads should be built to conform to the site grades; this will require a minimum amount of cut and fill. They should also be designed so that the side slopes and road grades are not excessively steep. Construction roads should not be constructed in areas which are wet or on highly erodible soils.

Spill Prevention and Response

Spills are a source of storm water contamination, and construction site spills are no exception. Spills can contaminate soil and water, can waste materials, and can result in potential health risks.

The following measures are appropriate for a spill prevention and response plan.

- *Materials management.* Store and handle materials to prevent spills. Tightly seal containers. Make sure all containers are clearly labeled. Stack containers neatly and securely.
- *Employee training.* Identify personnel responsible for responding to a spill of toxic or hazardous materials. Provide personnel with spill response training. Post the names of spill response personnel. Ensure that cleanup procedures are clearly posted.
- *Spill containment.* Contain any liquid. Stop the source of the spill. Reduce storm water contact if there is a spill.
- *Spill cleanup.* Cover the spill with absorbent material such as kitty litter or sawdust. Have cleanup materials readily available. Keep the spill area well ventilated. If necessary, use a private firm that specializes in spill cleanup.
- *Disposal.* Dispose of contaminated materials according to the manufacturer's instructions or state or local requirements.
- *Spill reporting.* Check the spill reporting requirements listed in the permit. Typically any spill should be reported.

Releases in excess of reportable quantities

The EPA construction general permit requires that the discharge of hazardous substances or oil from a site be prevented or minimized in accordance with the SWPPP. Furthermore, if a permitted discharge contains a hazardous substance or oil in an amount equal to or in excess of a reportable quantity (RQ) established under 40 CFR 110, 40 CFR 117, or 40 CFR 302, during a 24-h period, then the National Response Center (NRC) must be notified (dial 800-424-8802 or 202-426-2675 in the Washington, D.C., area). The NRC is charged with receiving reports of all chemical, radiological, oil, and biological releases regulated by the Clean Water Act. The NRC immediately relays reports to the appropriate state and federal on-scene coordinators. Depending on the type of release, severity, location, and receiving system (soil, air or water), additional local contacts may be notified (e.g., city fire departments or hazardous-material teams). Individual municipalities should contact their state or local response departments to request that they be provided with information when RQ releases occur to their storm sewer systems.

Only one permittee for a project needs to report an RQ release. The permittee with the most direct authority over the spill should make the report. Generally, this will be the permittee with day-to-day operational control of the construction project (e.g., the general contractor).

Also, within 14 calendar days of knowledge of the release, the SWPPP must be modified to include the date and description of the release, the circumstances leading to the release, responses to be employed for such releases, and measures to prevent the recurrence of such releases.

All necessary information related to RQ releases and the NRC is contained in the EPA construction general permit, and in 40 CFR Parts 110, 117, and 302. The National Oil and Hazardous Substances Pollution Contingency Plan [also known as the National Contingency Plan (NCP)], found at 40 CFR 300, provides additional information about the organizational structure and procedures for preparing for and responding to discharges of oil and releases of hazardous substances, pollutants, and contaminants. In addition to the NCP, regional contingency plans (RCPs) exist for every region, and area contingency plans (ACPs) may also exist. EPA regional offices should be contacted directly for copies of available materials. Additional information is available via the Internet at the following Web sites for the U.S. National Response Team (NRT) and the NRC: www.nrt.org and www.dot.gov/dotinfo/uscg/hq/nrc.

Sanitary and Septic Disposal

Almost all construction sites have temporary sanitary facilities for onsite personnel. There are three main types:

1. *Portable facilities* that store the sanitary wastes and must be emptied periodically. (See Fig. 14.9.) Domestic waste haulers should be contracted to regularly remove the sanitary and septic wastes and to maintain the facilities in good working order. This will prevent overloading of the system, which could allow discharges to storm water runoff. The state may have a licensing program for waste haulers. If so, only licensed haulers should be used; if not, a reputable hauler should be chosen.

2. *Facilities that employ septic systems* for treatment and disposal of sewage. Wastes should be treated to an appropriate level before discharge. Malfunctioning septic tanks may be a more significant surface runoff pollution problem than a groundwater problem. This is because a malfunctioning septic system is less likely to cause groundwater contamination where a bacterial mat in the soil retards the downward movement of wastewater. Surface contamination is caused by clogged or impermeable soils or when stopped-up or collapsed pipe forces untreated wastewater to the surface. Surface contamination can vary in degree from occasional damp patches on the surface to constant pooling or runoff of wastewater. These discharges have high bacteria, nitrate, and nutrient levels and can contain a variety of household chemicals.

3. *Facilities that discharge to a sanitary sewer system.* Facilities should be properly hooked into the sanitary sewer system to prevent illicit discharges.

Untreated, raw sewage or seepage should never be discharged or buried on the site. Sanitary or septic wastes that are generated on the site should be treated or disposed of in accordance with state or local requirements. Many states have regulations concerning *on-site disposal systems* (OSDSs) or discharges to sanitary sewers. Localities often have ordinances which deal with the proper management of sanitary and septic wastes. In addition, if sewage is being discharged to the sanitary sewer, the local publicly owned treatment work (POTW) should be contacted because it may have certain requirements as well.

Contact the proper authorities prior to the development of the SWPPP for the information needed to demonstrate compliance with the appropriate regulations.

Non–Storm Water Discharges

NPDES storm water permits for construction activities typically prohibit most non–storm water discharges. Permits will state that all discharges covered by the permit must be composed entirely of storm water. However, permits may list some non–storm water discharges

Figure 14.9 Portable sanitary facility for construction personnel.

that, when combined with storm water discharges, may be authorized by the permit. The EPA construction general permit only authorizes the discharge of the following nonstorm waters, and only when such discharges are identified in the storm water pollution prevention plan and appropriate controls are included:

- Fire flows from firefighting activities. The EPA does not require that flows from firefighting activities be identified in the SWPPP because of the emergency nature of such discharges and because of the unpredictability of their occurrence. If practicable, however, the discharger must take action to mitigate the impacts of firefighting runoff on receiving water quality.
- Vehicle wash water if detergents are not used. The type of vehicle wash water that may be discharged under the EPA construction general permit results only from the removal of sediment from trucks and other vehicles. Since detergents are not needed for this type of washing, they are not allowed in the wash water discharge.
- Dust control runoff in accordance with permit conditions.
- Potable water sources including waterline flushings.
- Uncontaminated groundwaters from dewatering of excavated areas. Discharge associated with the dewatering of trenches is the same type of water contemplated by the term *groundwater dewatering*. This discharge is authorized by the permit, provided that adequate controls are used to minimize the discharge of pollutants, as described later in this chapter.
- Foundation and footing drain discharges from subsurface drainage systems, where flows are not contaminated with materials such as solvents.
- Natural groundwater discharges such as springs, riparian habitats, and wetlands.
- Irrigation tailwater. Irrigation water discharged during seeding and planting practices.
- Building wash water. New construction exterior building washdown discharges, without detergents or other contaminants.
- Pavement wash waters from dust control and general housekeeping practices, where spills or leaks of toxic or hazardous materials have not occurred and where detergents are not used.
- Air-conditioning condensate (but not including cooling water from cooling towers, heat exchangers, or other sources).

During the construction process, the nonstorm waters listed above are authorized for discharge either alone or when commingled with storm water. The EPA can request individual permit applications for such discharges where appropriate. If a discharge is authorized under another NPDES permit, then that discharge may be mixed with storm water discharges (an allowable non–storm water discharge) authorized under a construction general permit.

The allowable non–storm water discharges should be identified in the SWPPP. *For each of the discharges except emergency firefighting discharges, practices or controls that will be used to prevent pollution from these discharges should be described in detail.* The following general practices should be considered to prevent pollution from these discharges:

- All downslope site sedimentation and erosion controls should be in place prior to the discharge.
- Discharges with sediment loads should be discharged so that sediment pollution is minimized. These discharges include dewatering operation discharges and discharges from sediment traps and basins.
- Discharge should be directed only to areas that are stabilized to minimize erosion (e.g., buffer zones, vegetated filter strips, inlet and outlet protection, level spreaders). Do not discharge non–storm water flows onto disturbed areas.
- Discharges with sediment should be directed to pass through a sediment-filtering device. Sediment-filtering devices include sediment traps, basins, silt fences, vegetated filter strips, sump pits, and sediment tanks.

At some locations, acid and alkaline solutions from exposed soil or rock units may be a problem. The control of these pollutants involves good site planning and preconstruction geological surveys. Sealing fractures in the bedrock with grout and bentonite will often reduce the amount of acid or alkaline seepage. The remaining runoff may have to be chemically neutralized.

Dewatering discharges

Dewatering is the method used to remove and discharge excess water from a construction site. The most common procedure used is to pump water out of excavated areas, sediment basins, and sediment traps. Dewatering may also include methods used to lower the groundwater table to provide a stabilized area for construction.

Dewatering discharges usually have a very high sediment content; therefore, sediment control should be provided before the discharge enters a receiving water.

Water remaining in excavated areas may be eliminated by dewatering so that construction can proceed on schedule. Sediment traps and basins are often used to remove sediment from dewatering of excavation areas.

Dewatering may be used to remove accumulated water and sediments from sediment traps and basins to ensure their effectiveness throughout the entire project. At the end of the project, dewatering of sediment traps and basins is appropriate prior to removal of the last sediment control measures. Filtering should also be provided when discharge results from dewatering a sediment trap or basin. Methods of filtering the discharge from dewatering a sediment trap or basin include the following:

1. *Sump pit.* This is the preferred method of sediment control for dewatering discharges. The following section provides details.
2. *Floating suction hose.* This is an apparatus which allows clean water at the surface to be pumped out before the hose sinks low enough to pick up sediment-laden water.
3. *Standpipe.* A vertical pipe is attached to the base of the sediment basin riser with slits to control inflow and wrapping of filter fabric to aid in filtering sediments.
4. *Sediment tank.* This is a portable metal tank designed to provide the storage volume needed for some sediment to settle out of the discharge, even for locations in which there is no suitable space to construct a sediment basin, sediment trap, or sump pit. Figure 14.10 illustrates a sediment tank. Sediment tanks may be connected in series, and they should provide a storage volume adequate for about 10 min of flow at the maximum inflow rate. In other words, if the maximum rate into the sediment tank is 100 gal/min (0.223 ft^3/s or 0.0063 m^3/s), then the required volume of the sediment tank is 60,000 gal (8000 ft^3 or 227 m^3).

Sump pits

A *sump pit* is a temporary hole or pit placed so that it can collect water from sediment traps and basins or excavations. In the center of the pit is a slotted or perforated standpipe surrounded by stone. Water that collects in the pit flows through the gravel into the standpipe and is pumped out to a filtering device or, in some cases, directly to a receiving water.

A sump pit may be used to dewater a sediment trap or basin, or it may be used during construction when water collects in an excavation. The number of sump pits and their location will depend on the individual site and any state or local requirements. The standpipe should have

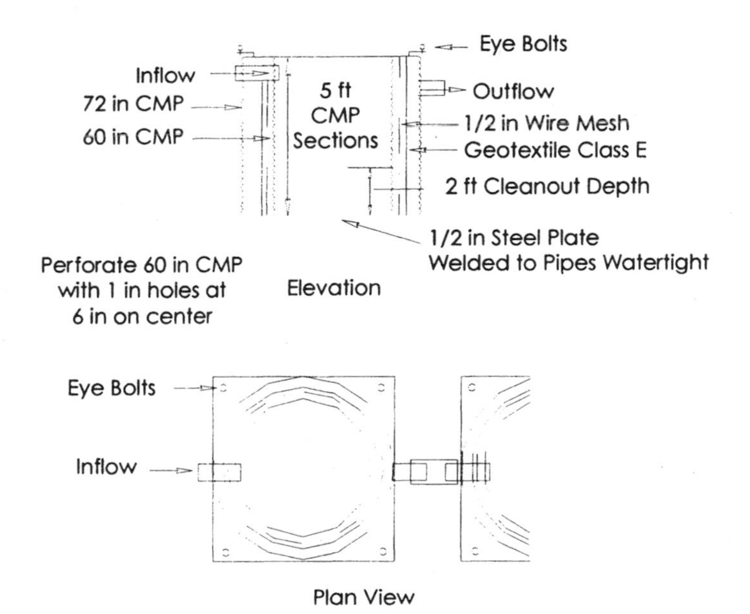

Figure 14.10 Sediment tank for dewatering discharges.

holes in it to allow water to flow in and should be extended at least 1 ft (0.3 m) over the top of the pit, to prevent overflows into the top of the standpipe. If the sump pit is to discharge directly into a receiving water, then the standpipe should be wrapped in filter fabric before the pit is backfilled with stone.

Figure 14.11 illustrates a typical sump pit. The estimated cost of a sump pit ranges from \$500 to about \$7000, depending upon soil conditions and other factors.

Non–storm water discharges not authorized by a general permit

There are three choices for handling non–storm water discharges which are not allowed by the NPDES storm water discharge permit:

1. Eliminate the source of the discharge.
2. Apply for a separate NPDES permit for the discharge.
3. Direct the discharge to a sanitary sewer system, after checking with the operator of the sewer system to see whether the material in question can be discharged into the sanitary sewer.

For discharges not covered by a construction general permit (such as industrial process wastewater or process wastewater mixed with storm water), the discharger must submit the appropriate application forms

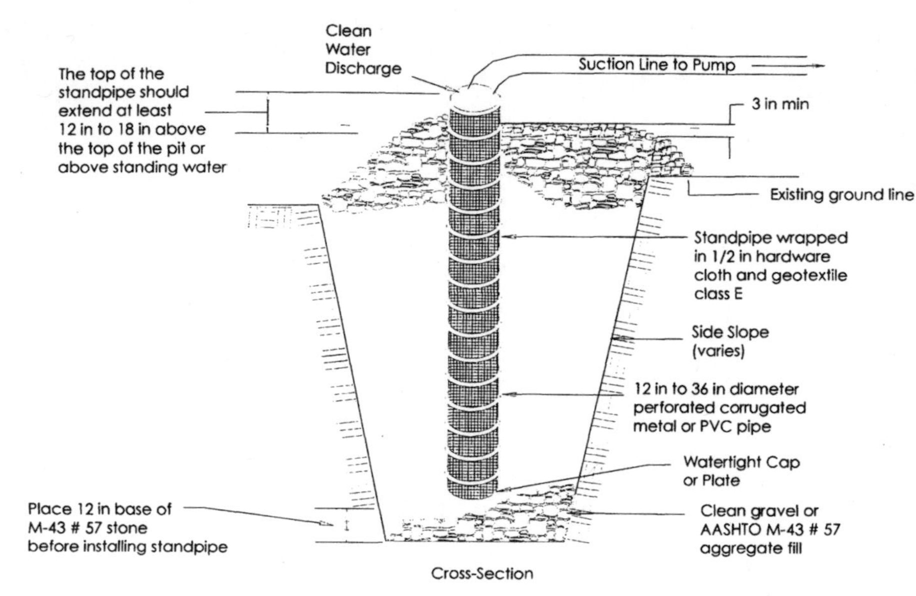

Figure 14.11 Sump pit for dewatering discharges (*Environmental Protection Agency, 1992*).

(Forms 1 and 2C) to obtain permit coverage or discontinue the discharge. "Allowable" non–storm water discharges cannot be authorized under this permit, unless they are directly related to and originate from a construction site or dedicated support activity site. For example, a pressure-washing company could not apply for coverage under the EPA construction general permit to broadly cover the discharges from all its pressure-washing of pavements, buildings, or vehicles, because these activities would not necessarily be associated with a construction site. Only those activities directly supporting a construction activity may be covered under a construction general permit.

References

Cosper, S. D., W. H. Hallenbeck, and G. R. Brenniman, 1993. *Construction and Demolition Waste: Generation, Regulation, Practices, Processing, and Policies,* Public Service Report OSWM-12, Office of Solid Waste Management, University of Illinois at Chicago, January.

Environmental Protection Agency, 1992. *Storm Water Management for Construction Activities: Developing Pollution Prevention Plans and Best Management Practices,* EPA Report 832/R-92/005, September.

Harris County/Harris County Flood Control District/City of Houston, 1992. *Storm Water Management Handbook for Construction Activities,* September 17.

von Stein, Edward L. and George M. Savage, 1994. "Current Practices and Applications in Construction and Demolition Debris Recycling," *Resource Recovery,* vol. 13, no. 4, pp. 85–94, April.

Appendix A

Storm Water Permit Requirements for Each State and Territory

This appendix provides specific storm water permitting requirements for each state and territory of the United States. Rather than describe all the requirements for each state, however, this chapter highlights only those areas in which a particular state's requirements differ from the EPA requirements. Therefore, it is most beneficial to review this material after you have already become familiar with the EPA requirements for storm water discharges from industrial facilities (as described in Chaps. 5 through 9) or construction sites (as described in Chaps. 10 through 14).

Role of the States in NPDES Permitting

The Clean Water Act of 1972 provided for the transfer of some NPDES permitting authority from the federal government to the states. Subsequent amendments to the Clean Water Act (CWA) have allowed and encouraged states to accept greater roles in NPDES permitting. For example, states could not administer the program as it applied to federally owned or operated facilities before 1977, when Congress amended Section 313 of the CWA to allow states to have this authority (Jessup, 1990).

The Federal Water Quality Act of 1987 also encouraged the transfer of authority to the states, because for the first time Congress authorized the EPA to transfer partial NPDES permitting authority to states. In other words, a state could assume authority for a major category of discharges or a major component of the NPDES program, even though that state was not prepared to fully enforce all other aspects of the NPDES permit program. Congress also allowed for phased program approval over a 5-year period (Jessup, 1990).

As a result of the consistent transfer of NPDES permitting authority from the EPA to the states since 1972, over 40 states are now authorized to manage their own NPDES programs. Some of these nonauthorized states are working toward becoming authorized. The federal government continues to administer the NPDES permit program on most Native American lands in all states.

Alabama

Alabama Department of Environmental Management
Water Quality Division
1751 Congressman W.L. Dickinson Drive
Montgomery, AL 36109-2608
Mailing address: P.O. Box 301463; Montgomery, AL 36130-1463
Jim Coles, industrial: (334) 271-7936
Richard Hulcher, mining: (334) 213-4312
Dennis Harrison, municipal: (334) 271-7801
Fax: (334) 279-3051

Alabama is an NPDES-delegated state with general permitting authority. The Alabama Department of Environmental Management (ADEM) has issued 21 general permits for storm water discharges from different types of industries. Alabama currently has no plans to adopt EPA's multisector general permit. The following ADEM branches are involved in the administration of these permits:

ADEM industrial branch. Asphalt/concrete industries; lumber and wood products industries (not including wood-preserving facilities); portland cement and concrete industries; primary metals, metals fabrication, and metals finishing industries; transportation terminals and warehousing activities; food and related substances industries; landfills (does not include landfills that discharge leachate); manufacturing of paint and related products; salvage and recycling activities; plastics and rubber industries; stone, glass, and clay industries; textile industries; cooling and boiler water; offshore facilities (facilities usually associated with the oil and gas industry); activities involved in the handling and storage of petroleum products; hydropower facilities; hydrostatic test waters; and uncontaminated storm water.

ADEM mining and nonpoint source section. Inactive or abandoned mining sites; nonmetallic and non–coal mining activities; and construction sites and other land-disturbing activities.

To be covered by one of the general permits, a Notice of Intent (NOI), a permit fee, and proof of public notification of the permit application are required (Thompson Publishing Group, August 1996).

The following Alabama municipal entities must obtain storm water discharge permits for their municipal separate storm sewer systems under phase I: Adamsville, Alabaster, Bessemer, Birmingham, Brighton, Brookside, Chickasaw, Creola, Daphne, Fairfield, Fairhope, Fultondale, Gardendale, Graysville, Helena, Homewood, Hoover, Hueytown, Huntsville, Indian Springs, Irondale, Leeds, Lipscomb, Madison, Maytown, Midfield, Mobile, Montgomery, Moody, Mountain Brook, Mulga, Pelham, Pleasant Grove, Prichard, Saraland, Satsuma, Tarrant, Trussville, Vestavia Hills, Baldwin County, Jefferson County (was listed in regulation; however, population dropped below 100,000 in 1990 census), Mobile County [unincorporated areas defined as beginning at the mouth of the South Fork Deer River extending west to the southwest (SW) corner, section 18, township 6 south, range 2 west, then north to northwest (NW) corner, section 6 township 2 south, range 2 west, then east to the Mobile County line, then south along the county line to U.S. Highway 90 bridge], Shelby County (all unincorporated areas of Shelby County within the drainage basin of the Cahaba River upstream of the confluence of Shoal Creek and the Cahaba River), St. Clair County (unincorporated areas of St. Clair County within the drainage basin of the Cahaba River), and the Alabama Highway Department [60 FR 50804 (Sept. 29, 1995)].

Alaska

Alaska Department of Environmental Conservation
410 East Willoughby Avenue
Juneau, AK 99801
Kenwyn George: (907) 465-5313, fax (907) 465-5274

U.S. EPA Region X
1200 6th Avenue
Seattle, WA 98101
Joe Wallace: (206) 553-8399, fax (206) 553-1280
E-mail: Wallace.Joe@epamail.epa.gov

Alaska does not have general permitting authority. The storm water program in Alaska is handled by the EPA region X office in Seattle. Therefore, the EPA construction general permit and multisector industrial permit are applicable within Alaska. However, the state has certain additional conditions which apply to storm water discharges.

Copies of the Notice of Intent (NOI), storm water pollution prevention plan (SWPPP), Discharge Monitoring Reports (DMRs), Reportable

Quantity (RQ) releases, and Notice of Termination (NOT) must be sent to the appropriate regional office of the Alaska Department of Environmental Conservation (ADEC) as well as to the EPA region X office (Personal correspondence, Kenwyn George, ADEC, February 1998).

Construction NOIs in Alaska must describe the activities on-site, the whole number of acres to be disturbed, the primary pollutants expected to be generated, and the type of treatment to be provided.

Industrial permittees should submit a copy of their inventory of exposed materials and records of spills and leaks to the appropriate ADEC regional office by the deadline for preparing the storm water pollution prevention plan.

Active construction sites must be inspected monthly in arid areas and weekly in other areas. Sites that have been stabilized must continue to be inspected monthly until a NOT is submitted for the site (Thompson Publishing Group, December 1994).

Under phase I, the city of Anchorage, the University of Alaska, and the Alaska Department of Transportation must obtain storm water discharge permits for their municipal separate storm sewer systems [60 FR 50804 (Sept. 29, 1995)].

American Samoa

American Samoa Environmental Protection Agency
Office of Governor
American Samoa Government
Pago Pago, AS 96799

U.S. EPA Region IX
75 Hawthorne Street
San Francisco, CA 94105
Eugene Bromley, Storm Water Coordinator: (415) 744-1906,
fax (415) 744-1235
E-mail: Bromley.Eugene@epamail.epa.gov

The territory of American Samoa does not have NPDES permitting authority. The storm water program in American Samoa is handled by the EPA region IX office in San Francisco. Therefore, the EPA construction general permit and multisector industrial permit are applicable within American Samoa.

A copy of the NOI as well as the storm water pollution prevention plan must be submitted to the American Samoa Protection Agency [57 FR 44464 (Sept. 25, 1992)].

Arizona

Arizona Department of Environmental Quality
3033 North Central Avenue
Phoenix, AZ 85012
Robert Wilson: (602) 207-4574, fax: (602) 207-4674

U.S. EPA Region IX
75 Hawthorne Street
San Francisco, CA 94105
Eugene Bromley: (415) 744-1906, fax (415) 744-1235
E-mail: Bromley.Eugene@epamail.gov

Arizona does not have NPDES permitting authority. The storm water program in Arizona is handled by the EPA region IX office in San Francisco. Dischargers in Arizona may seek coverage under the EPA baseline permit for construction activities or the multisector general permit for industrial activities. However, the state does have certain additional conditions which apply to storm water discharges.

The NOI and NOT must be sent to the Arizona Department of Environmental Quality (ADEQ) as well as to the EPA. This requirement applies to both the industrial as well as the construction NOI and NOT. A valid NOI submitted to the state of Arizona shall include the well registration number if storm water associated with industrial activity is discharged to a dry well or an injection well.

Facilities with Section 313 water priority chemicals must use appropriate measures to minimize discharges from liquid storage areas where storm water contacts tanks, containers, vessels, or equipment used to handle such materials. Appropriate measures may include secondary containment structures holding at least the volume of the largest single tank plus adequate freeboard to handle a 25-year, 24-h storm event; strong spill contingency and integrity testing plans; and/or equivalent measures (Thompson Publishing Group, March 1995).

Under phase I, Glendale, Mesa, Phoenix, Scottsdale, Tempe, Tucson, Pima County, and the Arizona Department of Transportation must obtain storm water discharge permits for their municipal separate storm sewer systems [60 FR 50804 (Sept. 29, 1995)].

Arkansas

Arkansas Department of Pollution Control and Ecology
8001 National Drive
P.O. Box 8913
Little Rock, AK 72219-8913
Jim Floyd: (501) 682-0627, fax (501) 682-0767
E-mail: floyd@adeq.state.ar.us

Arkansas is a delegated NPDES state with general permitting authority. Two general permits are available for storm water discharges: one for construction sites and one for industrial facilities. The state has no plans to adopt the EPA's model multisector general permits at this time (personal correspondence, Jim Floyd, ADPCE, June 1997).

Currently, the requirements of these Arkansas general permits are very similar to those for the baseline EPA general permit, except that the Arkansas industrial permit does not require semiannual sampling for any industrial discharges. However, Arkansas requires that some additional parameters be tested in the discharges of certain types of industries (Thompson Publishing Group, March 1995):

- *Primary metal industries:* 5-day biochemical oxygen demand (BOD).
- *Land disposal units, incinerators, boilers, and industrial furnaces:* Ammonia (as N) and nitrate plus nitrite nitrogen. (These facilities are not required to sample for total Kjeldahl nitrogen.)
- *Wood treatment facilities:* 5-day BOD.

Arkansas requests that industries submit their storm water pollution prevention plans. Industrial facilities passing annual biomonitoring 2 years in a row are exempt from monitoring from the rest of the period. A facility failing annual biomonitoring 2 years in a row requires quarterly biomonitoring; however, if it then passes 2 quarters in a row, it is exempt from further monitoring requirements. A facility failing quarterly biomonitoring 4 quarters in a row requires an individual permit.

Construction sites of 5 acres or more are required to obtain a storm water permit. Construction dischargers must submit their storm water pollution prevention plans along with their NOI. A $100 annual fee is assessed. Construction sites must post their permit number and must comply with the Endangered Species Act before the permit is issued.

Under phase I, the city of Little Rock must obtain storm water discharge permits for its municipal separate storm sewer system [60 FR 50804 (Sept. 29, 1995)].

California

California State Water Resources Control Board
Division of Water Quality
Attention: Storm Water Permit Unit
P.O. Box 1977
Sacramento, CA 95812-1977
Bruce Fujimoto or Leo Cosentini: (916) 657-1110, fax (916) 657-1011
Internet Web site: www.swrcb.ca.gov

California is a delegated NPDES state with general permitting authority. The state has issued two general permits for storm water discharges: one for the construction sites and one for industrial facilities. California has no current plans to adopt EPA's multisector model general permit. Although the state does not accept group applications for storm water permits, state regulations do allow for the formation of group monitoring plans.

California requires a storm water discharge permit for any industrial operation that is considered a primary activity even though the operation may be considered "auxiliary" to the facility. California also requires storm water discharge permits for runoff from large, flat, impervious surfaces, as this is considered a point source discharge rather than sheet flow. New industrial facilities must submit a Notice of Intent 14 days prior to the start of industrial activity. Industrial storm water pollution prevention plans must include a map extending approximately 0.25 mile beyond the facilities' boundaries, illustrating the general topography of the area and the facilities' discharge points into the local municipal separate storm sewer system. California also specifies that at least two dry-season (June through September) observations of the site be made for the presence of non–storm water discharges, monthly visual observations be made during the wet seasons (October through May), and annual certification be obtained of compliance with the storm water pollution prevention plan. Best management practices (BMPs) are required for authorized non–storm water discharge (California Industrial Activities Storm Water General Permit, April 1997). With some exceptions, industrial permittees are required to sample at least two storm events producing a significant storm water discharge each year, and these samples should be analyzed for the following parameters:

- Total suspended solids, pH, specific conductance, and total organic carbon (TOC), or oil and grease as an alternative to total organic carbon
- Toxic chemicals and other pollutants likely to be present in a facility's storm water discharge

Industrial facilities which can certify that no materials are exposed to storm water and can comply with a regional water board certification as well as a local agency certification program are eligible for sampling and analysis exemptions. Facility operators who have sampled six storm events are eligible for reduced sampling. Facilities subject to federal storm water effluent guidelines may have additional monitoring requirements.

In California, construction storm water discharge permits are issued to the owner of the construction site. Construction site permittees must certify annually that their construction activities are in compliance with the state construction activity general permit and the provisions of their SWPPP (Thompson Publishing Group, June 1997).

Under phase I, the following California municipal entities must obtain storm water discharge permits for their municipal separate storm sewer systems: Agoura Hills, Alameda, Albany, Alhambra, Anaheim, Arcadia, Artesia, Atherton, Azusa, Bakersfield, Baldwin Park, Bell, Bellflower, Bell Gardens, Belmont, Berkeley, Beverly Hills, Big Bear Lake, Bradbury, Brentwood, Brisbane, Burbank, Burlingame, Camarillo, Campbell, Carlsbad, Carson, Cerritos, Chula Vista, Claremont, Clayton, Colma, Commerce, Compton, Concord, Contra Costa County (15 cities), Coronado, Covina, Cudahy, Culver City, Cupertino, Daly City, Del Mar, Diamond Bar, Downey, Duarte, Dublin, East Palo Alto, El Cajon, El Monte, El Segundo, Emeryville, Encinitas, Escondido, Fairfield, Fillmore, Folsom, Foster City, Fremont, Fresno, Fullerton, Galt, Gardena, Garden Grove, Gilroy, Glendale, Glendora, Half Moon Bay, Hawaiian Gardens, Hawthorne, Hayward, Hermosa Beach, Hidden Hills, Hillsborough, Huntington Beach, Huntington Park, Imperial Beach, Industry, Inglewood, Irvine, Irwindale, La Canada Flintridge, Laguna Beach, Lake Tahoe Basin (one city), Lakewood, La Mesa, La Mirada, La Palma, La Puente, La Verne, Lawndale, Lemon Grove, Livermore, Lomita, Long Beach, Los Alamitos, Los Altos, Los Altos Hills, Los Angeles, Los Gatos, Lynwood, Manhattan Beach, Maywood, Menlo Park, Millbrae, Milpitas, Modesto, Monrovia, Montebello, Montery Park, Monte Sereno, Moorpark, Moreno Valley, Mountain View, National City, Newark, Norwalk, Oakland, Oceanside, Ojai, Ontario, Orange, Orange County (17 cities), Oxnard, Pacifica, Palo Alto, Palos Verdes Estates, Paramount, Pasadena, Pico Rivera, Piedmont, Pleasanton, Pomona, Port Hueneme, Poway, Rancho Cucamonga, Rancho Palos Verdes, Redondo Beach, Redwood City, Riverside, Riverside County (10 cities), Rolling Hills, Rolling Hills Estates, Rosemead, Sacramento, Salinas, San Bernardino, San Bernardino County (13 cities), San Bruno, San Carlos, San Diego, San Dimas, San Fernando, San Gabriel, San Jose, San Leandro, San Marcos, San Marino, San Mateo, Santa Ana, Santa Clara, Santa Clarita, Santa Fe Springs, Santa Monica, Santa Paula, Santa Rosa, Santee, Saratoga, Seal Beach, Sierra Madre, Signal Hill, Simi Valley, Solana Beach, South El Monte, South Gate, South Pasadena, South San Francisco, Stockton, Suisun City, Sunnyvale, Temple City, Thousand Oaks, Torrance, Union City, Vallejo, Vernon, Vista, Walnut, West Covina, West Hollywood, Westlake Village, Whittier, Woodside; Alameda County, Contra Costa County, Kern County, Lake Tahoe Basin (two counties), Los Angeles County, Orange County, Riverside County, Sacramento County, San Bernardino

County, San Diego County, San Mateo County, Santa Clara County, Ventura County; the Alameda County Flood Control District, zone 7 of the Alameda County Flood Control District, the California Department of Transportation, the Coachella Valley area, the Contra Costa County Flood Control District, the Orange County Flood Control Distict, the Riverside Flood Control District, the San Bernardino Flood Control District, the San Diego Unified Port District, and the Santa Clara Valley Water District [60 FR 50804 (Sept. 29, 1995)].

Colorado

Colorado Department of Public Health and Environment
Water Quality Control Division
4300 Cherry Creek Drive South
Denver, CO 80222-1530
Kathy Dolan: (303) 692-3596, fax (303) 782-0390

Colorado is a delegated NPDES state with general permitting authority. At this time, Colorado does not plan to adopt EPA's multisector general permit. Seven general permits are available for storm water discharges from various categories of industry: heavy industry; light industry; construction activity; metal mining operations; recycling industry; sand and gravel mining and processing (and other nonmetallic minerals except fuels); and coal mining facilities. Permit requirements are similar to those for the EPA baseline general permits. However, Colorado may impose various specific requirements for each of the seven general permits listed above (Thompson Publishing Group, January 1997).

Under phase I, the Colorado municipal entities of Aurora, Colorado Springs, Denver, Englewood, Lakewood, Arapahoe County, and the Colorado Department of Transportation must obtain storm water discharge permits for their municipal separate storm sewer systems [60 FR 50804 (Sept. 29, 1995)].

Connecticut

Connecticut Department of Environmental Protection
Water Management Bureau
79 Elm Street
Hartford, CT 06106-5127
Chris Stone: (860) 424-3850, fax (860) 424-7074

Connecticut is a delegated NPDES state with general permitting authority. The state has issued three general permits for storm water discharges: construction sites, industrial activities, and commercial facilities.

The commercial general permit covers any facility with an SIC code of 5XXX or 7XXX, where X represents any digit, or has 5 acres or more of impervious surface. General permit applications must include an 8½-in by 11-in copy of the applicable section of a U.S. Geological Survey quadrangle map illustrating the boundaries of the site, and must be certified by a professional engineer registered to practice in the state (personal correspondence, Chris Stone, CDEP, June 1997).

At least 15 days' notice is required before construction can begin under the general permit for construction activities. The construction permit authorizes the discharges of construction site storm water runoff and excavation dewatering discharges only. The construction general permit requires the installation of a sediment basin for all disturbed areas of 5 acres or more draining to a common discharge point. The basin must have a volume of 134 yd^3/acre of drainage area. This is approximately the same as the EPA baseline general permit requirement of 3600 ft^3/acre. The Connecticut construction permit requires that permanent storm water control measures be installed that will remove 80% of the total suspended solids from the storm water discharge after construction ends. Permanent velocity dissipation devices are also required (Thompson Publishing Group, February 1995).

Any industrial site discharging within a designated buffer zone of a nonfreshwater tidal wetland must discharge through a system designed to store the storm water runoff volume generated by 1 in of rainfall on the site. Industrial dischargers must construct impermeable containment structures which will hold at least the volume of the largest chemical container present, or 10% of the total volume of all containers, in areas where chemicals are stored. Industrial dischargers must perform annual sampling for the following constituents:

- Total oil and grease
- pH
- Chemical oxygen demand
- Total suspended solids
- Total phosphorus
- Total Kjeldahl nitrogen
- Nitrate as nitrogen
- Fecal coliform
- Total copper
- Total zinc

- Total lead
- Other pollutants listed in existing NPDES permits or effluent limitation guidelines for the facility

All these sampling parameters have target levels. If a facility samples below these levels for 2 consecutive years, then no sampling is required for the next 2 years. If small businesses (under 25 employees) or municipalities sample below these levels for 1 year, then no sampling is required for the next 2 years (personal correspondence, Chris Stone, CDEP, June 1997).

Additional requirements for acute toxicity testing apply under certain conditions (Thompson Publishing Group, February 1995).

Under phase I, the city of Stamford must obtain a storm water discharge permit for its municipal separate storm sewer system [60 FR 50804 (Sept. 29, 1995)].

Delaware

Delaware Dept. of Natural Resources and Environmental Control
Division of Water Resources, Water Pollution Control Branch
89 Kings Highway
P.O. Box 1401
Dover, DE 19903
Chuck Schadel: (302) 739-5731, fax (302) 739-3491

Delaware is an NPDES-delegated state with general permitting authority. The state has issued a general permit for storm water discharges with two parts: part 1 for industrial activities and part 2 for construction activities. The state is developing industry-specific permit provisions for the following industries:

- Ready-mixed concrete industry (SIC code 3273)
- Asphalt manufacturing and related activities (SIC codes 2951, 2950)
- Automotive salvaging and related activities (SIC codes 5015, 5093)
- Chemicals and allied products manufacturing (SIC code 28)
- Scrap and waste recycling and related activities
- Marinas (SIC code 4493)
- Air transportation maintenance and deicing activities
- Rail yard maintenance and related activities (SIC code 40)
- Automotive transportation maintenance—motor freight transportation and warehousing maintenance activities, including U.S. Postal Service transportation maintenance (SIC codes 41, 42, 43)

- Food and related product processing activities (SIC code 40)
- Metals manufacturing and associated industries (SIC codes 34, 35, 36, 37, 38)
- Wood products
- Concrete, stone, ceramics products, and related activities

At least 180 days' advance notice is required before permit coverage can be obtained for an industrial facility. Storm water plans must be developed within 90 days from industrial activity commencement, and implementation of the plan must begin within 90 days of the receipt of the NOI by the appropriate state department. Sampling is required only for facilities handling SARA Title III, Section 313 water priority chemicals.

Delaware is developing its own multisector permits, using the EPA program as a model. It is likely that the Delaware multisector permits will be somewhat more flexible than the EPA model, allowing more time for compliance and having less stringent monitoring requirements (Thompson Publishing Group, February 1997).

For construction activities, the NOI must be submitted through the approved state agency, conservation district, or municipality. This agency will review and approve the NOI and the related sediment and storm water management plan (SSMP) . The state requires approved sediment control plans for all construction activities disturbing at least 5000 ft^2. For larger developments, no more than 20 acres may be disturbed at one time. Permanent storm water management measures are required to remove at least 80% of the total suspended solids from post-construction discharges and must be designed for the 2-year, 10-year, and 100-year storm events. One employee from each construction site must attend a 3.5-h training course (Thompson Publishing Group, February 1997).

Under phase I, the Delaware municipal entities of Arden, Ardencroft, Ardentown, Bellefonte, Delaware City, Elsmere, Middletown, Newark, New Castle, Newport, Odessa, Townsend, Wilmington; New Castle County; and the Delaware Department of Transportation must obtain storm water discharge permits for their municipal separate storm sewer systems [60 FR 50804 (Sept. 29, 1995)].

District of Columbia

DC Department of Consumer and Regulatory Affairs
2100 Martin Luther King, Jr. Avenue S.E.
Washington, DC 20020
James Collier: (202) 645-6601 ext. 3040

U.S. EPA Region III
841 Chestnut Building
Philadelphia, PA 19107
Elaine Harbold: (215) 566-5744, fax (215) 566-2301

The District of Columbia does not have NPDES permitting authority. Storm water permitting is performed through the EPA region III office in Philadelphia. Facilities in the District of Columbia may apply for coverage under EPA's multisector general permit. All NOIs, NOTs, monitoring reports, and industrial storm water management plans must be submitted to the District of Columbia's Department of Consumer and Regulatory Affairs for review and approval. The provisions of the EPA construction general permit apply (Thompson Publishing Group, August 1996).

Under phase I, the city of Washington must obtain a storm water discharge permit for its municipal separate storm sewer system [60 FR 50804 (Sept. 29, 1995)].

Florida

Florida Department of Environmental Protection
Stormwater/Nonpoint Source Management Division
Twin Towers Office Building
2600 Blair Stone Road (MS 3570)
Tallahassee, FL 32399-2400
Eric Livingston: (850) 921-9915
John Cox, industrial permits: (850) 921-9383, fax (904) 921-5217

U.S. EPA Region IV
Atlanta Federal Center
100 Alabama Street, SW
Atlanta, GA 30303
Floyd Wellborn: (404) 562-9296, fax (404) 562-9224
E-mail: Wellborn.Floyd@epamail.epa.gov

Florida is authorized to manage NPDES permitting. However, the state will not assume authority of the storm water program until May 2000. Until then, EPA's region IV office in Atlanta will continue to issue all individual industrial permits, municipal separate storm sewer system permits, as well as construction and multisector general permits. EPA region IV has proposed to issue a state-specific construction permit for Florida only. The permit is very similar to the existing EPA construction general permit, except that it requires a permit for unpaved roads with a total surface area greater than 5 acres [62 FR 18605 (Apr. 16, 1997)].

Florida has the authority to issue wastewater permits. For some facilities, one permit may be obtained from the Florida Department of Environmental Protection (FDEP) that addresses both wastewater and storm water discharges, at the request of the permittee. The permittee also has the option to obtain a wastewater permit from FDEP and a separate storm water permit from EPA region IV (personal correspondence, Daryll Joyner, FDEP, June 1997).

Construction site operators must first receive their Florida State Stormwater or Environmental Resource Permit from the Florida Department of Environmental Protection or the appropriate water management district before submitting a Notice of Intent. They also must submit copies of the NOI and the construction SWPPP to the agency which issues their storm water permit (personal correspondence, Eric Livingston, FDEP, February 1998). Permanent storm water management controls are required to achieve 80% to 95% removal of total suspended solids from postconstruction discharges. Velocity dissipation devices at discharge outfalls and along the length of any outfall channel are also required. Construction site inspections are required as described in the EPA baseline general permit except that the minimum depth of rainfall required for an additional inspection is 0.25 in rather than the EPA-specified 0.5 in (Thompson Publishing Group, July 1995).

Under phase I, the following municipal entities must obtain storm water discharge permits for their municipal separate storm sewer systems: Apopka, Atlantic Beach, Atlantis, Auburndale, Bal Harbour, Bartow, Bay Harbor Islands, Bay Lake, Belleair, Belleair Beach, Belleair Bluffs, Belle Glade, Belle Isle, Boca Raton, Boynton Beach, Briny Breezes, Century, Clearwater, Cloud Lake, Coconut Creek, Cooper City, Coral Gables, Coral Springs, Dania, Davenport, Davie, Deerfield Beach, Delray Beach, Dundee, Dunedin, Eagle Lake, Eatonville, Edgewood, Fort Lauderdale, Fort Meade, Frostproof, Glen Ridge, Golden Beach, Golf, Golfview, Greenacres City, Gulfport, Gulf Stream, Haines City, Hallandale, Haverhill, Hialeah, Hialeah Gardens, Highland Beach, Highland Park, Hillcrest Heights, Hollywood, Homestead, Hypoluxo, Indian Creek, Indian Rocks Beach, Jacksonville Beach, Jacksonville, Juno Beach, Jupiter, Jupiter Inlet Colony, Key Biscanyne, Kenneth City, Lake Alfred, Lake Buena Vista, Lake Clarke Shores, Lake Hamilton, Lakeland, Lake Park, Lake Wales, Lake Worth, Lantana, Largo, Lauderdale-by-the-Sea, Lauderdale Lakes, Lauderhill, Lighthouse Point, Longboat Key, Madeira Beach, Maitland, Manalapan, Mangonia Park, Margate, Medley, Miami, Miami Beach, Miami Shores, Miami Springs, Miramar, Mulberry, Neptune Beach, North Bay Village, North

Lauderdale, North Miami, North Miami Beach, North Palm Beach, North Port, North Redington Beach, Oakland Park, Ocean Ridge, Ocoee, Oldsmar, Opa-locka, Orlando, Pahokee, Palm Beach, Palm Beach Gardens, Palm Beach Shores, Palm Springs, Parkland, Pembroke Park, Pembroke Pines, Pennsuee, Pensacola, Pinellas Park, Plantation, Plant City, Polk City, Pompano Beach, Redington Beach, Redington Shores, Reedy Creek Improvement District, Riviera Beach, Royal Palm Beach, Safety Harbor, St. Petersburg Beach, St. Petersburg, Sarasota, Sea Ranch Lakes, Seminole, South Bay, South Miami, South Palm Beach, South Pasadena, Sunrise, Surfside, Sweetwater, Tallahassee, Tamarac, Tampa, Tarpon Springs, Temple Terrace, Tequesta, Treasure Island, Venice, West Miami, West Palm Beach, Wilton Manors, Winter Garden, Winter Haven, Winter Park; Broward County, Dade County, Escambia County, Hillsborough County, Lee County, Leon County, Manatee County, Orange County, Palm Beach County, Pasco County, Pinellas County, Polk County, Sarasota County, Seminole County; the Florida Department of Transportation, and the urban water control districts [60 FR 50804 (Sept. 29, 1995)].

Georgia

Georgia Department of Natural Resources
Environmental Protection Division
Floyd Towers, East Room 1070
205 Butler Street, S.E.
Atlanta, GA 30334
Andrew Zurow: (404) 656-4887
Joseph Krewer, municipal permits: (404) 656-4887
Fax (404) 657-7379

Georgia is an NPDES-delegated state with general permitting authority. At the present time, Georgia does not intend to adopt the EPA multisector general permit. The state has issued separate general permits for storm discharges from construction sites and from industrial facilities. An NOI must be submitted at least 72 h before construction activity begins. Georgia state law requires permits for all activities disturbing land areas of more than 1.1 acres, with certain exceptions. Specific requirements apply to the control of sediment from such activities.

For industrial dischargers, permit requirements are similar to those for the EPA baseline general permits, except that only annual sampling is required by certain classes of industrial discharges; no industries are required to perform semiannual storm water sampling. There are also differences in the sampling parameters for certain industries. Whole effluent toxicity testing is not required by the Georgia general permit, except where compelled by the director of Environmental Protection Division (EPD) (personal correspondence, Andrew Zurow, GDNR, June 1997). The storm water pollution prevention plan developed by an industrial facility must be implemented within 60 days of the start of industrial activity (Thompson Publishing Group, March 1996).

Under phase I, the following Georgia municipal entities must obtain storm water discharge permits for their municipal separate storm sewer systems: Acworth, Alpharetta, Atlanta, Augusta Urban District, Augusta Suburban District, Austell, Avondale Estates, Berkely Lake, Bloomingdale, Buford, Chamblee, Clarkston, College Park, Columbus Consolidated Government, Dacula, Decatur, Doraville, Duluth, East Point, Fairburn, Forest Park, Garden City, Grayson, Hapeville, Jonesboro, Kennesaw, Lake City, Lawrenceville, Lilburn, Lithonia, Lovejoy, Macon, Marietta, Morrow, Norcross, Palmetto, Payne, Pine Lake, Pooler, Port Wentworth, Powder Springs, Riverdale, Roswell, Savannah, Smyrna, Snellville, Stone Mountain, Sugar Hill, Suwanee, Thunderbolt, Tybee Island, Union City; Bibb County, Chatham County, Clayton County, Cobb County, DeKalb County, Fulton County, and Gwinnett County (personal correspondence, Georgia DNR, June 1997).

Guam

Guam Environmental Protection Agency
P.O. Box 22439 GMF
Barrigada, Guam 96921

U.S. EPA Region IX
75 Hawthorne Street
San Francisco, CA 94105
Eugene Bromley, Storm Water Coordinator: (415) 744-1906,
fax (415) 744-1235
E-mail: Bromley.Eugene@epamail.epa.gov

The territory of Guam does not have NPDES permitting authority. The storm water program in Guam is handled by the EPA region IX office in San Francisco. Therefore, the EPA construction general permit is applicable within Guam. The multisector general permit for industrial activities was extended to include Guam. However, copies of all Notice of Intent forms, storm water pollution prevention plans, and monitoring reports must be submitted to the Guam EPA [61 FR 50020 (Sept. 24, 1996)].

Hawaii

Hawaii State Department of Health
Clean Water Branch
P.O. Box 3378
Honolulu, HI 96801
Alec Wong: (808) 586-4309, fax (808) 586-4352

Hawaii is a delegated NPDES state with general permitting authority. It is likely that Hawaii will require facilities that apply for coverage under Hawaii's baseline industrial permit to comply with additional requirements from EPA's multisector model general permit. The state has issued separate general permits for storm water discharges from construction sites and from industrial facilities. The NOI for general permit coverage must be submitted at least 30 days before industrial or construction activities may begin (personal correspondence, Alec Wong, HSDH, June 1997). Storm water pollution control plans must be submitted to the Department of Health (DOH) within 120 days of obtaining industrial general permit coverage, and implementation must begin within 180 days of submission to the DOH. The construction NOI is quite comprehensive, containing most of the elements normally required for a storm water pollution prevention plan. The construction general permit does not authorize any non–storm water discharges; separate permit coverage is required for excavation dewatering discharges and other non–storm water discharges.

All industrial dischargers must monitor annually for the following:

- Storm water flow
- Five-day biochemical oxygen demand
- Chemical oxygen demand
- Total suspended solids
- Total phosphorus
- Total nitrogen
- Nitrate plus nitrite nitrogen
- Oil and grease
- pH
- Any water priority pollutant identified in the facility's storm water pollution control plan
- Pollutants limited under an individual NPDES permit issued to the facility for process water
- Pollutants listed in an effluent guideline to which the industry is subject

Industrial discharge monitoring reports must be submitted to the DOH for review (Thompson Publishing Group, October 1995).

Under phase I, Honolulu County and the Hawaii Department of Transportation must obtain storm water discharge permits for their municipal separate storm sewer systems [60 FR 50804 (Sept. 29, 1995)].

Idaho

Idaho Department of Environmental Quality
1410 North Hilton
Boise, ID 83706
Todd Maguire: (208)-373-0502, fax (208) 334-0276

U.S. EPA Region X
1200 6th Avenue
Seattle, WA 98101
Joe Wallace: (206) 553-8399, fax: (206) 553-1280
E-mail: Wallace.Joe@epamail.epa.gov

Idaho does not have NPDES permitting authority. Therefore, the EPA final general permits are applicable within Idaho and are administered by the EPA region X office in Seattle. Dischargers in Idaho may apply for coverage under an EPA baseline general permit for construction activity or under EPA's multisector general permit for industrial activity. In contrast to most other non-NPDES states, however, Idaho has not added conditions to the EPA general permits. Therefore, the EPA general permits represent the complete requirements within Idaho, except that discharges must meet the state's water quality standards for groundwater (Thompson Publishing Group, May 1997).

Under phase I, the Idaho municipal entities of Boise City, Garden City, the Ada County Highway District, Boise State University, Drainage District 3, and the Idaho Department of Transportation must obtain storm water discharge permits for their municipal separate storm sewer systems (personal correspondence, Todd Maquire, IDEQ, June 1997).

Illinois

Illinois Environmental Protection Agency
Division of Water Pollution Control
Permit Section
2200 Churchill Road
P.O. Box 19276
Springfield, IL 62794-9276
Ginger Wells: (217) 782-0610, fax (217) 782-9891

Illinois is a delegated NPDES state with general permitting authority. The Illinois Environmental Protection Agency (IEPA) does not plan to adopt the multisector general permit at this time. The state has issued separate general permits for storm water discharges from construction sites and from industrial facilities. The requirements of these permits are very similar to those for the EPA general baseline permits. However, the Illinois permit for industrial discharges requires the submittal of quantitative sampling data as part of the application process for industries identified as potential significant contributors of pollutants to waters of the United States (personal correspondence, Ginger Wells, IEPA, June 1997). The Illinois general storm water permit does not require any storm water sampling after permit coverage has been obtained. Industrial dischargers are also required to submit annual inspection reports to the state.

New industrial facilities must submit an NOI form at least 180 days prior to the initiation of storm water discharges from the new site. Existing facilities that were not covered by a general or site-specific NPDES industrial permit must submit appropriate application information following the general storm water permit issue date.

Effective September 30, 1997, the NOI for construction permit coverage must be submitted at least 30 days prior to the commencement of construction (personal correspondence, Ginger Wells, IEPA, June 1997).

Under phase I, the Illinois municipal entities of Rockford, Springfield, and the Illinois Department of Transportation must obtain storm water discharge permits for their municipal separate storm sewer systems (personal correspondence, Ginger Wells, IEPA, June 1997).

Indiana

Indiana Department of Environmental Management
Office of Water Management
100 N. Senate Avenue
P.O. Box 6015
Indianapolis, IN 46206-6015
Laura Bieberich: (317) 233-6725 and 232-8637, fax (317) 232-8637
E-mail: lbieb@opn.dem.state.in.us

Indiana is a delegated NPDES state with general permitting authority. At this time, Indiana does not intend to adopt EPA's multisector model general permit. The state has issued separate general permits for storm water discharges from construction sites and from industrial facilities. Permit requirements are similar to those for the EPA baseline general permits, except that all industrial facilities must sample one storm event before implementing the SWPPP and two additional storm events after implementation of the SWPPP. Thereafter, only semiannual visual inspections are required, unless the state determines that further sampling is needed. Industrial dischargers must submit annual reports to the Indiana Department of Environmental Management (IDEM). All construction site sedimentation and soil erosion plans must be reviewed by the appropriate soil and water conservation districts (Thompson Publishing Group, December 1995).

Under phase I, the city of Indianapolis must obtain a storm water discharge permit for its municipal separate storm sewer systems (personal correspondence, Laura Bieberich, IDEM, July 1997).

Iowa

Iowa Department of Natural Resources
Environmental Protection Division
502 East 9th Street
Des Moines, IA 50319-0034
Joe Griffin: (515) 281-7017, fax (515) 281-8895

Iowa is a delegated NPDES state with general permitting authority. Iowa does not plan to adopt EPA's multisector general permit, nor does it plan to issue its own version of the permit. Instead, Iowa anticipates drafting individual permits based on EPA's multisector permit for industrial storm water dischargers that have participated in EPA-approved group applications. The draft individual permit will include information on public notices as well as a 30-day public comment period before a final permit is issued. Currently, individual permits have been issued for 22 industrial groups (personal correspondence, Joe Griffin, IDNR, February 1998). Permit requirements are similar to those for the EPA baseline general permit. However, the requirements for the NOI differ somewhat. For example, the applicant must submit proof that a public notice of the application has been published in at least two newspapers. In addition, the advance-notice requirement for industrial dischargers is 24 h. There are also differences in the sampling requirements and parameters for various industries. The state also issues general permits for construction activity, and these requirements are similar to those of the EPA baseline general permit (Thompson Publishing Group, January 1997).

Under phase I, Cedar Rapids, Davenport, and Des Moines must obtain storm water discharge permits for their municipal separate storm sewer systems [60 FR 50804 (Sept. 29, 1995)].

Johnston Atoll, Territory of

U.S. EPA Region IX
75 Hawthorne Street
San Francisco, CA 94105
Eugene Bromley, Storm Water Coordinator: (415) 744-1906, fax (415) 744-1235
E-mail: Bromley.Eugene@epamail.epa.gov

The territory of Johnston Atoll does not have NPDES permitting authority. The storm water program on Johnston Atoll is handled by the EPA Region IX office in San Francisco. Therefore, the EPA construction general permit and multisector industrial permit are applicable for Johnston Atoll.

Kansas

Kansas Department of Health and Environment
Bureau of Water
Forbes Field, Building 283
Topeka, KS 66620
David Freise: (785) 296-5557

Kansas is an EPA-delegated state with general permitting authority. Kansas plans to issue a general permit for industrial activity; until that time, the Kansas Department of Health and Environment (KDHE) advises industrial storm water dischargers to consider complying with EPA's baseline general permits for industrial activity or multisector general permit. Facilities that have submitted an EPA NOI, participated in a group application, or sent an individual application to KDHE will be offered general permit coverage when Kansas completes its industrial general permit.

Kansas also issues general permits for construction activity, and these requirements are similar to the EPA general permit for construction activity (Thompson Publishing Group, January 1997; and David Freise, KDHE, June 1997).

Under phase I, Kansas City, Overland Park, Topeka, Wichita, the Fairfax Drainage District, and the Kaw Valley Drainage District must obtain storm water discharge permits for their municipal separate storm sewer systems [60 FR 50804 (Sept. 29, 1995)].

Kentucky

Kentucky Department of Environmental Protection
KPDES Branch, Division of Water
14 Reilly Road
Frankfort, KY 40601
Douglas Allgeier, industrial permits: (502) 564-3410
Herb Ray, municipal permits: (502) 564-3410
Fax: (502) 564-4245

Kentucky is an NPDES state with general permitting authority. Kentucky has no plans to adopt the multisector model general permit, but may incorporate some elements of the multisector model permit into its general permits. Kentucky currently has eight general permits covering various industrial categories, including construction sites disturbing 5 acres or more of land; primary metal industries; active and inactive landfills, land application sites, and open dumps that receive industrial wastes; wood treatment facilities using chlorophenolic formulations and/or creosote; wood treatment facilities using inorganic preservatives containing arsenic or chromium; facilities with coal pile runoff; oil and gas exploration and/or production facilities; and industrial facilities not subject to any other category. Permit requirements are similar to those for the EPA baseline general permits, except that no special containment controls are required for SARA Title III, Section 313 water priority chemicals. Industrial best management plans must be prepared within 180 days of coverage under the state general permit and must be implemented within 365 days. In addition, whole effluent toxicity testing is not required. There are also differences in the sampling requirements and parameters for certain industries (Thompson Publishing Group, December 1995).

Under phase I, the Kentucky municipal entities of Lexington-Fayette, Louisville, and Jefferson County must obtain storm water discharge permits for their municipal separate storm sewer systems [60 FR 50804 (Sept. 29, 1995)].

Louisiana

Louisiana Department of Environmental Quality
Office of Water Resources
7290 Bluebonnet Boulevard
P.O. Box 82215
Baton Rouge, LA 70884-2215
Darlene Bernard: (504) 765-0525, fax (504) 765-0635
E-mail: darlene_b@deq.state.la.us

Louisiana is a delegated NPDES state with general permitting authority. Permitting authority was granted in August 1996, and Louisiana continues to regulate storm water dischargers under separate general permits for industrial facilities and construction sites developed by EPA. The state currently has no plans to significantly change the EPA storm water program. The state does, however, have certain additional conditions which apply to storm water dischargers.

The state requires that industrial dischargers meet additional numeric limitations on certain storm water contaminants. All industrial dischargers must meet effluent limitations for total organic carbon (not to exceed a daily maximum of 50 mg/L) and oil and grease (not to exceed daily maximum of 15 mg/L). Oil and gas facilities must meet additional storm water effluent limitations for chemical oxygen demand and chlorides.

If there is the potential for water contamination or the discharge of excessive volumes of storm water, storm water dischargers in areas where solid or liquid industrial materials are stored may have to obtain Louisiana Pollution Discharge Elimination System (LPDES) permits. These permits may require such facilities to implement more stringent best-available technology controls or water quality controls. Additional permits may be required for oil and gas facilities, activities that alter state waterways and stream bottoms, or activities that impact coastal wetlands. These activities are regulated by various state offices, and the permits may contain additional regulations concerning storm water discharges (Thompson Publishing Group, February 1997).

Under phase I, Baton Rouge, New Orleans, Shreveport, East Baton Rouge Parish, Jefferson Parish, and the Louisiana Department of Transportation must obtain storm water discharge permits for their municipal separate storm sewer systems [60 FR 50804 (Sept. 29, 1995)].

Maine

Maine Department of Environmental Protection
Bureau of Land and Water Quality
State House Station 17
Augusta, ME 04333
Norm Marcotte: (207) 287-3901, fax (207) 287-7826

U.S. EPA Region I
JFK Federal Building
Boston, MA 02203
Thelma Hamilton: (617) 565-3569, fax (617) 565-4940
E-mail: Hamilton.Thelma@epamail.epa.gov

Maine does not have NPDES permitting authority. The EPA region I office in Boston issues individual, construction, and multisector general permits within the state. However, Maine has certain additional conditions which apply to storm water dischargers.

Maine specifies that discharges must not reduce the quality of receiving waters below the minimum requirements of its water quality classifications. Maine also requires that "facilities that must conduct whole effluent toxicity (WET) testing must base half of their freshwater vertebrate testing on *Salvelinus fontinalis* (brook trout) and half using EPA Region I's testing protocols for the fathead minnow." Additional state regulations may apply to many construction projects under the Natural Resources Protection Act, 38 Maine Revised Statutes Annotated (MRSA), Section 480 A-Q; the Site Location of Development Act, 38 MRSA Sections 481 to 490; or local shore land zoning ordinances developed under the Mandatory Shoreline Zoning Act, 38 MRSA, Section 435 (Thompson Publishing Group, May 1996).

Under phase I, none of the municipalities in Maine is required to obtain a municipal storm water discharge permit [60 FR 50804 (Sept. 29, 1995)].

Maryland

Maryland Department of the Environment
Water Managment Administration
2500 Broening Highway
Baltimore, MD 21224
Edward Gertler, industrial permits: (410) 631-3323
Brian Clevenger, municipal permits: (410) 631-3543
Lois McNamara, construction permits: (410) 631-3510
Fax: (410) 631-4883

Maryland is a delegated NPDES state with general permitting authority. The state has issued separate general permits for storm water discharges from construction sites and from industrial facilities. While Maryland has no plans to adopt EPA's multisector model general permit, the Maryland Department of the Environment (MDE) has issued eight industry-specific general permits for both storm water and non–storm water discharges from the following industrial activities: pipe and tank discharges, surface coal mines, mineral mines, seafood processors, animal wastes, marinas, oil terminals, and ground remediation of oil contamination. In addition, Maryland plans to issue permits regulating discharges from the following activities: swimming pools; facilities whose only discharge is cooling waters; car washes; and well development for activities involved in providing drinking water or performing groundwater monitoring (personal correspondence, Ed Gertler, MDE, June 1997).

The state has had controls on sediment discharges since 1970, and the construction general permit incorporates these controls. An erosion and sediment control plan must be submitted to a local soil conservation district for approval before construction permit coverage is obtained for construction projects disturbing 5000 ft^2 of earth or more. Additionally, the state storm water management law requires new development project designs to include best management practices to replicate predevelopment runoff conditions after construction. The construction general permit incorporates this as well (personal correspondence, Brian Clevenger, MDE, June 1997). Industrial permit requirements are similar to those for the EPA baseline general permit, except that fewer additional requirements are imposed on SARA Title III, Section 313 facilities. In addition, no sampling is required for industrial dischargers (Thompson Publishing Group, October 1995).

Under phase I, Baltimore, Anne Arundel County, Baltimore County, Carroll County, Charles County, Frederick County, Harford County, Howard County, Montgomery County, Prince George's County, and the Maryland State Highway Administration must obtain storm water discharge permits for their municipal separate storm sewer systems [60 FR 50804 (Sept. 29, 1995)].

Massachusetts

Massachusetts Department of Environmental Protection
Division of Watershed Management
627 Main St. 2nd floor
Worcester, MA 01608
Bryant Firmin: (508) 792-7470

U.S. EPA Region I
JFK Federal Building
Boston, MA 02203
Thelma Hamilton: (617) 565-3569, fax (617) 565-4940
E-mail: Hamilton.Thelma@epamail.epa.gov

Massachusetts does not have NPDES permitting authority. The EPA region I office in Boston is responsible for permitting storm water dischargers under the EPA industrial, construction, or multisector general permit. However, the state has certain additional conditions which apply to storm water dischargers to ensure that no state water quality standards are violated. Outfall pipes releasing new or increased storm water discharges to any outstanding resource water are subject to additional setback and BMP requirements. Dischargers to outstanding resource waters must submit a copy of their NOI to the Massachusetts Department of Environmental Protection. Construction projects and industrial storm water dischargers may be subject to additional requirements under the state Wetlands and Waterways Protection Act and the state Ocean Sanctuaries Act. Certain areas of the state are affected by the National Estuary Program (Thompson Publishing Group, May 1996).

Under phase I, Boston and Worcester must obtain storm water discharge permits for their municipal separate storm sewer systems [60 FR 50804 (Sept. 29, 1995)].

Michigan

Michigan Department of Natural Resources
Surface Water Quality Division
P.O Box 30273
Lansing, MI 48909-7773
Susan Benzie: (517) 335-4188, fax (517) 373-9958

Michigan has NPDES general permitting authority. The state has issued a general permit for industrial activity which is administered by nine district offices under the authority of the Department of Environmental Quality (DEQ). The state has issued a general permit for industrial activity. Construction activity is regulated separately by a state "permit by rule." No sampling is required for either industrial or construction dischargers, and storm water treatment and control measures must be under the direct supervision of a certified storm water operator (personal correspondence, Susan Benzie, MDEQ, June 1997). In order to obtain a construction permit by rule, a notice of coverage must be submitted before construction activity begins (Thompson Publishing Group, December 1994).

Under phase I, Ann Arbor, Flint, Grand Rapids, Sterling Heights, Warren, the University of Michigan, and the Michigan Department of Transportation must obtain storm water discharge permits for their municipal separate storm sewer systems [60 FR 50804 (Sept. 29, 1995)].

Midway Island, Territory of

U.S. EPA Region IX
75 Hawthorne Street
San Francisco, CA 94105
Eugene Bromley, Storm Water Coordinator: (415) 744-1906,
fax (415) 744-1235
E-mail: Bromley.Eugene@epamail.epa.gov

The territory of Midway Island does not have NPDES permitting authority. The storm water program on Midway Island is handled by the EPA region IX office in San Francisco. Therefore, the EPA construction general permit and multisector industrial permit are applicable for Midway Island.

Minnesota

Minnesota Pollution Control Agency
Water Quality Division
520 Lafayette Road
St. Paul, MN 55155
Gene Soderbeck, general and industrial permits: (612) 296-8280
David Sahli, municipal discharges: (612) 296-8722
Keith Cherryholmes, construction activities: (612) 296-6945
Fax: (612) 282-6247

Minnesota is a delegated NPDES state with general permitting authority. The state has issued separate general permits for storm water discharges from construction sites and from industrial facilities. While the Minnesota Pollution Control Agency (MPCA) plans to review the EPA multisector model general permit in assessing the environmental risks posed by various industrial groups, Minnesota does not plan to adopt the multisector permit for use within the state. The MPCA will continue issuing the industrial general permit for federal group applicants. Sand and gravel industries are subject to a "multimedia" general permit which covers air and water regulations (Thompson Publishing Group, May 1997).

The Minnesota construction general permit applies to both public and private land disturbing greater than 5 acres, with no exemption for small municipal entities. Additional differences in the construction general permit include the following: both the owner and general contractor are copermittees; maintenance schedules are established within the permit; and the conversion of 1 acre or more of pervious surface to impervious surface requires the construction of a permanent wet storm water retention pond designed to handle the runoff from the project's ultimate development (personal correspondence, Gene Soderbeck, MPCA, July 1997).

Under phase I, the Minnesota municipal entities of Minneapolis and St. Paul must obtain storm water discharge permits for their municipal separate storm sewer systems [60 FR 50804 (Sept. 29, 1995)] (personal correspondence, Gene Soderbeck, MPCA, July 1997).

Mississippi

Mississippi Department of Environmental Quality
Office of Pollution Control
General Permits Branch
P.O. Box 10385
2380 Highway 80 West
Jackson, MS 39289-0385
Jim Morris, industrial and municipal permits: (601) 961-5151
Kenneth LaFleur: (601) 961-5074
C. Adam Smith: (601) 961-5029
Fax: (601) 961-5703

Mississippi is a delegated NPDES state with general permitting authority. The state has issued separate general permits for storm water discharges from construction sites and from eight other classes of industrial facilities, including baseline; mining; coal pile; land disposal; oil and gas; primary metal; SARA Title III, Section 313 facilities; and wood treaters. While Mississippi has no current plan to adopt EPA's multisector model general permit, the state intends to review the multisector permit when making changes to its general permits. For construction sites and mining operations, NOIs must be submitted at least 30 days prior to commencement of activity; for other industrial sites, NOIs must be filed 60 days in advance of activity. Permit requirements are similar to those for the EPA baseline general permits, except that certain special provisions are included for each separate class of industry (Thompson Publishing Group, August 1996).

Under phase I, only the city of Jackson must obtain a storm water discharge permit for its municipal separate storm sewer system [60 FR 50804 (Sept. 29, 1995)].

Missouri

Missouri Department of Natural Resources
Water Pollution Control Program
205 Jefferson Street
P.O. Box 176
Jefferson City, MO 65102
Linda Vogt: (573) 526-5630, fax (573) 751-9396

Missouri is a delegated NPDES state with general permitting authority. Missouri has developed 35 final industrial general permits for storm water discharges addressing a variety of industrial activities. Missouri does not plan to adopt EPA's multisector general permit; however, as the state's general permits expire, the multisector permit may be examined as a source of additional information (Personal correspondence, Linda Vogt, MDNR, July 1997). Permit requirements are similar to those for the EPA baseline general permit, except that there are specific requirements for certain industries. Missouri does not require whole effluent toxicity testing for any of its general permits (Thompson Publishing Group, August 1996).

Under phase I, Independence, Kansas City, and Springfield must obtain storm water discharge permits for their municipal separate storm sewer systems [60 FR 50804 (Sept. 29, 1995)].

Montana

Montana Department of Environmental Quality
Water Protection Bureau
1520 East 6th Avenue, Metcalf Building
P.O. Box 200901
Helena, MT 59620-0901
Roxann Lincoln: (406) 444-2406 ext. 5338
Fax: (406) 444-1374

Montana is a delegated NPDES state with general permitting authority. Montana does not plan to adopt EPA's multisector general permit; however, the multisector permit will be examined as a source of additional information for use in implementation of the state's general permits. The state has issued separate general permits for storm water discharges from construction sites, from general industrial facilities, and from oil and gas and mining activities. Permit requirements are similar to those for the EPA baseline general permits, except that more information is required on the NOI form for industrial discharges. In addition, the NOI must be submitted at least 30 days before the discharge begins, and the discharger must receive approval before beginning to discharge. Similarly, construction dischargers must submit a *storm water erosion control plan* (SECP) at least 30 days before construction begins and must receive approval before commencing construction. The industrial permit also imposes additional sampling requirements for certain industrial groups. SARA Title III, Section 313 facilities are not subject to additional requirements in their storm water pollution prevention plans. Discharge monitoring reports and an annual certification of compliance must be submitted by all industrial dischargers to the Department of Environmental Quality (Thompson Publishing Group, August 1996).

Under phase I, none of the municipalities in Montana are required to obtain municipal storm water discharge permits [60 FR 50804 (Sept. 29, 1995)].

Nebraska

Nebraska Department of Environmental Quality
Water Quality Division
1200 N Street, Suite 400
P.O. Box 98922
Lincoln, NE 68509-8922
Jim Yeggy: (402) 471-2023, fax (402) 471-2909

Nebraska is an NPDES-delegated state with general permitting authority. Nebraska does not intend at this time to adopt EPA's multisector general permit. The state has issued separate general permits for storm water discharges from construction sites and from industrial facilities. Written authorization is required for all general permits. Permit requirements are similar to those for the EPA baseline general permit, except that routine monitoring is not required (Thompson Publishing Group, September 1996).

Under phase I, Lincoln and Omaha must obtain storm water discharge permits for their municipal separate storm sewer systems [60 FR 50804 (Sept. 29, 1995)].

Nevada

Nevada Department of Conservation and Natural Resources
Division of Environmental Protection
333 West Nye Lane, Room 138
Carson City, NV 89706-0866
Rob Saunders: (702) 687-4670 ext. 3149
Fax: (702) 687-5856

Nevada is a delegated NPDES state with general permitting authority. The state has issued separate general permits for storm water discharges from construction sites and from two classes of industrial facilities: metal mining and other industries. Nevada does not plan to adopt EPA's multisector model general permit. The Department of Conservation and Natural Resources (DCNR) will require group applicants to seek coverage under the state's industrial general permit (personal correspondence, Rob Saunders, NDCNR, June 1997). Permit requirements are similar to those for the EPA baseline general permits. However, no storm water sampling is required for any facilities except subchapter N mining facilities. Special requirements apply within the Lake Tahoe drainage basin, and state water quality standards vary widely throughout the state (Thompson Publishing Group, April 1995).

Under phase I, Henderson, Las Vegas, North Las Vegas, Reno, Sparks, Clark County, Washoe County, Clark County Flood Control

District, and the Nevada Department of Transportation must obtain storm water discharge permits for their municipal separate storm sewer systems [60 FR 50804 (Sept. 29, 1995)].

New Hampshire

New Hampshire Department of Environmental Services
Water Division
64 North Main St., Third Floor
Concord, NH 03301
Jeff Andrews: (603) 271-2984, fax (603) 271-7894

U.S. EPA Region I
JFK Federal Building
Boston, MA 02203
Thelma Hamilton: (617) 565-3569, fax (617) 565-4940
E-mail: Hamilton.Thelma@epamail.epa.gov

New Hampshire does not have NPDES permitting authority. The storm water permit program is administered by the EPA region I office in Boston. However, construction projects disturbing more than 100,000 ft^2 of land, or 50,000 ft^2 of land adjacent to surface waters, are required to obtain a "Significant Alteration of the Terrain" permit. In contrast to most other non-NPDES states, the state has not added conditions to the EPA general permits. The EPA general permits represent the complete requirements in New Hampshire (Thompson Publishing Group, May 1996).

Under phase I, none of the municipalities in New Hampshire is required to obtain a municipal storm water discharge permit [60 FR 50804 (Sept. 29, 1995)].

New Jersey

New Jersey Department of Environmental Protection
Bureau of Nonpoint Pollution Control
Division of Water Quality
401 East State Street
P.O. Box 029
Trenton, NJ 08625-0029
Ed Frankel: (609) 633-7021, fax (609) 984-2147
E-mail: efrankel@dep.state.nj.us
Web site: www.state.nj.us/dep/dwq/dwqhome.htm

New Jersey is a delegated NPDES state with general permitting authority. While New Jersey does not intend to adopt EPA's multisector general permit, the multisector permit may examined as a source of information as the state develops industry-specific and individual permits. Currently, the state's Department of Environmental Protection (DEP) is proposing to require discharge permits for all industrial sources of storm water, including nonpoint sources. The state has issued separate general permits for storm water discharges from construction sites greater than 5 acres and from industrial facilities (recently renewed until January 2002) as well as two industry-specific general permits:

1. Scrap metal processing and recycling facilities engaged in the dismantling of motor vehicles
2. Concrete product manufacturing (includes concrete aggregate)

Permit requirements differ from those for the EPA baseline general permits in several areas:

- Applicants are required to submit a Request for Authorization (RFA) to discharge. This is similar to the EPA Notice of Intent. The RFA must be submitted at least 30 days prior to the date of expected storm water discharge.
- Coverage under the New Jersey industrial general permit is limited to storm water that has no contact with significant materials.
- New Jersey does not regulate rooftop drainage, unless an industrial activity is conducted on the rooftop.
- Non–storm water discharges are not authorized by a general permit except where specifically mentioned.
- In addition, annual certification of compliance with the storm water pollution prevention plan must be submitted to the state (Thompson Publishing Group, April 1996).
- Construction requirements are based upon state requirements already in place prior to the EPA storm water discharge regulations. These requirements include sedimentation and soil erosion controls for most construction disturbing more than 5000 ft^2.

Under phase I, no municipalities in New Jersey are required to obtain storm water discharge permits for their municipal separate storm sewer systems [60 FR 50804 (Sept. 29, 1995)].

New Mexico

New Mexico Environment Department
Surface Water Quality Bureau
1190 St. Francis Drive

P.O. Box 26110
Santa Fe, NM 87502
Richard Powell: (505) 827-2798, fax (505) 827-0160
E-mail: Richard-Powell@nmenv.state.nm.us

U.S. EPA Region VI
1445 Ross Avenue, Suite 1200
Dallas, TX 75202-2733
Monica Burrell: (214) 665-7530, fax (214) 665-2191
E-mail: Burrell.Monica@epamail.epa.gov
Multisector general permit hotline: (800) 245-6510

New Mexico does not have NPDES permitting authority. The EPA region VI office in Dallas is responsible for permitting storm water dischargers under the EPA industrial, construction, and multisector general permits. However, the state has certain additional conditions that apply to storm water dischargers. New Mexico requires that all notifications be sent to the New Mexico Environmental Department as well as to the EPA.

Under phase I, within New Mexico, Albuquerque, the Albuquerque Metropolitan Flood Control Authority, and the New Mexico Department of Transportation must obtain storm water discharge permits for their municipal separate storm sewer systems [60 FR 50804 (Sept. 29, 1995)].

New York

New York Department of Environmental Conservation
Bureau of Water Regulation Programs
50 Wolf Road
Albany, NY 12233-3505
Ken Stevens: (518) 457-0624, fax (518) 485-7786
Internet Web site: http://www.dec.state.ny.us

New York is a delegated NPDES state with general permitting authority. New York does not plan to adopt the EPA multisector general permit; however, as the state's general permits expire, the multisector permit will be examined as a source of additional information. The state has issued separate general permits for storm water discharges from construction sites and from industrial facilities. Industrial permit requirements are similar to those for the EPA baseline general permit. However, the construction general permit is somewhat more stringent. For example, the construction general permit requires submittal of the SWPPP to applicable local regulatory agencies and requires that streams not be substantially degraded as a result of construction storm water discharges (Thompson Publishing Group, September 1996).

Under phase I, New York City, the Bronx, Brooklyn, Manhattan, Queens, and Staten Island must obtain storm water discharge permits for their municipal separate storm sewer systems [60 FR 50804 (Sept. 29, 1995)].

North Carolina

North Carolina Department of Environment and Natural Resources
Water Quality Section
512 N. Salisbury Street
P.O. Box 29535
Raleigh, NC 27626-0535
Bill Mills, industrial permits: (919) 733-5083, ext. 548
Jeanette Powell, municipal permits: (919) 733-5083, ext. 537
Fax: (919) 733-9919

North Carolina is a delegated NPDES state with general permitting authority. The state has issued 18 separate general permits for storm water discharges from various categories of industrial facilities, including construction activities. The state expects to issue an additional general permit covering general aviation airports in 1998. Most permit requirements are similar to those for the EPA multisector permit, except for some specific monitoring requirements and other miscellaneous differences (personal correspondence, Bill Mills, NCDENR, June 1997). Construction activities disturbing 1 acre or more of land may be required to comply with additional state regulations (Thompson Publishing Group, March 1997).

Under phase I, Charlotte, Durham, Greensboro, Raleigh, Winston-Salem, Fayetteville/Cumberland County, and the North Carolina Department of Transportation must obtain storm water discharge permits for storm water discharges from their jurisdictional area [60 FR 50804 (Sept. 29, 1995)].

North Dakota

North Dakota Department of Health
Division of Water Quality
1200 Missouri Avenue
P.O. Box 5520
Bismark, ND 58502-5520
Jim Collins: (701) 328-5242, fax (701) 328-5200
E-mail: ccmail.jcollins@ranch.state.nd.us
Internet: www.ehs.health.state.nd.us

North Dakota is a delegated NPDES state with general permitting authority. The state has issued separate general permits for storm water discharges from construction sites (expiring in September 1999), from industrial facilities (expiring in March 2000), and from mining activities (expiring in June 1999). Industrial storm water dischargers must submit a storm water pollution prevention plan at least 30 days prior to the start of industrial activity. Storm water pollution prevention plans for construction sites must be submitted along with the NOI at least 5 days prior to the start of construction (Thompson Publishing Group, September 1995).

Under phase I, none of the municipalities in North Dakota is required to obtain a municipal storm water discharge permit [60 FR 50804 (Sept. 29, 1995)].

Northern Mariana Islands, Commonwealth of the

Commonwealth of the Northern Mariana Islands
Division of Environmental Quality
P.O. Box 1304
Saipan, MP 96950

U.S. EPA Region IX
75 Hawthorne Street
San Francisco, CA 94105
Eugene Bromley, Storm Water Coordinator: (415) 744-1906, fax (415) 744-1235
E-mail: Bromley.Eugene@epamail.epa.gov

Ohio

Ohio Environmental Protection Agency
Division of Surface Water
1800 Watermark Dr.
P.O. Box 1049
Columbus, OH 43216-1049
Bob Phelps: (614) 644-2034, fax (614) 644-2329
Storm water hotline: (614) 644-2053

Ohio is a delegated NPDES state with general permitting authority. The state has issued separate general permits for storm water discharges from construction sites, industrial facilities, and group applicants. Ohio also has issued a general permit for coal strip mining that addresses both process wastewater and storm water discharges (personal correspondence, Bob Phelps, OEPA, June 1997). Industrial permit requirements are similar to those for the EPA baseline general permits. However, construction permit requirements differ significantly, because the Ohio permit requirements are based on previous state standards and regulations for sediment and erosion control. Major differences include timing requirements as to when stabilization must take place, specific requirements to protect state waters, and requirements for sediment basins for large projects. In addition, construction NOIs must be submitted at least 45 days prior to the start of construction. The owner is solely responsible for the NOI. For construction activity on small lots, a modified NOI must be submitted (Thompson Publishing Group, February 1997).

Under phase I, the Ohio municipal entities of Akron, Columbus, Dayton, and Toledo must obtain storm water discharge permits for their separate storm sewer systems. The state is seeking exemptions for Cincinnati and Cleveland. Permitting for the Ohio Department of Transportation has not been addressed (Personal correspondence, Bob Phelps, OEPA, 1997).

Oklahoma

Oklahoma Department of Environmental Quality
Water Quality Division
1000 N.E. 10th St.
Oklahoma City, OK 73117-1212
Dave Farrington: (405) 271-5205, ext. 118, fax (405) 271-7339
E-mail: Dave.Farrington@oklaosf.state.ok.us

Oklahoma has become an NPDES-authorized state. However, permitting and enforcement authority is being gradually phased in.

The state prohibits new point source discharges of storm water to waters designated by the state as *outstanding resource waters* and *scenic rivers* as well as certain other designations.

Under phase I, the Oklahoma municipal entities of Oklahoma City, Tulsa, the Oklahoma Turnpike Authority, and the Oklahoma Department of Transportation must obtain storm water discharge permits for their municipal separate storm sewer systems [60 FR 50804 (Sept. 29, 1995)] and (Personal correspondence, Dave Farrington, ODEQ, June 1997).

Oregon

Oregon Department of Environmental Quality
Water Quality Division
811 S.W. Sixth Ave.
Portland, OR 97204
Paul Keiran: (503) 229-5937, fax (503) 229-5359

Oregon is a delegated NPDES state with general permitting authority. Oregon does not plan to adopt the EPA multisector general permit. However, as the state's general permits expire, the multisector permit will be examined as a source of additional information. The state has issued 13 separate general permits for storm water discharges from various classes of industrial facilities, including construction. Permit requirements are similar to those for the EPA baseline general permits, except that the construction general permit requires the submittal of an *erosion control plan* (ECP) to the state at least 30 days before construction may begin. Other portions of the application must be submitted at least 90 days before construction may begin. All industrial permits require semiannual monitoring, and the parameters monitored vary for each class of industrial facility. Industrial dischargers may also be required to have their storm water pollution prevention plans prepared by a registered engineer or architect, depending upon the size of the facility (Thompson Publishing Group, September 1996).

Under phase I, the following Oregon municipal entities must obtain storm water discharge permits for municipal separate storm sewer systems: Banks, Barlow, Beaverton, Canby, Cornelius, Durham, Estacada, Eugene, Fairview, Forest Grove, Gaston, Gladstone, Gresham, Happy Valley, Hillsboro, Johnson City, King City, Lake Oswego, Milwaukie, Molalla, North Plains, Oregon City, Portland, Rivergrove, Salem, Sandy, Sherwood, Tigard, Tualatin, West Linn, Wilsonville; Clackamas County, Multnomah County, Washington County; the Oregon Department of Transportation; and the Port of Portland [60 FR 50804 (Sept. 29, 1995)].

Pennsylvania

Industrial permits:

Pennsylvania Department of Environmental Protection
Water Quality Protection Bureau
Division of Wastewater Management
400 Market Street, 11th Floor
Rachel Carson State Office Building
P.O. Box 8774
Harrisburg, PA 17105-8774
R. B. Patel: (717) 787-8184, fax (717) 772-5156

Construction permits:

Pennsylvania Department of Environmental Protection
Bureau of Land & Water Conservation
Division of Storm Water Management and Sediment Control
400 Market Street, 11th Floor
Rachel Carson State Office Building
P.O. Box 8555
Harrisburg, PA 17105-8555
Ken Marin: (717) 787-6827

Pennsylvania is a delegated NPDES state with general permitting authority. The state has issued separate general permits for storm water discharges from construction sites and from industrial facilities. Pennsylvania has no plans to adopt the multisector general permit. However, certain requirements of the multisector general permit have been incorporated into the industrial general permit. Permit requirements differ from those for the EPA baseline permit requirements in several respects. Industrial dischargers must submit a Notice of Intent to the appropriate regional office of the DEP at least 180 days before beginning to discharge. The state requires a *preparedness, prevention, and contingency (PPC) plan* for most dischargers. Industrial dischargers must prepare and implement their PPC plans prior to submitting the NOI. In addition, the Pennsylvania permit sets a limit of 7 mg/L for dissolved iron and a pH limit of 6.0 to 9.0. Sampling requirements differ from the EPA baseline permit requirements in certain circumstances. For construction dischargers, the PPC plan is required only when there is a potential for accidental pollution of the environment or public endangerment. However, erosion and sediment control plans are required for all construction permittees. Construction dischargers are required to submit the following in order to obtain coverage under the construction general permit: NOI; U.S. Geological Survey map illustrating the location of the project; proof of notification of appropriate municipal and county government; an NOI fee; documentation of compliance with water quality standards and effluent limits, including a completed erosion and control plan; and documentation of the project's eligibility under the general permit. The construction NOI must be submitted 30 days prior to the start of construction activity. Erosion control plans must be submitted to the local county conservation district for review. Individual storm water permits are required for projects disturbing 25 acres or more, except for projects occurring in phases (Thompson Publishing Group, August 1997).

Under phase I, Allentown, Philadelphia, and the Pennsylvania Department of Transportation must obtain storm water discharge permits for their municipal separate storm sewer systems [60 FR 50804 (Sept. 29, 1995)].

Puerto Rico, Commonwealth of

Puerto Rico Environmental Quality Board
431 Ponce de Leon Avenue
5th Floor, Office 527
P.O. Box 11488
Hato Rey, PR 00910
Wanda Garcia-Hernandez: (787) 767-8731

U.S. EPA Region II
Water Permits and Compliance Branch
290 Broadway, 25th Floor
New York, NY 10007-1866
Sergio Bosques, Storm Water Regional Coordinator: (212) 637-3717
Fax: (212) 637-3771
E-mail: Bosques.Sergio@epamail.epa.gov

Puerto Rico does not have NPDES permitting authority. The storm water program in Puerto Rico is handled by the EPA region II office in New York. However, the territory has certain additional conditions which apply to storm water dischargers. Construction and industrial permittees must submit the following information to the Puerto Rico Environmental Quality Board (EQB) as well as to the EPA Region II office:

- The number of storm water discharges associated with industrial activity at the site
- A drawing indicating the drainage area of each storm water outfall

Industrial dischargers must also indicate the sampling point at each outfall. Storm water pollution prevention plans, certification that the plans have been implemented according to the conditions of the permit, and discharge monitoring reports must also be submitted to the EQB. Puerto Rico requires EQB approval before construction of any treatment facility for waters composed entirely of storm water. The territory has certain sampling procedure requirements which vary somewhat from the EPA baseline requirements. In addition, water quality standards must not be compromised due to storm water discharges from regulated facilities (Thompson Publishing Group, November 1994).

Rhode Island

Rhode Island Department of Environmental Management
Office of Water Resources
Rhode Island Pollutant Discharge Elimination System (RIPDES) Program
235 Promenade Street
Providence, RI 02908-5767
Angelo S. Liberti: (401) 222-4700, ext. 7225
David J. Cluley: (401) 222-4700, ext. 7046
Fax: (401) 521-4230
Web site: www.state.ri.us/dem

Rhode Island is a delegated NPDES state with general permitting authority. The state issued separate general permits for storm water discharges from construction sites and from industrial facilities in March 1993. Both of these general permits expired in March 1998. In February 1998, the RIDEM finalized its 1998 general storm water permits with only minor modifications to the 1993 general permits. Within two years, RIDEM intends to review significant changes to the industrial general permit including the incorporation of portions of the EPA multisector general permit and the modification of monitoring requirements. The RIDEM plans to issue a permit modification to address these changes (personal correspondence, Angelo Liberti, RIDEM, March 1998). The requirements of the 1998 permits are similar to those of the EPA baseline general permits, except for some differences in sampling requirements and other miscellaneous variations. For example, no whole effluent toxicity (WET) testing is required under the state general permit. Industrial facilities discharging storm water to the subsurface are regulated under state Groundwater Protection Program and its federally delegated Underground Injection Control (UIC) Program (Thompson Publishing Group, June 1997); (Personal correspondence, Angelo Liberti, RIDEM, July 1997).

Under phase I, none of the municipalities in Rhode Island is required to obtain a municipal storm water discharge permit [60 FR 50804 (Sept. 29, 1995)]. The city of Providence, which would otherwise qualify for permit coverage, has requested a waiver based on the population served by its combined sewer system.

South Carolina

South Carolina Department of Health and Environmental Control
Bureau of Water Pollution Control
2600 Bull Street
Columbia, SC 29201
Arturo Ovalles, industrial permits: (803) 734-5308, fax (803) 734-5216
E-mail:
WPO#123#DHEC4005.COLUMB35#C#0VALLEAR#125#@gm.state.sc.us

South Carolina is a delegated NPDES state with general permitting authority. The state has issued separate general permits for storm water discharges from construction sites and from industrial facilities. South Carolina has decided not to adopt EPA's multisector general permit (MSGP) at present, but it may adopt portions later. The state strongly encourages applicants to implement the best management practices in the MSGP. The state requires group applicants to apply for coverage under the state industrial general permit (personal correspondence, Arturo Ovalles, SCDHEC, June 1997). Permit requirements are similar to those for the EPA baseline general permit (Thompson Publishing Group, June 1997).

Construction activities disturbing 1 acre or more are regulated under the South Carolina Stormwater Management and Sediment Reduction Act of 1991. For these smaller construction activities, a storm water pollution prevention plan must be prepared and submitted to the South Carolina Department of Health and Environmental Control (DHEC) or the appropriate local entity for approval (Thompson Publishing Group, June 1997).

Under phase I, Greenville County, Richland County, and the Harbor of Charleston must obtain storm water discharge permits for their municipal separate storm sewer systems [60 FR 50804 (Sept. 29, 1995)].

South Dakota

South Dakota Department of Environment and Natural Resources
Division of Environmental Services
Joe Floss Building
523 East Capitol
Pierre, SD 57501-3181
Norma Job: (605) 773-4040, fax (605) 773-6035
E-mail: NormaJ@denr.state.sd.us
Storm water information hotline: (800) 737-8676

South Dakota is a delegated NPDES state with general permitting authority. The state has issued separate general permits for storm water discharges from construction sites and from industrial facilities. However, the state is in the process of rewriting the industrial general permit. South Dakota anticipates requiring industrial dischargers to submit an NOI prior to the start of the industrial activity, as well as eliminating certain monitoring requirements. The state has also issued "multimedia" general permits which incorporate storm water discharges along with air emissions for three industrial categories: concrete batch plants, rock crushers, and asphalt plants (personal correspondence, Norma Job, SDDES, 1997).

The construction general permit mirrors the EPA general permit, except that NOIs must be submitted 15 days before construction begins, and the certification of applicant form must be signed and notarized (personal correspondence, Norma Job, SDDENR, June 1997); (Thompson Publishing Group, May 1997).

Under phase I, Sioux Falls must obtain a storm water discharge permit for its municipal separate storm sewer system [60 FR 50804 (Sept. 29, 1995)].

Tennessee

Tennessee Department of Environment and Conservation
Division of Water Pollution Control
L&C Annex, 6th Floor
401 Church Street
Nashville, TN 37243-1534
Robert Haley, III: (615) 532-0669, fax (615) 532-0503
E-mail: rhaley@mail.state.tn.us

Tennessee is a delegated NPDES state with general permitting authority. The state has issued a construction general permit as well as a multisector general permit that is very similar to EPA's multisector model general permit. However, Notices of Intent (NOIs) for industrial facilities must be submitted at least 5 days prior to storm water discharge. Cutoff concentrations for various monitoring parameters may vary somewhat from those of the EPA's multisector permit. The appropriate field office of the Division of Water Pollution Control must be notified upon completion of a storm water pollution prevention plan. In addition, the Tennessee multisector general permit does not provide coverage for metal mining facilities (Tennessee Storm Water Multi-Sector General Permit Fact Sheet, March 31, 1997).

Under phase I, the Tennessee municipal entities of Barlett, Belle Meade, Berry Hill, Chattanooga, Collierville, East Ridge, Forest Hills, Germantown, Goodlettsville, Knoxville, Lakewood, Memphis, Nashville-Davidson, Oak Hill, Red Bank Ridgetop, and the Tennessee Department of Transportation must obtain storm water discharge permits for their municipal separate storm sewer systems [60 FR 50804 (Sept. 29, 1995)].

Texas

Texas Natural Resource Conservation Commission
Water Quality Division
Mail Stop 148
P.O. Box 13087
Austin, TX 78711-3087
Stephen Ligon, industrial permits: (512) 239-4527
E-mail: sligon@smtpgate.tnrcc.state.tx.us
Firoj Vahora, municipal permits: (512) 239-4540
E:mail: fvahora@smtpgate.tnrcc.state.tx.us
Fax: (512) 239-4430

U.S. EPA Region VI
1445 Ross Avenue, Suite 1200
Dallas, TX 75202-2733
Monica Burrell: (214) 665-7530, fax (214) 665-2191

E-mail: Burrell.Monica@epamail.epa.gov
Multisector general permit hotline: (800) 245-6510

Texas does not have NPDES permitting authority. However, the state has applied for NPDES delegation of the storm water program. Currently, storm water permitting for Texas is handled by the EPA region VI office in Dallas. The state has certain additional conditions which apply to storm water discharges.

Texas has placed numeric effluent limitations on arsenic, barium, cadmium, chromium, copper, lead, manganese, mercury, nickel, selenium, silver, and zinc, for both inland and tidal waters.

Whole effluent toxicity (WET) testing is required for all wood treatment facilities in Texas. The WET test must be performed using freshwater test organisms, and the WET testing must indicate a survival rate of greater than 50% over 24 h in undiluted grab samples of storm water effluent. Additional WET testing must be conducted if more than 10% of the control organisms die (Thompson Publishing Group, October 1996).

Under phase I, Abilene, Amarillo, Arlington, Austin, Beaumont, Corpus Christi, Dallas, El Paso, Fort Worth, Garland, Houston, Irving, Laredo, Lubbock, Mesquite, Pasadena, Plano, San Antonio, Waco, Harris County, the Harris County Flood Control District, the University of Texas at Austin, the University of Texas at Arlington, and Texas A&M University at Corpus Christi must obtain storm water discharge permits for their municipal separate storm sewer systems. The Texas Department of Transportation is a copermittee on municipal permits [60 FR 50804 (Sept. 29, 1995)] (personal correspondence, Monica Burrell, USEPA, June 1997).

Utah

Utah Department of Environmental Quality
Division of Water Quality
288 North 1460 West
P.O. Box 144870
Salt Lake City, UT 84114-4870
Harry Campell: (801) 538-6923, fax (801) 538-6016

Utah is a delegated NPDES state with general permitting authority. The state has issued separate general permits for storm water discharges from construction sites and from industrial facilities, and both permits expired September 30, 1997. Utah plans to reissue the construction general permit; however, the industrial general permit will be replaced with a state multisector general permit modeled after EPA's multisector permit. The state's multisector permit will likely contain fewer details pertaining to what needs to be done for the Endangered Species Act and the Historic Preservation Act, but it will mention that the permit cannot cause problems relating to the two acts (personal correspondence, Harry Campbell, UDEQ, June 1997). Permit requirements are similar to those for the EPA baseline general permits, except that the state regulates discharges to groundwater. Utah also does not regularly require whole effluent toxicity tests of industrial facilities (Thompson Publishing Group, May 1997).

Under phase I, Salt Lake City, Salt Lake County, and the Utah Department of Transportation must obtain storm water discharge permits for their municipal separate storm sewer systems [60 FR 50804 (Sept. 29, 1995)].

Vermont

Vermont Department of Environmental Conservation
Permits, Compliance, and Protection Division
103 S. Main Street
Waterbury, VT 05676
Brian Kooiker: (802) 241-3822, fax (802) 244-5141

Vermont is a delegated NPDES state with general permitting authority. Vermont has prepared separate draft permits for construction sites and industrial activity. The state does not plan to adopt the EPA multisector general permit; however, the multisector permit may be used as a source of additional information as Vermont finalizes the industrial general permit. Permit requirements are similar to those for the EPA baseline general permits, except that no monitoring is likely to be required for low-risk industrial facilities. Construction activities involving new residential, commercial, and subdivision projects may be required to implement postdevelopment hydraulic controls (Thompson Publishing Group, July 1996).

Under phase I, none of the municipalities in Vermont must obtain a storm water discharge permit for its municipal separate storm sewer system [60 FR 50804 (Sept. 29, 1995)].

Virginia

Virginia Department of Environmental Quality
Water Division
629 E. Main Street
P.O. Box 10009
Richmond, VA 23240-0009
Burton Tuxford: (804) 698-4086, fax (804) 698-4032

Virginia is a delegated NPDES state with general permitting authority. The state has issued four separate general permits for storm water discharges from construction sites (EPA category 10) and from some classes of industrial facilities (EPA categories 2, 5, 6, 7, 8, and 11). Industrial dischargers must submit a *general permit registration statement* (GPRS) and permit fee at least 30 days prior to commencement of industrial activity. Construction dischargers must submit a GPRS and permit fee at least 14 days prior to the start of construction activity. The monitoring requirements of the current Virginia industrial permits mirror those of EPA's baseline general permit, except that Virginia does not require any facility with a general storm water permit to submit discharge monitoring reports until permit reapplication (personal correspondence, Burton Tuxford, VDEQ, June 1997); (Thompson Publishing Group, March 1997).

Under phase I, Chesapeake, Hampton, Newport News, Norfolk, Portsmouth, Virginia Beach, Arlington County, Chesterfield County, Fairfax County, Henrico County, and Prince William County must obtain storm water discharge permits for their municipal separate storm sewer systems [60 FR 50804 (Sept. 29, 1995)] (personal correspondence, Burton Tuxford, VDEQ, June 1997).

Virgin Islands of the United States

Virgin Islands Department of Planning and Natural Resources
Environmental Protection Division
1118 Watergut Homes, Christiansted
St. Croix, VI 00820-4433
(809) 773-0565

For St. Thomas:

Virgin Islands Department of Planning and Natural Resources
Wheatley Center II
St. Thomas, VI 00802
Phyllis Hall: (340) 777-4577, fax (340) 774-5416

U.S. Environmental Protection Agency
1118 Watergut Homes, Christiansted
St. Croix, VI 00820-5065

The Virgin Islands of the United States is a delegated territory. It issues NPDES permits through the Territorial Pollutant Discharge Elimination System. These are essentially the same as the EPA baseline general permits.

Wake Island, Territory of

U.S. EPA Region IX
75 Hawthorne Street
San Francisco, CA 94105
Eugene Bromley, Storm Water Coordinator: (415) 744-1906,
fax (415) 744-1235
E-mail: Bromley.Eugene@epamail.epa.gov

The territory of Wake Island does not have NPDES permitting authority. The storm water program here is handled by the EPA region IX office in San Francisco. Thus, the EPA construction general permit and multisector industrial permit are applicable for Wake Island.

Washington

Washington Department of Ecology
Office of Water Programs
300 Desmond Drive
P.O. Box 47600
Olympia, WA 98504-7600
Mike Templeton: (360) 407-6295, fax (360) 407-6426

Washington is a delegated NPDES state with general permitting authority. The state has issued a general permit for storm water discharges from construction sites of 5 acres or more. The state has also issued a general permit covering most other types of industrial facilities.

Construction site owners must submit an NOI at least 38 days before the start of construction activity if they will discharge storm water from the site to a surface water of the state. Operators of industrial facilities in categories listed in the general permit must submit an NOI at least 38 days before the start of industrial activity if they will discharge storm water to a surface water of the state. No storm water sampling is required in either permit, but industrial discharges must conduct semiannual inspections.

Storm water requirements for boatyards and for sand and gravel and similar mining facilities are covered under general permits for process wastewater and storm water for those industries. Facilities in certain industrial categories must obtain individual permits covering both process wastewater and storm water, e.g., shipyards and wood treaters (personal correspondence, Mike Templeton, WDOE, July 1997).

Under phase I, the Washington municipal entities of Seattle, Tacoma, Clark County, King County, Pierce County, Snohomish County, Spokane County, and the Washington Department of

Transportation must obtain storm water discharge permits for their municipal separate storm sewer systems [60 FR 50804 (Sept. 29, 1995)].

West Virginia

West Virginia Department of Commerce
Division of Environmental Protection, Office of Water Resources
Engineering Branch
1201 Greenbrier St.
Charleston, WV 25311
Arthur Vickers: (304) 558-8855
Pravin Sangami: (304) 558-4086
Fax: (304) 558-5903

West Virginia is a delegated NPDES state with general permitting authority. The state has issued general permits for storm water discharges from construction sites and from industrial facilities. Permit coverage is required for construction activities that disturb at least 3 acres. West Virginia is also in the process of developing a state multisector general permit based on EPA's model multisector permit. Industrial permit requirements are similar to those for the EPA baseline general permits, except for variations in sampling requirements. Construction dischargers must apply for permit coverage at least 30 days before construction begins. The state also maintains a voluntary permit program for construction sites disturbing at least 1 acre but less than 3 acres (Thompson Publishing Group, June 1997).

Under phase I, none of the municipalities in West Virginia is required to obtain a municipal storm water discharge permit [60 FR 50804 (Sept. 29, 1995)].

Wisconsin

Wisconsin Department of Natural Resources
Bureau of Wastewater Management
101 S. Webster
P.O. Box 7921 WT/2
Madison, WI 53707-7921
Percy Mather: (608) 266-9263, fax (608) 267-2800
E-mail: mathep@dnr.state.wi.us
Web site: http:\\www.dnr.state.wi.us\eq\wq\nps

Wisconsin is a delegated NPDES state with general permitting authority. Dischargers in Wisconsin may apply for coverage under the state's separate industrial, construction, or municipal storm water discharge permits. Industrial general permits are issued under a three-tier system that groups industries by type and by how likely outdoor operations are to contaminate storm water. Construction dischargers must submit an NOI at least 14 days before construction begins. In addition to large and medium MS4s, storm water discharge permits are required for the following:

- Any municipality in a Great Lakes "area of concern" where storm water quality has been identified as a serious concern
- Municipalities with a population of more than 50,000 that are located within a watershed which has a nonpoint source control plan through the state's priority watershed program
- Communities surrounding a municipality with a storm water permit if the storm water runoff systems meet the criteria in the state's administrative code (personal correspondence, Percy Mather, WDNR, June 1997).

Under phase I, the municipalities of Allouez, Ashwaubenon, De Pere, Eau Claire, Green Bay, Madison, Marinette, Milwaukee, Racine, Sheboygan, Superior, Waukesha and West Allis must obtain municipal storm water discharge permits. The major University of Wisconsin campuses and Wisconsin's Department of Transportation must also obtain permits or equivalent controls, respectively, where they operate separate storm sewer systems in these cities (personal correspondence, Percy Mather, WDNR, June 1997); (Thompson Publishing Group, February 1995) [60 FR 50804 (Sept. 29, 1995)].

Wyoming

Wyoming Department of Environmental Quality
Herschler Building, 4th Floor
122 West 25th Street
Cheyenne, WY 82002
Barbara L. Sahl: (307) 777-7570, fax (307) 777-5973
E-mail: bsahl@missc.state.wy.us
Web site: http:\\deq.state.wy.us

Wyoming is a delegated NPDES state with general permitting authority. The state has issued separate general permits for storm water discharges from construction sites and from industrial facilities. Currently, Wyoming does not plan to adopt EPA's multisector general permit. The coverage and requirements of the industrial storm water permit are similar to those for the EPA baseline general permit for discharges from industrial activities. According to the Wyoming Department of

Environmental Quality (DEQ), the purpose of the industrial general permit is to prevent or minimize pollution to state waters. Therefore, only minimal runoff controls may be required in desert areas, but extensive controls may be required for discharges in areas of steep terrain or near sensitive waters. Best management practices or other controls must be designed so that discharges meet state water quality standards. Most dischargers are not required to perform storm water sampling, only on-site inspections. Certain timber, concrete, gypsum, metal mining, mineral mining, scrap recycling, and waste recycling facilities are required to sample their storm water discharge quarterly during the second year and (possibly) fourth year of their permit (personal correspondence, Barb Sahl, WDEQ, February 1998).

Construction sites with runoff that affects bodies of water which are perennially wet must not increase turbidity by more than 10 to 15 turbidity units above background levels. Other discharges are not subject to the turbidity limitation, but are prohibited from discharging sediments that degrade the aesthetics or habitat characteristics of the water body. Storm water management controls must be designed to remove 80% of the total suspended solids from all flows exceeding predevelopment levels. If this goal is not attainable, the permittee must supply justification of why particular management measures are not implemented. Other requirements are similar to those for the EPA baseline general permit, except that the frequency of inspection may be reduced to once per quarter during seasonal shutdowns and after completion of construction but before the site has returned to "approximate preconstruction conditions" (Thompson Publishing Group, May 1995).

Under phase I, none of the municipalities in Wyoming is required to obtain a storm water discharge permit for its municipal separate storm sewer system [60 FR 50804 (Sept. 29, 1995)].

References

Environmental Protection Agency, 1996. *Storm Water List of Contacts,* October 31.

Jessup, Deborah Hitchcock, 1990. *Guide to State Environmental Programs,* 2d ed., Bureau of National Affairs, Rockville, MD.

Thompson Publishing Group, 1997. *Stormwater Permit Manual,* Washington, updated monthly.

Government Printing Office: *Federal Register,* Office of the Federal Register, National Archives and Records Administration, Washington.

Appendix

B

Glossary of Storm Water Terms and Acronyms

ABEL EPA's computer model for analyzing a violator's ability to pay a civil penalty.

Acute toxicity The ability of a substance to cause poisonous effects resulting in severe biological harm or death soon after a single exposure or dose. Also, any severe poisonous effect resulting from a single short-term exposure to a toxic substance. (*See* Chronic toxicity.)

Administrative order A legal document signed by the EPA directing an individual, business, or other entity to take corrective action or refrain from an activity. It describes the violations and actions to be taken, and it can be enforced in court. Such orders may be issued, e.g., as a result of an administrative complaint whereby the respondent is ordered to pay a penalty for violations of a statute.

Administrative procedures act A law that spells out procedures and requirements related to the promulgation of regulations.

APA Administrative procedures act.

Arsenic (As) A heavy metal that can accumulate in the environment and is highly toxic. (*See* Heavy metals.)

BACT Best available control technology.

BAP Benefits analysis program.

BASINS (better assessment science integrating point and nonpoint sources) A computer-run tool that contains an assessment and planning component that allows users to organize and display geographic information for selected watersheds. It also contains a modeling component to examine impacts of pollutant loadings from point and nonpoint sources and to characterize the overall condition of specific watersheds.

BAT Best available technology.

BCPCT Best conventional pollutant control technology.

BCT Best control technology.

BEN EPA's computer model for analyzing a violator's economic gain from not complying with the law.

Beryllium (Be) An airborne metal hazardous to human health when inhaled. It is discharged by machine shops, ceramic and propellant plants, and foundries.

Best management practices (BMPs) Methods that have been determined to be the most effective, practical means of preventing or reducing pollution from storm water runoff. These include schedules of activities, prohibitions of practices, maintenance procedures, and other management practices. BMPs also include treatment requirements, operating procedures, and practices to control plant site runoff, spillage or leaks, sludge or waste disposal, or drainage from raw material storage.

Biochemical oxygen demand (BOD) A measure of the amount of oxygen consumed in the biological processes that break down organic matter in water. The greater the BOD, the greater the degree of pollution.

Biological oxygen demand (BOD) An indirect measure of the concentration of biologically degradable material present in organic wastes. It usually reflects the amount of oxygen consumed in 5 days by biological processes breaking down organic waste.

BMP Best management practice.

BOD Biochemical oxygen demand or biological oxygen demand.

BOD5 The amount of dissolved oxygen consumed in 5 days by biological processes breaking down organic matter.

Buffer strips (buffer zones) *See* Filter strip.

Bypass The intentional diversion of waste streams from any portion of a treatment facility.

By-product Material, other than the principal product, generated as a consequence of an industrial process.

Cadmium (Cd) A heavy-metal element that accumulates in the environment. (*See* Heavy metals.)

Cap A layer of clay or other impermeable material installed over the top of a closed landfill to prevent entry of rainwater and to minimize leachate.

Cells In solid waste disposal, holes where waste is dumped, compacted, and covered with layers of dirt daily.

CERCLA Comprehensive Environmental Response, Compensation, and Liability Act of 1980 (Public Law 96-510), which established a program to mitigate releases of hazardous waste from inactive hazardous waste sites that endanger public health and the environment.

CFR Code of Federal Regulations

Chemical oxygen demand (COD) A measure of the oxygen required to oxidize all compounds, both organic and inorganic, in water.

Chromium (Cr) A heavy metal that can accumulate in the environment. (*See* Heavy metals.)

Chronic effect An adverse effect on a human or an animal in which symptoms recur frequently or develop slowly over a long time.

Chronic toxicity The capacity of a substance to cause long-term poisonous human health effects. (*See* Acute toxicity.)

Clean Water Act (CWA) The Clean Water Act (formerly referred to as the Federal Water Pollution Control Act or Federal Water Pollution Control Act Amendments of 1972), Public Law 92-500, as amended by Public Law 96-483 and Public Law 97-117, 33 U.S.C. 1251 et seq. The Clean Water Act (CWA) contains a number of provisions to restore and maintain the quality of the nation's water resources. One of these provisions is Section 303(d), which establishes the total maximum daily load program.

Cleanup Action taken to deal with a release or threat of release of a hazardous substance that could affect humans and/or the environment. The term *cleanup* is sometimes used interchangeably with the terms *remedial action, removal action, response action,* or *corrective action.*

Closure The procedure a landfill operator must follow when a landfill reaches its legal capacity for solid waste: ceasing acceptance of solid waste and placing a cap on the landfill site.

Coal pile runoff The rainfall runoff from or through any coal storage pile.

Coastal zone Lands and waters adjacent to the coast that exert an influence on the uses of the sea and its ecology, or whose uses and ecology are affected by the sea.

Coastal Zone Act Reauthorization Amendments of 1990 (CZARA) Section 6217 of CZARA requires states to have federally approved programs to implement and enforce nonpoint source controls. This provision applies only to the coastal zones of the 29 coastal states and territories that have approved Coastal Zone Management Act (CZMA) programs.

COD Chemical oxygen demand.

Colocated industrial activity A facility that has industrial activities being conducted on site that are described under more than one of the sectors in the EPA multisector general permit. Facilities with colocated industrial activities shall comply with all applicable monitoring and pollution prevention plan requirements of each sector in which a colocated industrial activity is described.

Combined sewer overflow Discharge of a mixture of storm water and domestic waste when the flow capacity of a sewer system is exceeded during rainstorms.

Combined sewers A sewer system that carries both sewage and storm water runoff. Normally, its entire flow goes to a waste treatment plant, but during a heavy storm, the volume of water may be so great as to cause overflows of untreated mixtures of storm water and sewage into receiving waters. Storm water runoff may also carry toxic chemicals from industrial areas or streets into the sewer system.

Commencement of construction The initial disturbance of soils associated with clearing, grading, or excavating activities or other construction activities.

Commercial treatment and disposal facilities Facilities that receive, on a commercial basis, any produced hazardous waste (not their own) and treat or dispose of those wastes as a service to the generators. Such facilities treating and/or disposing of exclusively residential hazardous wastes are not included in this definition.

Compliance monitoring Collection and evaluation of data, including self-monitoring reports, and verification to show whether pollutant concentrations and loads contained in permitted discharges are in compliance with the limits and conditions specified in the permit.

Compliance schedule A negotiated agreement between a pollution source and a government agency that specifies dates and procedures by which a source will reduce emissions and thereby comply with a regulation.

Composite sample A series of water samples taken over a given time and weighted by flow rate or time.

Consent decree A legal document, approved by a judge, that formalizes an agreement reached between EPA and dischargers through which the dischargers will cease or correct actions or processes that are polluting the environment; or otherwise comply with EPA-initiated regulatory enforcement actions. The consent decree describes the actions dischargers will take and may be subject to a public comment period.

Construction and demolition waste Waste building materials, dredging materials, tree stumps, and rubble resulting from construction, remodeling, repair, and demolition of homes, commercial buildings, and other structures and pavements. May contain lead, asbestos, or other hazardous substances.

Contaminant Any physical, chemical, biological, or radiological substance or matter that has an adverse effect on air, water, or soil.

Contamination Introduction into water, air, and soil of microorganisms, chemicals, toxic substances, wastes, or wastewater in a concentration that makes the medium unfit for its next intended use. Also applies to surfaces of objects and buildings, and various household and agricultural use products.

Contingency plan A document setting out an organized, planned, and coordinated course of action to be followed in case of a fire, explosion, or other accident that releases toxic chemicals, hazardous waste, or radioactive materials that threaten human health or the environment.

Continuous sample A flow of water from a particular place in a plant to the location where samples are collected for testing; may be used to obtain grab or composite samples.

Conventional pollutants Statutorily listed pollutants understood well by scientists. These may be in the form of organic waste, sediment, acid, bacteria, viruses, nutrients, oil and grease, or heat.

Cover material Soil used to cover compacted solid waste in a sanitary landfill.

CWA Clean Water Act.

CZARA The Coastal Zone Act Reauthorization Amendments of 1990 (Public Law 101-508).

CZMA Coastal Zone Management Act.

Dedicated portable asphalt plant A portable asphalt plant that is located on or contiguous to a construction site and that provides asphalt only to the construction site that the plant is located on or adjacent to. The term *dedicated portable asphalt plant* does not include facilities that are subject to the asphalt emulsion effluent limitation guideline at 40 CFR 443.

Dedicated portable concrete plant A portable concrete plant that is located on or contiguous to a construction site and that provides concrete only to the construction site that the plant is located on or adjacent to.

Deep-well injection Deposition of raw or treated, filtered hazardous waste by pumping it into deep wells, where it is contained in the pores of permeable subsurface rock.

Delegated state A state (or other government entity such as a tribal government) that has received authority to administer an environmental regulatory program in lieu of a federal counterpart. As used in connection with the NPDES program, the term does not connote any transfer of federal authority to a state.

Design capacity The average daily flow that a treatment plant or other facility is designed to accommodate.

Dewater 1. Remove or separate a portion of the water in a sludge or slurry to dry the sludge so it can be handled and disposed. 2. Remove or drain the water from a tank or trench.

Dike A low wall that can act as a barrier to prevent a spill from spreading.

Direct discharger A municipal or industrial facility which introduces pollution through a defined conveyance or system such as outlet pipes; a point source.

Direct filtration A method of treating water which consists of the addition of coagulant chemicals, flash mixing, coagulation, minimal flocculation, and filtration. Sedimentation is not used.

Direct runoff Water that flows over the ground surface or through the ground directly into streams, rivers, and lakes.

Director The regional administrator of the Environmental Protection Agency or an authorized representative.

Discharge Flow of surface water in a stream or canal, or the outflow of groundwater from a flowing artesian well, ditch, or spring.

Dissolved oxygen (DO) The amount of oxygen that is dissolved in water. It also refers to a measure of the amount of oxygen available for biochemical activity in a water body, and as an indicator of the quality of that water.

DMR Discharge Monitoring Report.

Drainage basin The area of land that drains water, sediment, and dissolved materials to a common outlet at some point along a stream channel. Also called a watershed.

Dump A site used to dispose of solid waste without environmental controls.

Effluent Wastewater (either treated or untreated) that flows out of a treatment plant, sewer, or industrial outfall. Generally refers to wastes discharged into surface waters.

Effluent guidelines Technical EPA documents which set effluent limitations for given industries and pollutants.

Effluent limitations Restrictions established by a state or EPA on quantities, rates, and concentrations in wastewater discharges.

Effluent standard *See* Effluent limitations.

Endangered species Animals, birds, fish, plants, or other living organisms threatened with extinction by artificial or natural changes in their environment. Requirements for declaring a species endangered are contained in the Endangered Species Act.

Endangerment assessment A study to determine the nature and extent of contamination at a site on the National Priorities List and the risks posed to public health or the environment. EPA or the state conducts the study when a legal action is to be taken to direct potentially responsible parties to clean up a site or pay for it. An endangerment assessment supplements a remedial investigation.

Enforceable requirements Conditions or limitations in permits issued under the Clean Water Act, Section 402 or 404, that, if violated, could result in the issuance of a compliance order or initiation of a civil or criminal action under federal or applicable state laws. If a permit has not been issued, the term includes any requirement which, in the regional administrator' s judgment, would be included in the permit when issued. Where no permit applies, the term includes any requirement which the regional administrator determines is necessary for the best practical waste treatment technology to meet applicable criteria.

Enforcement EPA, state, or local legal actions to obtain compliance with environmental laws, rules, regulations, or agreements and/or obtain penalties or criminal sanctions for violations. Enforcement procedures may vary, depending on the requirements of different environmental laws and related implementing regulations.

Environmental audit An independent assessment of the current status of a party's compliance with applicable environmental requirements or of a party's environmental compliance policies, practices, and controls.

EPCRA Emergency Planning and Community Right-to-Know Act of 1986, another name for Title III of SARA. Title III was created as a freestanding law establishing requirements for federal, state, and local governments and industry regarding emergency planning and reporting of hazardous and toxic chemicals.

Ephemeral stream A stream that flows only in direct response to precipitation in the immediate watershed or in response to the melting of a cover of snow and ice, and which has a channel bottom that is always above the local water table [30 CFR 701.5] (Hawley and Parsons, 1980).

Erosion The wearing away of the land surface by running water, waves, or moving ice and wind, or by such processes as mass wasting and corrosion (solution and other chemical processes). The term *geologic erosion* refers to natural erosion processes occurring over long (geologic) time spans. *Accelerated erosion* generically refers to erosion in excess of what is presumed or estimated to be naturally occurring levels, and which is a direct result of human activities (e.g., farming, residential or industrial development, road building, or logging) (Hawley and Parsons, 1980).

ESA Endangered Species Act of 1973 (Public Law 93-205), which regulates a wide range of activities affecting plants and animals designated as endangered or threatened.

Feedlot A confined area for the controlled feeding of animals. It tends to concentrate large amounts of animal waste that cannot be absorbed by the soil and, hence, may be carried to nearby streams or lakes by rainfall runoff.

Filter strip Strip or area of vegetation used for removing sediment, organic matter, and other pollutants from runoff and wastewater.

Final stabilization The condition that is reached when all soil-disturbing activities at a construction site have been completed, and a uniform perennial vegetative cover with a density of 70% of the cover for unpaved areas and areas not covered by permanent structures has been established, or equivalent permanent stabilization measures (such as the use of riprap, gabions, or geotextiles) have been employed.

Flow-weighted composite sample A composite sample consisting of a mixture of aliquots collected at a constant time interval, where the volume of each aliquot is proportional to the flow rate of the discharge.

Flume A natural or artificial channel that diverts water.

FR *Federal Register.*

General permit A permit applicable to a class or category of dischargers.

Grab sample A single sample collected at a particular time and place that represents the composition of the water only at that time and place.

Grassed waterway Natural or constructed watercourse or outlet that is shaped or graded and established in suitable vegetation for the disposal of runoff water without erosion.

Ground cover Plants grown to keep soil from eroding.

Gully A small channel with steep sides cut by running water and through which water ordinarily runs only after a rain or icemelt or snowmelt. The distinction between a gully and a rill is one of depth. A gully generally is an obstacle to wheeled vehicles and is too deep to be obliterated by ordinary tillage; a rill is less deep and can be smoothed over by ordinary tillage (Hawley and Parsons, 1980).

Gully erosion Severe erosion in which trenches are cut to a depth greater than 30 cm (about 1 ft).

Hazardous waste A by-product of society that can pose a substantial or potential hazard to human health or the environment when improperly managed. It possesses at least one of four characteristics (ignitability, corrosivity, reactivity, or toxicity), or appears on special EPA lists.

Hazardous waste landfill An excavated or engineered site where hazardous waste is deposited and covered.

Heavy metals Metallic elements with high atomic weights, e.g., mercury, chromium, cadmium, arsenic, and lead. They can damage living things at low concentrations and tend to accumulate in the food chain.

Hydrology The science dealing with the properties, distribution, and circulation of water.

Impoundment A structure or depression, either naturally formed or artificially built, that holds water, sediment, slurry or other liquids or semiliquids [30 CFR 701.5]

Incinerator A furnace for burning waste under controlled conditions.

Injection well A well into which fluids are injected for purposes such as waste disposal, improving the recovery of crude oil, or solution mining.

Intermittent stream A stream, or reach of a stream, that does not flow year-round and that flows only when it receives base flow solely during wet periods, or it receives groundwater discharge or protracted contributions from melting snow or other erratic surface and shallow subsurface sources (Hawley and Parsons, 1980) 1. A stream or reach of a stream that drains a watershed of at least 1 mi^2. 2. A stream or reach of a stream that is below the local water table for at least some part of the year, and obtains its flow from both surface runoff and groundwater discharge [30 CFR 701.5].

Irrigation return flow Surface and subsurface water which leaves the field following application of irrigation water.

Kjeldahl nitrogen [total Kjeldahl nitrogen (TKN)] A measure of the ammonia and organic nitrogen present in a water sample.

Land application Discharge of wastewater onto the ground for treatment or reuse.

Land application unit An area where wastes are applied to or incorporated into the soil surface (excluding manure-spreading operations) for treatment or disposal.

Landfill An area of land or an excavation in which wastes are placed for permanent disposal, and which is not a land application unit, surface impoundment, injection well, or waste pile.

Large and medium-size municipal separate storm sewer systems All municipal separate storm sewers that are (1) located in an incorporated place (city) with a population of 100,000 or more as determined by the latest Decennial Census by the Bureau of Census (these cities are listed in Appendices F and G of 40 CFR Part 122); or (2) located in the counties with unincorporated urbanized populations of 100,000 or more, except municipal separate storm sewers that are located in the incorporated places, townships, or towns within such counties (these counties are listed in Appendices H and I of 40 CFR Part 122); or (3) owned or operated by a municipality other than those described in paragraph (1) or (2) and that are designated by the Director as part of the large or medium-size municipal separate storm sewer system.

Leachate Water that collects contaminants as it trickles through wastes, pesticides or fertilizers. Leaching may occur in farming areas, feedlots, and landfills, and it may result in hazardous substances entering surface water, groundwater, or soil.

Lead (Pb) A heavy metal that is hazardous to health if breathed or swallowed. Its use in gasoline, paints, and plumbing compounds has been sharply restricted or eliminated by federal laws and regulations. (*See* Heavy metals.)

Lift In a sanitary landfill, a compacted layer of solid waste and the top layer of cover material.

Load allocation (LA) The portion of a receiving water's loading capacity that is attributed either to one of its existing or future nonpoint sources of pollution or to natural background sources. Load allocations are best estimates of the loading, which may range from reasonably accurate estimates to gross allotments, depending on the availability of data and appropriate techniques for predicting the loading. Wherever possible, natural and nonpoint source loads should be distinguished [40 CFR 130.2(g)].

Marsh Periodically wet or continually flooded area with the surface not deeply submerged. It is covered dominantly with sedges, cattails, rushes, or other hydrophytic plants (Soil Science Society of America, 1987).

Mercury (Hg) A heavy metal that can accumulate in the environment and is highly toxic if breathed or swallowed. (*See* Heavy metals.)

Monitoring Periodic or continuous surveillance or testing to determine the level of compliance with statutory requirements and/or pollutant levels in various media or in humans, plants, and animals.

MS4 Municipal separate storm sewer system.

Mudflat A relatively level area of fine-grained material (e.g., silt) along a shore (as in a sheltered estuary) or around an island, alternately covered and uncovered by the tide or covered by shallow water, and barren of vegetation (Bates and Jackson, 1987).

Municipal discharge Discharge of effluent from wastewater treatment plants that receive wastewater from households, commercial establishments, and industries in the coastal drainage basin. Combined sewer and separate storm overflows are included in this category.

Municipal separate storm sewer system (MS4) A conveyance or system of conveyances (including roads with drainage systems, municipal streets, catch basins, curbs, gutters, ditches, artificial channels, or storm drains) that is owned or operated by a state, city, town, borough, county, parish, district, association, or other public body which is designed or used for collecting or conveying storm water. Any public drainage system owned by a nonfederal government agency could be considered to be a *municipal* system. A *separate* storm sewer system is one that is designed to carry only wet-weather flows (as opposed to a combined sewer, which carries sewage as well as wet-weather flows).

National Pollutant Discharge Elimination System (NPDES) The national program for issuing, modifying, revoking and reissuing, terminating, monitoring, and enforcing permits, and imposing and enforcing pretreatment requirements, under Sections 307, 402, 318, and 405 of the Clean Water Act. The CWA prohibits discharge of pollutants into waters of the United States unless a special permit is issued by EPA, a state, or, where delegated, a tribal government on an Indian reservation.

National Response Center The federal operations center, operated by the U.S. Coast Guard 24 hours per day, that receives notifications of all releases of oil and hazardous substances into the environment, and that evaluates all reports and notifies the appropriate agency.

Natural Resources Defense Council (NRDC) An environmental group active in water quality issues.

Navigable waters Traditionally, waters sufficiently deep and wide for navigation by all or specified vessels. Such waters in the United States come under federal jurisdiction and are protected by certain provisions of the Clean Water Act.

New source Any stationary source built or modified after publication of final or proposed regulations that prescribe a given standard of performance.

New Source Performance Standards (NSPSs) Uniform national EPA air emission and water effluent standards which limit the amount of pollution allowed from new sources or from modified existing sources.

NHPA National Historic Preservation Act of 1966 (Public Law 89-665), which establishes a program and system of regulations intended to preserve historic places.

Nitrate Plant nutrient and inorganic fertilizer that enters water supply sources from septic systems, animal feedlots, agricultural fertilizers, manure, industrial wastewaters, sanitary landfills, and garbage dumps.

Nitrite 1. An intermediate in the process of nitrification. 2. Nitrous oxide salts used in food preservation.

Nitrogen (N) An essential chemical food element that can contribute to the eutrophication of lakes and other water bodies. Various nitrogen compounds including nitrates, nitrites, ammonia, and organic nitrogen are among the "conventional pollutants" present in many storm water discharges.

NOI Notice of Intent to be covered by a general permit.

Noncontact cooling water Water used for cooling which does not come into direct contact with any raw material, product, by-product, or waste.

Nonconventional pollutant Any pollutant not statutorily listed or that is poorly understood by the scientific community.

Non–point source Diffuse pollution sources (i.e., without a single point of origin or not introduced into a receiving stream from a specific outlet). The pollutants generally are carried off the land by storm water. Nonpoint sources can be divided into activities related to either land or water use including failing septic tanks, improper animal-keeping practices, forest practices, and urban and rural runoff.

NOT Notice of Termination of general permit coverage.

Notice of Deficiency An EPA request to a facility owner or operator for additional information before a preliminary decision on a permit application can be made.

NRDC Natural Resources Defense Council.

Open dump An uncovered site used for disposal of waste without environmental controls. (*See* Dump.)

Outfall The place where effluent is discharged into receiving waters.

Overburden Rock and soil cleared away before mining.

Parshall flume Device used to measure the flow of water in an open channel.

Parts per billion (ppb)/parts per million (ppm) Units commonly used to express contamination ratios, as in establishing the maximum permissible amount of a contaminant in water, land, or air.

Perennial stream A stream that flows continuously during all the calendar year as a result of groundwater discharge or surface runoff [30 CFR 701.5]. The water in a perennial stream is generally lower than the water table adjacent to the region adjoining the stream (Bates and Jackson, 1987).

Permit An authorization, license, or equivalent control document issued by EPA or an approved state agency to implement the requirements of an environmental regulation; e.g., a permit to operate a wastewater treatment plant or to operate a facility that may generate harmful emissions.

pH An expression of the intensity of the basic or acidic condition of a liquid. The pH may range from 0 to 14, where 0 is the most acid and 7 is neutral. Natural waters usually have a pH of between 6.5 and 8.5.

Phosphorus (P) An essential chemical food element that can contribute to the eutrophication of lakes and other water bodies. Increased phosphorus levels result from discharge of phosphorus-containing materials into surface waters. Phosphorus is one of the "conventional pollutants" present in many storm water discharges.

Playa The usually dry and nearly level lake plain that occupies the lowest parts of closed depressions, such as those occurring on intermontane basin floors. Temporary flooding occurs primarily in response to precipitation runoff events. Playa deposits are fine-grained, and may or may not have high water table and saline conditions (Hawley and Parsons, 1980).

Playa lake A shallow, intermittent lake in a arid or semiarid region, covering or occupying a playa in the wet season but drying up in summer; an ephemeral lake that upon evaporation leaves or forms a playa (Bates and Jackson, 1987).

Point source Any discernible, confined, and discrete conveyance, including but not limited to any pipe, ditch, channel, tunnel, conduit, well, discrete fissure, container, rolling stock, concentrated animal feeding operation, landfill leachate collection system, vessel or other floating craft from which pollutants are or may be discharged. This term does not include return flows from irrigated agriculture or agricultural storm water runoff.

Pollutant Generally, any substance introduced into the environment that adversely affects the usefulness of a resource.

Pollution Generally, the presence of matter or energy whose nature, location, or quantity produces undesired environmental effects. Under the Clean Water Act, e.g., the term is defined as the human-made or human-induced alteration of the physical, biological, chemical, and radiological integrity of water.

Pollution prevention The active process of identifying areas, processes, and activities which create excessive waste by-products for the purpose of substitution, alteration, or elimination of the process to prevent waste generation.

Pond 1. A natural body of standing freshwater occupying a small surface depression, usually smaller than a lake and larger than a pool. 2. A small artificial body of water, used as a source of water (Bates and Jackson, 1987).

Pothole (geomorphology) A generic, imprecise term for any pot-shaped pit or hole (Bates and Jackson, 1987). Glacial geology: A term applied to a small pit depression (1 to 15 m deep), generally circular or elliptical, occurring in an outwash plain, a recessional moraine, or a till plain. Lake: A shallow depression, generally less than 10 ha in area, occurring on disintegration moraines and commonly containing an intermittent or seasonal pond or marsh (Bates and Jackson, 1987).

POTW Publicly owned treatment work.

Prairie potholes Marshlike ponds that have formed in shallow basins caused by glaciation in the Dakotas, Iowa, and the Canadian prairies.

Pretreatment Processes used to reduce, eliminate, or alter the nature of wastewater pollutants from nondomestic sources before the pollutants are discharged into publicly owned treatment works.

Primacy Having the primary responsibility for administrating and enforcing regulations.

Process wastewater Any water that comes into contact with any raw material, product, by-product, or waste.

RCRA Resource Conservation and Recovery Act of 1976 (Public Law 94-580), which requires a regulatory system for the generation, treatment, storage, and disposal of hazardous wastes.

Receiving waters A river, lake, ocean, stream or other watercourse into which wastewater or treated effluent is discharged.

Release Any spilling, leaking, pumping, pouring, emitting, emptying, discharging, injecting, escaping, leaching, dumping, or disposing into the environment of a hazardous or toxic chemical or extremely hazardous substance.

Reportable quantity (RQ) Quantity of a hazardous substance that triggers reports under CERCLA. If a substance exceeds its RQ, the release must be reported to the National Response Center, the State Emergency Commission Response (SERC), and community emergency coordinators for areas likely to be affected.

Rill A small channel eroded into the soil surface by runoff. It can easily be smoothed out and obliterated by normal tillage.

Risk assessment Qualitative and quantitative evaluation of the risk posed to human health and/or the environment by the actual or potential presence and/or use of specific pollutants.

Runoff That part of precipitation, snowmelt, or irrigation water that runs off the land into streams or other surface water. It can carry pollutants from the air and land into receiving waters.

Runoff coefficient The fraction of total rainfall that will appear at the conveyance as runoff.

Sanitary landfill *See* landfill.

Sanitary sewers Underground pipes that carry off only domestic or industrial waste, not storm water.

SARA Superfund Amendments and Reauthorization Act of 1986 (Public Law 99-499), amended CERCLA.

SDWA Safe Drinking Water Act Amendments of 1996 (Public Law 104-182), which require states to establish source water assessment programs to delineate source water protection areas, to inventory significant contaminants in these areas, and to determine the susceptibility of each public water supply to contamination.

Section 313 water priority chemical A chemical or chemical categories which (1) are listed at 40 CFR 372.65 pursuant to Section 313 of the Emergency Planning and Community Right-to-Know Act (EPCRA) (also known as Title III of the Superfund Amendments and Reauthorization Act of 1986); (2) are present at or above threshold levels at a facility subject to EPCRA Section 313 reporting requirements; and (3) meet at least one of the following criteria: (i) Are listed in Appendix D of 40 CFR 122 on Table II (organic priority pollutants), Table III (certain metals, cyanides, and phenols), or Table V (certain toxic pollutants and hazardous substances); (ii) are listed as a hazardous substance pursuant to Section 311(b)(2)(A) of the CWA at 40 CFR 116.4; or (iii) are pollutants for which EPA has published acute or chronic water quality criteria.

Sediment Material, both mineral and organic, that is in suspension, is being transported, or has been moved from its site of origin by water, wind, ice, or mass wasting and has come to rest on the earth's surface either above or below sea level (Hawley and Parsons, 1980). Sediment piles up in reservoirs, rivers, and harbors, destroying fish and wildlife habitat and clouding the water so that sunlight cannot reach aquatic plants. Careless farming, mining, and building activities will expose sediment materials, allowing them to wash off the land after rainfall.

Sedimentation Settling out of solids from wastewater by gravity during treatment.

Sedimentation tanks Wastewater tanks in which floating wastes are skimmed off and settled solids are removed for disposal.

Significant materials Materials that must be considered in preparing a SWPPP. These include, but are not limited to, raw materials; fuels; materials such as solvents, detergents, and plastic pellets; finished materials such as metallic products; raw materials used in food processing or production; hazardous substances designated under Section 101(14) of CERCLA; any chemical the facility is required to report pursuant to EPCRA Section 313; fertilizers; pesticides; and waste products such as ashes, slag, and sludge that have the potential to be released with storm water discharges.

Significant spills Include, but are not limited to, releases of oil or hazardous substances in excess of quantities that are reportable under Section 311 of CWA [see 40 CFR 110.10 and 117.21] or Section 102 of the Comprehensive Environmental Response, Compensation, and Liability Act (CERCLA) [see 40 CFR 302.4]. Significant spills may also include releases of oil or hazardous substances that are not in excess of reporting requirements and releases of materials that are not classified as oil or a hazardous substance.

Silt Sedimentary materials composed of fine or intermediate-sized mineral particles.

Silviculture Management of forestland for timber.

Slough 1. A small marsh, especially a marshy area lying in a local, shallow, closed depression on a piece of dry land, as on the prairie of the midwest. 2. A term used, especially in the Mississippi Valley, to refer to a creek or sluggish body of water in a tidal flat, floodplain, or coastal marshland. 3. A sluggish channel of water, such as a side channel of a river, in which water flows slowly through low, swampy ground, as along the Columbia River, or a section of an abandoned river channel which may contain stagnant water and occurs in a flood plain or delta. 4. An area of soft, miry, muddy, or waterlogged ground, a place of deep mud (Bates and Jackson, 1987).

Sludge A semisolid residue from any of a number of air or water treatment processes; can be a hazardous waste.

SMCRA Surface Mining Control and Reclamation Act of 1977 (Public Law 95-87), which regulates surface coal mining operations and the acquisition and reclamation of abandoned mines.

Soil conditioner An organic material such as humus or compost that helps soil absorb water, build a bacterial community, and take up mineral nutrients.

Spill prevention control and countermeasures (SPCC) Plan covering the release of hazardous substances as defined in the Clean Water Act.

Storm sewer A system of pipes (separate from sanitary sewers) that carries only water runoff from buildings and land surfaces.

Storm water Storm water runoff, snowmelt runoff, and surface runoff and drainage.

Storm water associated with industrial activity The discharge from any conveyance which is used for collecting and conveying storm water and which is directly related to manufacturing, processing, or raw materials storage areas at an industrial plant. The term does not include discharges from facilities or activities excluded from the NPDES program. For the categories of industries identified in paragraphs (i) through (x) of this definition, the term includes, but is not limited to, storm water discharges from industrial plant yards; immediate access roads and rail lines used or traveled by carriers of raw materials, manufactured products, waste material, or by-products used or created by the facility; material handling sites; refuse sites; sites used for the application or disposal of process wastewaters [as defined at 40 CFR 401]; sites used for the storage and maintenance of material handling equipment; sites used for residual treatment, storage, or disposal; shipping and receiving areas; manufacturing buildings; storage areas (including tank farms) for raw materials, and intermediate and finished products; and areas where industrial activity has taken place in the past and significant materials remain and are exposed to storm water. For the categories of industries identified in paragraph (xi) of this definition, the term includes only storm water discharges from all areas (except access roads and rail lines) listed in the previous sentence, where material handling equipment or activities, raw materials, intermediate products, final products, waste materials, by-products, or industrial machinery is exposed to storm water. For the purposes of this paragraph, material handling activities include the storage, loading, and unloading; transportation; or conveyance of any raw material, intermediate product, finished product, by-product, or waste product. The term excludes areas located on plant lands separate from the plant's industrial activities, such as office buildings and accompanying parking lots, as long as the drainage from the excluded areas is not mixed with storm water drained from the above-described areas. Industrial facilities [including industrial facilities that are federally, state, or municipally owned or operated that meet the description of the facilities listed in paragraphs (i) to (xi) of this definition] include those facilities designated under 122.26(a)(1)(v). The following categories of facilities are considered to be engaging in "industrial activity" for purposes of this subsection:

i. Facilities subject to storm water effluent limitations guidelines, new source performance standards, or toxic pollutant effluent standards under 40 CFR subchapter N [except facilities with toxic pollutant effluent standards which are exempted under category (xi) of this definition].
ii. Facilities classified as Standard Industrial Classifications 24 (except 2434), 26 (except 265 and 267), 28 (except 283), 29, 311, 32 (except 323), 33, 3441, and 373.
iii. Facilities classified as Standard Industrial Classifications 10 through 14 (mineral industry) including active or inactive mining operations [except for areas of coal mining operations no longer meeting the definition of a reclamation area under 40 CFR 434.11(1) because the performance bond issued to the facility by the appropriate SMCRA authority has been released, or except for areas of non–coal mining operations which have been released from applicable state or federal reclamation requirements after December 17, 1990] and oil and gas exploration, production, processing, or treatment operations, or transmission facilities that discharge storm water contaminated by contact with, or that has come into contact with, any overburden, raw material, intermediate products, finished products, by-products, or waste products located on the site of such operations; inactive mining operations are mining sites that are not being actively mined, but which have an identifiable owner/operator.
iv. Hazardous waste treatment, storage, or disposal facilities, including those that are operating under interim status or a permit under Subtitle C of RCRA.
v. Landfills, land application sites, and open dumps that have received any industrial waste (waste that is received from any of the facilities described under this subsection), including those that are subject to regulation under Subtitle D of RCRA.
vi. Facilities involved in the recycling of materials, including metal scrapyards, battery reclaimers, salvage yards, and automobile junkyards, including but limited to those classified as Standard Industrial Classifications 5015 and 5093.
vii. Steam electric power generating facilities, including coal handling sites.

viii. Transportation facilities classified as Standard Industrial Classifications 40, 41, 42 (except 4221 to 4225), 43, 44, 45, and 5171 which have vehicle maintenance shops, equipment-cleaning operations, or airport deicing operations. Only those portions of the facility that either are involved in vehicle maintenance (including vehicle rehabilitation, mechanical repairs, painting, fueling, and lubrication), equipment-cleaning operations, or airport deicing operations, or are otherwise identified under paragraphs (i) to (vii) or (ix) to (xi) of this subsection are associated with industrial activity.

ix. Treatment works treating domestic sewage or any other sewage sludge or wastewater treatment device or system, used in the storage treatment, recycling, and reclamation of municipal or domestic sewage, including land dedicated to the disposal of sewage sludge that is located within the confines of the facility, with a design flow of 1.0 Mgal/day or more, or required to have an approved pretreatment program under 40 CFR 403. Not included are farmlands, domestic gardens, or lands used for sludge management where sludge is beneficially reused and which are not physically located in the confines of the facility, or areas that are in compliance with 40 CFR 503.

x. Construction activity including clearing, grading, and excavation activities except operations that result in the disturbance of less than 5 acres of total land area which are not part of a larger common plan of development or sale.

xi. Facilities under Standard Industrial Classifications 20, 21, 22, 23, 2434, 25, 265, 267, 27, 283, 285, 30, 31 (except 311), 323, 34 (except 3441), 35, 36, 37 (except 373), 38, 39, and 4221 to 4225 [and which are not otherwise included within categories (i) to (x)].

Sump A pit or tank that catches liquid runoff for drainage or disposal.

Surface runoff Precipitation, snowmelt, or irrigation in excess of what can infiltrate the soil surface and be stored in small surface depressions; a major transporter of nonpoint source pollutants.

Surface water All water naturally open to the atmosphere (rivers, lakes, reservoirs, ponds, streams, impoundments, seas, estuaries, etc.) and all springs, wells, or other collectors directly influenced by surface water.

Suspended solids Small particles of solid pollutants that float on the surface of, or are suspended in, a sample liquid. They resist removal by conventional means. Total suspended solids (TSS) is a measure of the suspended solids in a sample liquid, determined by tests for total suspended nonfilterable solids.

Swale 1. A slight, open depression which lacks a defined channel that can funnel overland or subsurface flow into a drainageway. 2. A shallow depression in an undulating ground moraine due to uneven glacial deposition. 3. A long, narrow, generally shallow, troughlike depression between two beach ridges, and aligned roughly parallel to the coastline (Bates and Jackson, 1987).

Tailings Residue of raw material or waste separated out during the processing of crops or mineral ores.

Tailwater The runoff of irrigation water from the lower end of an irrigated field.

Tidal flat An extensive, nearly horizontal, marshy, or barren tract of land that is alternately covered and uncovered by the tide, and consists of unconsolidated sediment (mostly clays, silts, and/or sand). It may form the top surface of a deltaic deposit (Bates and Jackson, 1987).

Time-weighted composite A composite sample consisting of a mixture of equal-volume aliquots collected at a constant time interval.

TKN Total Kjeldahl nitrogen (*see* Kjeldahl nitrogen).

Total maximum daily load (TMDL) The sum of the individual waste load allocations (WLAs) for point sources, load allocations (LAs) for nonpoint sources and natural background, and a margin of safety (MOS). TMDLs can be expressed in terms of mass per time, toxicity, or other appropriate measure that relates to a state's water quality standard.

Toxic chemical Any chemical listed in EPA rules as "toxic chemicals subject to Section 313 of the Emergency Planning and Community Right-to-Know Act of 1986."

Toxic pollutants Materials that cause death, disease, or birth defects in organisms that ingest or absorb them. The quantities and exposures necessary to cause these effects can vary widely.

Toxic substance A chemical or mixture that may present an unreasonable risk of injury to health or the environment.

Toxicity assessment Characterization of the toxicological properties and effects of a chemical, with special emphasis on establishment of dose-response characteristics.

Toxicity testing Biological testing (usually with an invertebrate, fish, or small mammal) to determine the adverse effects of a compound or effluent.

Treatment Any method, technique, or process designed to remove solids and/or pollutants from solid waste, waste streams, effluents, and air emissions.

Treatment plant A structure built to treat wastewater before discharging it into the environment.

Treatment, storage, and disposal (TSD) facility Site where a hazardous substance is treated, stored, or disposed of. TSD facilities are regulated by EPA and states under RCRA.

TSDF Treatment, storage, and disposal facility.

TSS Total suspended solids (*see* Suspended solids).

Upset An exceptional incident in which there is unintentional and temporary noncompliance with the numeric effluent limitations of a permit because of factors beyond the reasonable control of the permittee. An upset does not include noncompliance to the extent caused by operational error, improperly designed treatment facilities, inadequate treatment facilities, lack of preventive maintenance, or careless or improper operation.

Urban runoff Storm water from city streets and adjacent domestic or commercial properties that carries pollutants of various kinds into the sewer systems and receiving waters.

Vegetative controls Storm water pollution control practices that involve vegetative cover to reduce erosion and minimize discharge of pollutants.

Waste load allocation (WLA) The portion of a receiving water's loading capacity that is allocated to one of its existing or future point sources of pollution. WLAs constitute a type of water quality-based effluent limitation [40 CFR 130.2(h)].

Waste pile Any noncontainerized accumulation of solid, nonflowing waste that is used for treatment or storage.

Water quality–based limitations Effluent limitations applied to dischargers when mere technology-based limitations would cause violations of water quality standards. Usually applied to discharges into small streams.

Water quality criteria Levels of water quality expected to render a body of water suitable for its designated use. Criteria are based on specific levels of pollutants that would make the water harmful if used for drinking, swimming, farming, fish production, or industrial processes.

Water quality standards State-adopted and EPA-approved ambient standards for water bodies. The standards prescribe the use of the water body and establish the water quality criteria that must be met to protect designated uses.

Water table The upper surface of a zone of saturation, where the body of groundwater is not confined by an overlying impermeable zone [30 CFR 701.5].

Waters of the United States:

(a) All waters which are currently used, were used in the past, or may be susceptible to use in interstate or foreign commerce, including all waters which are subject to the ebb and flow of the tide;
(b) All interstate waters, including interstate "wetlands."
(c) All other waters such as intrastate lakes, rivers, streams (including intermittent streams), mudflats, sandflats, wetlands, sloughs, prairie potholes, wet meadows, playa lakes, or natural ponds, the use, degradation, or destruction of which would affect or could affect interstate or foreign commerce including any such waters.
    1. Which are or could be used by interstate or foreign travelers for recreational or other purposes
    2. From which fish or shellfish are or could be taken and sold in interstate or foreign commerce
    3. Which are used or could be used for industrial purposes by industries in interstate commerce

(d) All impoundments of waters otherwise defined as waters of the United States under this definition.
(e) Tributaries of waters identified in paragraphs (a) through (d) of this definition;
(f) The territorial sea.
(g) Wetlands adjacent to waters (other than waters that are themselves wetlands) identified in paragraphs (a) through (f) of this definition.

Waste treatment systems, including treatment ponds or lagoons designed to meet the requirements of CWA, are not waters of the United States.

Watershed The land area that drains into a stream. The watershed for a major river may encompass a number of smaller watersheds that ultimately combine at a common delivery point.

Watershed-based trading Trading arrangements among point source dischargers, nonpoint sources, and indirect dischargers in which the "buyers" purchase pollutant reductions at a lower cost than they would spend to achieve the reductions themselves. Sellers provide pollutant reductions and may receive compensation. The total pollution reduction, however, must be the same as or greater than what would be achieved if no trade occurred.

Weir A notched wall or plate placed in an open channel to measure the flow of water.

Wet meadows Grasslands with soil that is waterlogged after precipitation events.

Wetland An area that is saturated by surface or groundwater with vegetation adapted for life under those soil conditions, such as swamps, bogs, fens, marshes, and estuaries.

References

Bates, R. L., and J. A. Jackson (eds.), 1987. *Glossary of Geology,* 3d ed., American Geological Institute, Alexandria, VA.
Soil Science Society of America, 1987. *Glossary of Soil Science Terms,* Madison, WI.
Hawley, J. W., and R. B. Parsons, 1980. *Glossary of Selected Geomorphic and Geologic Terms.* Mimeo. USDA Soil Conservation Service, West National Technical Center, Portland, OR.

Appendix

C

Units, Abbreviations, and Conversions

This appendix lists all metric (SI) and U.S. Customary System (USCS) measurement units used in this book and gives the abbreviation and appropriate conversion constants for each.

American unit	Abbreviation	Conversion(s)
foot	ft	1 ft = 0.3048 m
Inch	in	1 in = 2.54 cm = 25.4 mm
Yard	yd	1 yd = 3 ft = 0.9147 m
pound (mass)	lb	1 lb = 2.205 kg
Ton	ton	1 ton = 2000 lb = 0.907 metric ton (t)
Gallon (U.S.)	gal	1 gal = 0.1337 ft^3 = 3.785 L
Gallons per minute	gal/min	1000 gal/min = 2.23 ft^3/s
cubic feet per second	ft^3/s	1 ft^3/s = 0.02832 m^3/s
Acre	acre	1 acre = 0.4047 ha
Square foot	ft^2	1 ft^2 = 0.09290 m^2
Square yard	yd^2	1 yd^2 = 0.8361 m^2
cubic foot	ft^3	1 ft^3 = 0.02832 m^3
cubic yard	yd^3	1 yd^3 = 0.765 m^3

SI unit	Abbreviation	Conversion(s)
Meter	m	1 m = 3.281 ft = 1.0937 yd
Centimeter	cm	1 cm = 0.3937 in
Millimeter	mm	1 mm = 0.03937 in
Kilogram	kg	1 kg = 2.2046 lb
Metric ton	mt	1 mt = 2205 lb = 1.1025 tn
Liter	L	1 L = 0.2642 gal
Cubic meters per second	m^3/s	1 m^3/s = 35.33 ft^3
Hectare	ha	1 ha = 10,000 m^2 = 2.471 acres
Square meter	m^2	1 m^2 = 10.76 ft^2 = 1549.44 in^2
Cubic meter	m^3	1 m^3 = 35.30 ft^3 = 1.31 yd^3

Index

ABOUT THE AUTHOR

Roy D. Dodson is the president of Dodson & Associates, Inc., a water-resources engineering firm in Houston, Texas. Mr. Dodson has extensive experience in the planning, design, and analysis of water resources systems for flood control, water supply, and water quality applications. Since 1986, he has presented training seminars for the American Society of Civil Engineers and others. The seminar topics include storm water pollution control, flood plain management, and computer modeling of hydrology and hydraulics.

Mr. Dodson graduated from Texas Tech University and Stanford University with degrees in civil engineering. In his graduate work at Stanford Mr. Dodson specialized in water resources infrastructure planning and management. Mr. Dodson is registered as a Professional Engineer in several states. He is also a registered Professional Hydrologist and a Certified Flood Plain Manager.